Rasa Shāstra

The Āyurvedic Art of Medicine Making and Alchemy

Revised & Expanded Edition

Andrew Mason

AEON

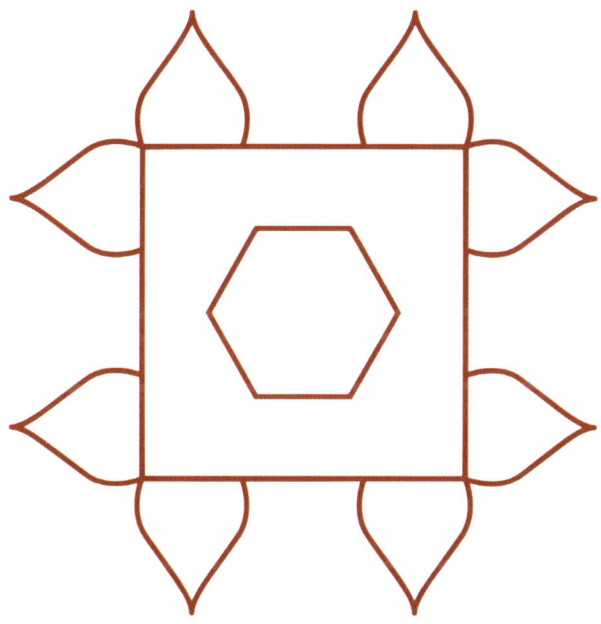

Dedicated to my wife Atsuko and daughter Himiko

First published 2024 by
Aeon Books Ltd
118 Finchley Road
London NW3 5HT

British Library Cataloguing in Publication Data

A C.I.P. for this book is available from the British Library

ISBN-978-1-80152-113-0

Printed in Great Britain by Bell & Bain Ltd, Glasgow

www.aeonbooks.co.uk

Contents

Foreword by Vaidya Ātreya Smith ix
Acknowledgements xiii
Notes on revised and expanded edition xv
Disclaimer xix

Sri Lanka 1
1/ Introduction 1
2/ Return to Sri Lanka 5
3/ Practical 8
4/ Journey to the factory 9
5/ Setting up shop 13
6/ Contents 15
7/ Final result 16
8/ Return 19

Part I 25

Setting the scene 29
Sagar Manthan (Churning the Milky Ocean) 29

Section 1 Overview of Āyurveda 37
1.1/ Ancient technology 37
1.2/ An early encounter with Āyurveda 40
1.3/ Rasa Shāstra 43
1.4/ Health and longevity in Āyurveda 45
1.5/ Ṣaḍkiryakalas (six stages of disease) 48
1.6/ Ojas and Sapta Dhātu 55
1.7/ The six actions of taste 59
1.8/ Ṣaḍupakarmas (six catagories of therapeutics) 63
1.9/ Pañcakarma (five purification therapies) 67
1.10/ Patients unsuited for Saṁśodhana 87

Section 2 Metals and metal-working 91
2.1/ Early metallurgy 91
2.2/ Loha (metal) 95
2.3/ Reduction and conversion of metal 99
2.4/ Metallic immune booster 107

Section 3 Visha (toxins) 109
3.1/ Origins of Visha 109
3.2/ Signs of Visha 117
3.3/ Visha as medicine 118
3.4/ Visha and caste 118
3.5/ Snake venom 121

Part II: Workshop, equipment, method and apparatus 125

Section 4 Siting of Rasashala 127
4.1/ Siting of the workshop 127
4.2/ Sarpa (snakes) 130
4.3/ Modern interpretations 134
4.4/ Celestial considerations 136
4.5/ Final note 138

Section 5 Selection of apprentices 141
5.1/ Alchemist and apprentice 141
5.2/ Shat-Samudrik Shāstra 143
5.3/ Hastā Rekha Shāstra 144
5.4/ Shodhana (purification) 152

Section 6 Preparation of medicines 157
6.1/ Testing of alchemical preparations 157
6.2/ Mercury (a special case) 159
6.3/ Khalwa Yantra (pestle and mortar) 160
6.4/ Man vs. machine – automation in Rasa Shāstra 166
6.5/ Mediums used for Bhavana 168
6.6/ Mārana (calcination) 173
6.7/ Puṭa paka (temperature) 176
6.8/ Pisti (Agnitapta Bhasma) 182
6.9/ Anupāna (vehicle of delivery) 183

Part III: Materials, formula and processing 187

Rasa materials 189
Quick reference guide 189
Dhātu – subcategories of iron 198

Section 7 Mercury 201
7.1/ Use of mercury-based medicines 201
7.2/ Origins of mercury and its impurities 204
7.3/ Transmutation of base metals 207
7.4/ Mercury – planet and metal 214
7.5/ Ancient sources of mercury 217
7.6/ Common purification methods for mercury 219
7.7/ Kajjali (HgS) black sulphide of mercury 222
7.8/ Rasa Parpati (HgS) 225
7.9/ Makara Dwaja (HgS) (red sulphide of mercury) 227
7.10/ Extraction of mercury from cinnabar 231
7.11/ Hiṅgula (cinnabar) 232
7.12/ Summing up and dangers of mercury 241

Section 8 Minerals 247
8.1/ Use of mineral-based medicines 247
8.2/ Sulphur 248
8.3/ Bitumen 256
8.4/ Arsenic trisulphide 262
8.5/ Rasa Maanikya 266
8.6/ Lavaṇavarga (types of salt) 270

Section 9 Metals 279
9.1/ Use of metal-based medicines 279
9.2/ Copper 280
9.3/ Tin 284
9.4/ Zinc 288

Section 10 Gemstones 293
10.1/ The origins of gemstones 293
10.2/ Red agate (Akika) 294
10.3/ Blue sapphire (Nilama) 297
10.4/ Diamond (Hiraka) 300

Section 11 Animal products ... 305
11.1/ Use of animal products as medicines ... 305
11.2/ Deer horn ... 306
11.3/ Pearl ... 309
11.4/ Peacock feather ... 318
11.5/ Ghee (clarified butter) ... 323

Section 12 Plants ... 333
12.1/ Use of plant-based medicines ... 333
12.2/ Dhattura ... 333
12.3/ Aconite ... 338
12.4/ Bhallātaka (bhilawan nut) ... 343
12.5/ Langali (*Gloriosa superba Linn*) ... 349
12.6/ Guggulu (resin) ... 353

Section 14 A living tradition of herbo-mineral-metalic medicines ... 361
14.1/ Rasa formulae ... 361
14.2/ Rasa Shāstra in therapeutic application ... 383
14.3/ Anti-cancer herbs used by the Niripola clinic ... 384

Appendix: Catalogue of materials and their use ... 387
Bibliography ... 453
Resources ... 459
Glossary ... 461
Index ... 479
About the author ... 495

Foreword

The book you are now holding in your hands is a gem. Desire is a necessary ingredient to obtain any precious stone, and the gem you hold is no exception, since its author had a burning desire to devote his life to the pursuit of ancient knowledge (guarded to this day by a select few) and to master this knowledge to the point that it can be shared with the general population. This understanding has been purified slowly by the fire of knowledge that is born from experience and hard knocks; true clarity of the ancient science of alchemy has found expression in these pages.

It is a difficult task to learn the fabled skills that most of us believe to be nothing more than a myth. Equally difficult were the cultural differences that the author struggled to understand, adapt to, overcome and eventually embrace as his own during his prolonged apprenticeship. True knowledge of any subject is not easy to come by, and reading this book shows me how fully the author has imbibed the art and science of alchemy. This is exhibited by his clear explanations of a methodology that is intentionally kept in obscurity.

The origins of Indian history are obscured by time. Oral tradition in India tells us that the basis of Indian culture is presented in the *Vedas* – four groups of verses or hymns that are named as *Rig Veda, Yajur Veda, Sama Veda* and *Atharva Veda*. They are known as *Apaurusheya*, meaning that they have not evolved from a human mind; rather, they are conceived by the *Atman*, or divine mind. They are eternal and have no beginning or end. Together they form what is called the *Sanatana Dharma*, or the 'eternal truth'. Thus, the Vedas cannot be limited historically to any period of history or pre-history as understood by those historians who formed the history of India in the European colonial era.

My teacher told me that, according to the Brahmin tradition, the current Vedic era is more than 40,000 years old. It is important to understand that India is a culture based on oral, not written tradition. The history of Āyurveda, traditional Indian medicine, is shrouded in mystery because ancient Indian culture felt that writing was an inferior method of record keeping. Many Indians who follow this tradition are well known for their ability to memorise 40,000 to 200,000 *sutras* (verses or hymns) and recite them flawlessly with perfect Sanskrit intonation and metre. A master of this art can not only do this, but can also recite them backwards, flawlessly, from the end of the text to the beginning!

Because of this view of learning and data recording, very little has been written about ancient India by these Indians themselves. Unfortunately, most Western scholars try to date Indian and Āyurvedic texts according to written Sanskrit. This is a major flaw when trying to understand the historical origins of an oral tradition. Writing is a relatively new development in India, and hence it limits the Veda, Āyurveda and other traditional knowledge to the last few thousand years. We do, however, know a great deal about ancient India from the Chinese and Persians, who were great writers and historians. Thus, most of what we have in recorded history about ancient India and Āyurvedic medicine comes from foreign merchants, or scholars, not all of whom were necessarily doctors. Judging from the records of the Chinese we can surmise that Āyurveda began to be formulated into public healthcare prior to 3000 BCE. This is supported by archaeological discoveries in the last decade of cities in the Indus Valley region that are more than 7000 years old and have indoor plumbing, attached bathrooms, and sewer systems to evacuate waste.

The history of medicine is clear in that the most important discovery for the health of humanity has not been vaccines, medicines or alchemy, but public sanitation. Therefore, it is illogical to assume that the public health of large, ancient cities was an arbitrary manifestation of luck. Moreover, it is clearly stated in the oldest Āyurvedic text, the *Caraka Saṃhitā*, that Āyurveda came into being due to the apprehension of the *Rishi* (wise men) of the period who saw the grouping of people into city environments as a major concern for general health and welfare due to disease epidemics. Āyurveda developed from the physical and sociological needs of humanity in that time and place.

The alchemical tradition in India's Āyurvedic medicine is called *Rasa Shāstra*. While there is a radical difference in modern historians' view of Āyurveda and that of the oral tradition that was passed down over several thousand years, the development of alchemy is historically well documented. The author explains this beautifully, and the relations between the ancient cultures is intriguing – bordering on passionate. Rasa Shāstra takes indigestible substances, such as metals and minerals, and transforms them into substances that are easy for the human body to assimilate.

This is the first book to clearly present Rasa Shāstra to Westerners. It is also one of the most enlightening books on ancient alchemy ever published. This book will be of great interest to students of Āyurveda. However, its interest is not limited to those passionate about traditional medicine, as it is a captivating medical text regardless of one's orientation in medical practice. Additionally, the general public will find this book a fascinating study of history and a cultural journey worthy of the most experienced antique traveller – I myself have thoroughly enjoyed travelling through time while reading this book and recommend it to everyone interested in the timeless knowledge of alchemy.

Vaidya Ātreya Smith
Director of several schools for Āyurveda in Europe, and author of six books on Āyurvedic medicine

Acknowledgements

The author would like to extend his appreciation to the following people, without whom this book could not have been written.

My sincere thanks go to:
Douglas Whyte, Udaya Dandunnage, Don Gunasena Ranatunga, Priyanta Senanayake, Alex Florshultz, Sebastian Hirsch, Heiner Fruehauf, Andrew and Laura Shakeshaft, Danny Cavanagh, Jonathan Edwards, Dr Partap Chauhan, Carl Peters-Bond, Fred Hecktil and Biopharm (UK), Sebastian Pole and the Mansuva Clinic (Colombo) and their factory facility in Dompe, Sri Lanka 2004–2005.

With special thanks to:
Dominik and Dagmar Wujastyk, Dr Ghanashyam Marda, Dr Janaki Perera, Dr Rukman Jayasinghe, Dr Mauroof Athique, Meulin Athique, Dr Venkata Narayana Joshi, Ātreya Smith, Kavirāja H.S. Sharma, Dr Claudia Welch, Dr Robert Svoboda, Dr Palitha Serasinghe, Dr Gamage Jayawardhane, Dr Arambepola Gishanti, Dr Chandralatha Samarasekera, Philip Weeks, Hakim Salim Khan, Dr Tasnim Munshi, Dr Kamal Serasinghe, Alexandra Thornton and Oliver Rathbone (Aeon Publishing).

Notes on revised and expanded edition

In 2014, I authored *Rasa Shāstra: The Hidden Art of Medical Alchemy*, a work that emerged a decade after my initial journey to Sri Lanka. In the period of studying prior to writing I accumulated numerous notes, drawings, and ideas, all of which served as the foundation for this introductory book, tailored for a Western audience seeking accessible information about an ancient and captivating system of medicine.

Even after the book's publication, my engagement with Indian alchemy remained ceaseless. I maintained a profound fascination and an insatiable thirst for further knowledge, consistently striving to enhance my skills in the preparation of these medicines.

The original book encompassed various themes, including travel experiences, thoughts, and dialogues that intertwined with the alchemical arts. Notably, a significant portion delved into Vedic astrology, known as Jyotish. This inclusion aimed to highlight the interconnectedness of alchemy and astrology, which has often been overlooked in our commercialised era. Subsequently, I began writing a second book, *Jyotish: The Art of Vedic Astrology*, exclusively dedicated to Jyotish, a project that brought me immense joy.

All along, I harboured the intention to revisit my first book, updating it with the knowledge accumulated during the intervening years and streamlining the focus to concentrate solely on alchemical material. However, nearly a decade passed before this aspiration could be realised.

What are the key differences between the 2014 and 2023 texts?

Primarily, the textual changes involve reformatting tables and reevaluating the processing and utilisation of materials. These improvements stem from my personal experiences and enhanced understanding, which naturally result from the learning process.

The section on Pārada now includes a comprehensive account of purification processes. This became possible through my involvement in the *Ayuryog Project*, where I had the opportunity to recreate numerous ancient processes for academic research. Upgrading the book to an all-colour format will also prove immensely beneficial to readers, since adding colour enhances the comprehension of explanations and indulges the visual sense.

Another valuable section, Essentials in the pharmacy, helps set the scene and stock the shelves (so to speak) in our alchemical pharmacy. This can be particularly challenging in Rasa Shāstra due to the numerous acceptable alternative methods. Additionally, regional variations and individual preferences exist, though the latter should be cautiously approached. Through my work in various Rasa pharmacies, I quickly learned about seasonal variation and the sudden unavailability of ingredients, which can leave one in a predicament. For instance, during one particular copper-processing task, I required an elephant foot yam, typically available year-round but most abundant in late summer. Unfortunately, my locale had none in stock and no future expectations. Fortunately, one of the resident doctors had two unharvested yams from his small allotment, which allowed work to continue with minimal interruption. However, I later discovered that alternative options such as *Limonia acidissima* (wood apple) are much more seasonably available and so aptly demonstrate that degrees in hindsight are two a penny.

Part III, Materials, which I consider one of the most crucially updated parts of this book, now includes a comprehensive section on Lavaṇa, the most critical types of salt used in Rasa Shāstra. Despite being a relatively abundant and humble material, the significance of salt in alchemy cannot be underestimated. Additionally, I have included Grtha, (clarified butter), in both its regular and medicated forms. Like salt, Grtha is an essential component in alchemical processing, along with all Pancagavya (bovine drugs). Both Grtha and Lavaṇa can be considered the cornerstones of purification in the alchemical laboratory.

In this same section, I have also added Langali, an Upavisha[1] plant. The remarkable potential of this exquisite lily, oddly unnoticed during my time in Asia, was likely due to only encountering its dried tuber in pharmacy drawers, an uninspiring sight – it resembles a sweet potato and is often mistaken for one, however consuming Langali is highly toxic and potentially fatal. While I had some knowledge of this plant, seeing it in the flesh rather than this root form created a deeper connection for me. Oddly, it was not in Asia that it

1 *Upavisha* are semi-toxic plants, often used in connection with Pārada (mercury)

crossed my path, but in my own locale, as a visit to a local sub-tropical garden centre had this exquiste vine growing around the main entrance and looking very healthy and prosperous, happily thriving in a cool northerly climate and outdoors! Further evaluation and potential use of this plant have just started to hint at its importance and pharmacological potency.

Needless to say, the entire Materials section has also been enriched and enlivened with vivid photographs and useful additional information.

And so, I am therefore delighted to present this *Revised and Expanded Edition*, which essentially embodies the book I had envisioned writing prior to 2014. While I still wholeheartedly endorse my original version, I firmly believe that this book will gratify those seeking a deeper connection with the elixirs and essences of Indian alchemy.

Andrew Mason, 2024

Disclaimer

The material in this book is for reference work only; it is not intended to be used to treat, diagnose or prescribe. The information herein must not be considered a substitute for professional guidance or consultations with a duly licensed healthcare professional.

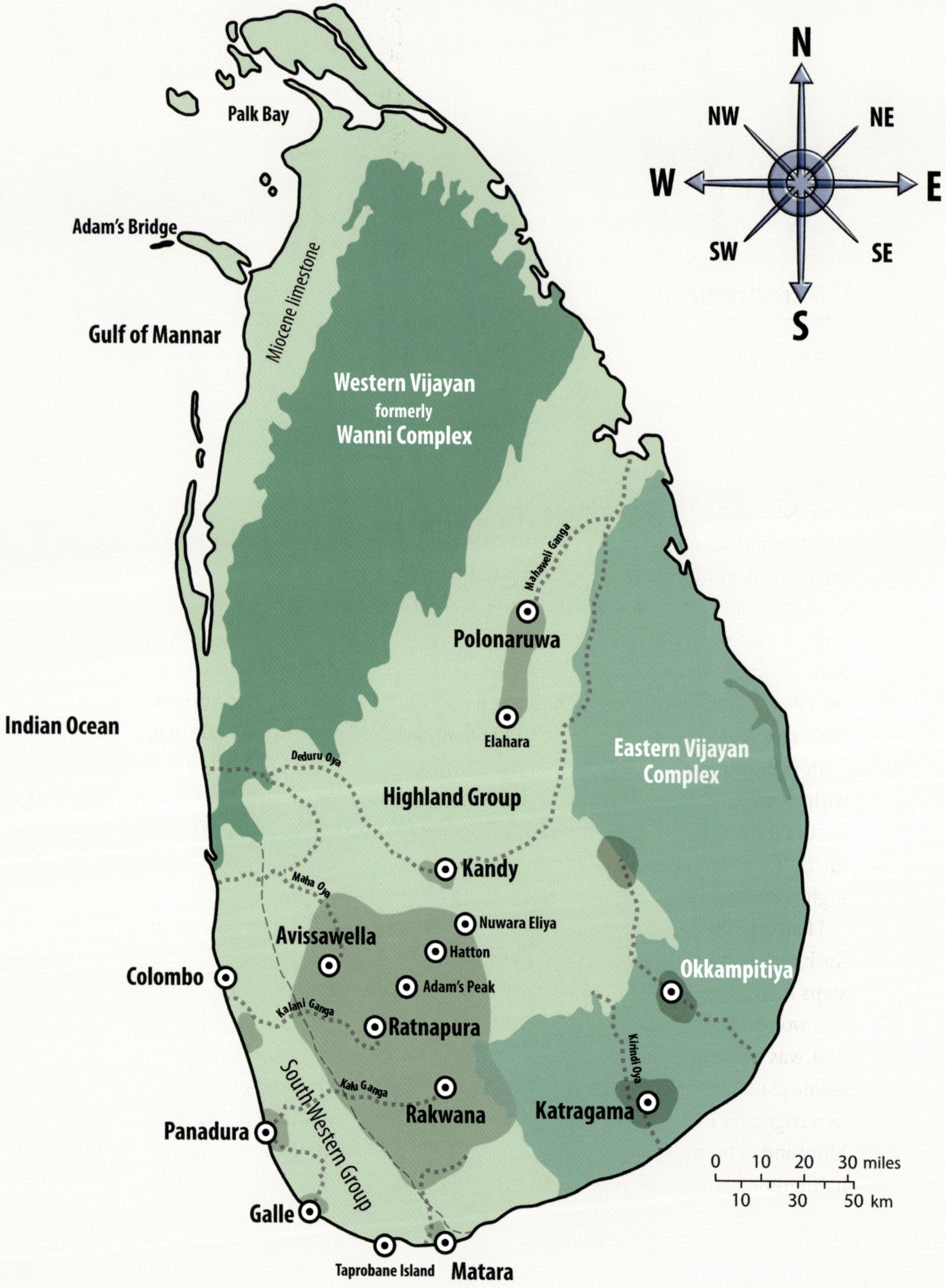
N
NW
NE
W
E
SW
SE
S
Palk Bay
Adam's Bridge
Gulf of Mannar
Miocene limestone
Western Vijayan
formerly
Wanni Complex
Mahaweli Ganga
Polonaruwa
Elahara
Indian Ocean
Eastern Vijayan
Complex
Deduru Oya
Highland Group
Kandy
Maha Oya
Nuwara Eliya
Avissawella
Hatton
Colombo
Adam's Peak
Okkampitiya
Kalani Ganga
Ratnapura
Kirindi Oya
South-Western Group
Kalu Ganga
Rakwana
Katragama
Panadura
0 10 20 30 miles
10 30 50 km
Galle
Taprobane Island
Matara

Sri Lanka

1/ Introduction

> 'Among all types of physician viz. Varuna, Indra, it is Rudra (Lord Shiva), the originator of Rasa Shāstra, whom has been awarded prime position in Rigveda and Shuklayajurveda.'
>
> *Rig Veda*

I awoke around 5:00am to the sounds of dawn prayer at the local mosque; Fajar was around 5:45. Although it was still dark outside, the cacophony of the early morning traffic was growing louder and seemingly nearer. Traffic in Colombo never actually sleeps; it just grows a little more subdued before dawn, as if taking a breath before starting a renewed roar of horns, clattering and shouting. All of this hustle and bustle is intermixed with Buddhist chants on PA systems, played at ear-splitting volumes, adding to the background daily chaos. It sounds unbearable, but after a week, you cease to hear any of it.

In the dim light of the room, I searched for my wristwatch and peered with bleary eyes at the dial. Just past 5:00am, so a little time to prepare myself for the trauma of the coming day. Even at this hour, the air temperature was uncomfortably hot and oppressive, but it was really just limbering up for the high heat of midday.

I had arrived in Sri Lanka in late December 2004 and still remembered how the blast of hot air hit me as I stepped off the plane. I hadn't even made it three steps forward before I began to feel the sensation of swimming in humidity and wondered if I could withstand it for any prolonged period.

It was now mid-2005, actually the 2nd of June, and up to this point, the whole plan had veered so wildly from its original course that I no longer tried to navigate in any meaningful direction. Sri Lanka is a country of unpredictability and extremes; you really do go five steps forward and then end up going half a mile backwards to get to where you need to be. It's also a country of

Far left, map of Sri Lanka

great possibility, spectacular delays and sometimes terrible disappointments but, on the whole, it's a country worthy of exploration.

Sitting now in the relative comfort of my home in the UK writing these words, it's easy to mentally dip back into those days, recalling the many high and low points in minute detail. This day stands out particularly from my time abroad – it was to be one of my more memorable days, seeing the culmination of twelve weeks of hard work preparing just a single alchemical remedy. I had been reading about it for a number of years, and was now about to complete it. The name of the preparation was *Makara Dwaja*,[2] a formulation much favoured in Rasa Shāstra. Getting to this point had been no easy ride, but it had also been an authentic experience, from its inception to its near completion.

I had arrived in Sri Lanka to begin my studies with a local doctor who specialised in Rasa Shāstra; I had met this doctor one year previously and watched her preparing some remedies for the local Āyurvedic hospital. I had been so intrigued that I had asked if she was prepared to tutor me in this subject. Aside from giving me a quizzical look, her reply had been, "How much time can you give?" This obviously implied that tuition could be lengthy or that I looked like a student who might slow the pace after a few days.

At that juncture, I had been about to return to the UK, this initial visit to Sri Lanka having served only as a little reconnaissance to see if I liked the place and could work there for an extended period. I'd already made it to the end of the third year of a four-year part-time college course back home in London. This course had offered students a chance to spend some time in an Indian Āyurvedic facility as part of their final years training, and somehow (never being one to do things the conventional way) I had found myself asking for a placement in Sri Lanka.

First, I'd planned this short trip, no more than a fortnight in Sri Lanka. This was just to get a feel for the place before committing to a longer stay. Regular internship offered by the London-based college was three months. In some way

2 In the Ramayana, the Makaradwaja is described as a hominid/reptile fusion that guards the gates of the underworld. The Macaroon was also a mythical creature fashioned from the head and body of a crocodile, with the tail of a fish. Crocodiles are creatures of great longevity, often portrayed wallowing in primordial waters.

the idea of visiting an island made things seem less intimidating. This in part turned out to be true, but also this deviation from the normal college protocol presented a whole set of new problems and ultimately, opportunites.

It had gone well. I had covered a lot of ground in a short two-week period, connecting with some excellent doctors and seeing lots of interesting things. It was only a few days until my return to the UK and now this new avenue had opened before me and it just felt right. My instinctive reply to her question was, "How long do you need?"

Above, crucible for cooking Nāga Pashana (serpentine)

I left that day with a buzzing head, plans already starting to take shape. I had already decided to complete my Āyurveda training and perhaps quit my job and take a year out, possibly in Sri Lanka or India. The idea of taking time out was not such a concern; it was just that the timing had come upon me unexpectedly. I had felt that something would present itself to me in Sri Lanka, but had not figured it would reveal itself in this manner. I found myself just saying yes, and agreeing to come back in 2004 and start studying. Two weeks of getting the feel for a country and then almost two days before I leave, I find what I came for – I guess that's the way it goes.

Here I should briefly explain why I made this rash decision, though I hope that as the book unfolds, the fascination of the medical system of Āyurveda will become self-evident. Āyurveda had been my primary introduction to India's indigenous medical system, a tradition of several thousand years by Western chronology. As I immersed myself more deeply into this study, it quickly became apparent that there was so much more to this practice than met the eye. One of the most interesting aspects of the work was its emphasis on the implementation of a daily routine to *avoid* disease. This part of the science was exceptionally well developed, emphasising right thought, right action and right diet. Āyurveda also takes positive steps to slow the natural ageing process, seeking to limit its decline through a process known as *Rasāyana*, literally meaning 'to follow the pathway of rejuvenation' – a return to the vigours of one's youth, so to speak.

This use of rejuvenating medicines/techniques represents only one of Eight Branches on the holistic tree of the Āyurvedic science. More on that later on, but it was on this branch that I discovered alchemical medicines. After a little research, this area of Āyurveda started to look very interesting, for here was something quite extraordinary, quietly tucked away in amongst the other healing modalities. I knew at that point that many remedies prepared as Rasāyana drugs contained some mercurial component, but I kept telling myself it must be a mistake or that the use of this metal was overstated. I wondered if it was just a synonym for something else. I was bemused, as most would be, that a dangerous substance like mercury could be allowed anywhere near a pharmacy (unless packed into a thermometer), and yet there it was, consistently mentioned in formulations. I decided to pursue this and made a mental note to ask about it on my travels to Sri Lanka.

After spending two weeks watching several practitioners prepare medicated oils, herbal compounds and other local brews, I had almost given up hope of finding anything that resembled alchemy. During those first two weeks, I'd been lodging at a private hospital and pharmacy. This had its own manufacturing facility, but not locally, so it required a little travelling to witness the sight of traditional medicines being prepared. However, less than a mile from where I was lodging was a large government-sponsored Āyurvedic hospital, and it was to be here that I finally found what I was looking for. While making a quick visit to the main hospital, my guide introduced me to the aforementioned doctor who specialised in Rasa medicines – so I had finally hit the jackpot, a few days prior to departure.

I left Sri Lanka early in the New Year, talking about start dates in July 2004. We'd decided to keep the study arrangements flexible. Still, after returning home, I began to feel really positive about committing to this adventure and almost immediately began to execute my plans.

Once home, I started buying as many books I could lay my hands on regarding Rasa Shāstra, spending many hours picking my way through numerous accounts describing outlandish procedures that made little or no sense. I immediately got the feeling it would not be easy to replicate any of this work without the benefit of some practical experience under the tutelage of an expert. Written information was all well and good, but so many questions came to mind whilst reading – things which were probably self-evident to the writers and so not included, but which were vital for a student.

Among the many formulae, I came across two that, although looking fairly similar, appeared to have different applications. Both were essentially a form of mercuric sulphide, heated at high temperature, eventually rendering a red crystalline powder. One was called Rasa Sindoora[3] and the other Makara Dwaja. The latter I remember seeing mentioned in a book by Dr Robert Svoboda called *Āyurveda: Life, Health and Longevity*,[4] in which he'd written a great introduction to the use of mineral-metal-based medicines. In one particular section called 'Compounds of Mercury and Sulphur', he had outlined the basic procedures of manufacture and their preferred uses.

I had read and enjoyed the book in the late nineties, but back then, most of that information seemed too far removed from the day-to-day practical remedies (in those days just getting basic herbal compounds was difficult). However, it had stuck in the back of my mind as worthy material to return to at some point if and when resources became available. With the prospects of return and study looking imminent, I decided that when I began, I should prioritise these two remedies, obtaining the necessary ingredients upon arrival, and then work on them during my study periods.

2/ Return to Sri Lanka

The following eleven months turned out to be a complete nightmare – escaping the clutches of the UK was not as easy as I had anticipated. Wrapping up affairs at home, negotiating a student visa, making Embassy trips and extricating myself from a business partnership all proved much more frustrating and time-consuming than I had imagined. Then, when it seemed like escape was impossible, doors miraculously opened, and insurmountable problems

3 *Rasa Sindoora* (mercuric sulphide) is chemically (HgS).

4 Svoboda, R.E. (1992) *Āyurveda: Life, Health and Longevity*. London. Penguin Arkana.

vanished. Visas arrived, money moved, papers got signed, and there I was, sitting on a plane at Heathrow about to take off – unbelievable!

In a dream-like state, I arrived in Sri Lanka on the 24th of December 2004, unpacked, made some phone calls and quickly settled into the same accommodation in Colombo. I spent Christmas day in pretty much the same stupor, trying to get my head around what I had just done before the 26th of December brought a world-headlining Tsunami that hit the island's south-eastern coastline, throwing everything into turmoil and a national emergency. Luckily the capital remained largely unaffected, most of the damage occurring along the southerly coastline. It was something of a shock, to say the least, as this event just added to the chaos and uncertainty. Undaunted, I set about trying to track down my contacts and pick up where we left off while trying to fend off concerned phone calls advising that I take the next flight home. However, this was not an option now – hastily retreating from home had meant I'd burnt my bridges and was now stuck here for a while, for better or worse.

After a week or so the island seemed to calm slightly, and things began to feel more relaxed. I eventually tracked down my previous tutor, and almost immediately things started to go awry. I should mention that Sri Lanka is a place where 'yes' can mean 'no', and 'no' can mean 'maybe'. It then appeared that what I had taken to be a 'yes' ten months earlier had, in fact, been a 'maybe', as the doctor concerned now no longer seemed to be sure about the original arrangement. This new revelation called for some extremely fast thinking and skilful negotiation as I tried to salvage a ship that was quickly sinking. I managed to speak with another doctor there with whom I'd spent a little time the previous December; I quickly explained my predicament: my tutor had agreed to teach theory and practical, but she was now no longer in a position to teach the latter. I explained to my contact that, without practical teaching, the study would not, I felt, be worthwhile. He considered the problem and asked for a little time to see what could be arranged. Having worked with him before, I knew he would come up with something and relaxed a little, but I still felt anxious about my next move.

The next day he called back and said he might have an answer; he talked about a trip we'd made in December 2003 to a research institute and one of the doctors I'd met there. Of course my mind was blank, but I stuck with the conversation. He mentioned a female doctor whom I had spoken with at the institute and explained that she specialised in Āyurvedic Surgery[5] and Rasa Shāstra. I then spoke with her at some length, and she agreed to meet me to discuss what might be done to remedy my predicament. He told me she would drop by the next day around lunchtime and to be ready to meet her. I hung up and stood for a while, thinking over what he'd just told me – could I really be this lucky?

The next day, as arranged, she arrived. I vaguely recognised her, and after we spoke, she reminded me of my visit to the research institute the previous year. Thinking back, I could not remember what I had been doing then, but I was amazed I'd not tuned into her and her connection to Rasa Shāstra. Looking back, I just see the whole thing as bizarre synchronicity, as if events had already been set in motion a year before I'd even decided to return and study.

Having the good fortune to cross her path again and in an hour of need, I asked if she might consider instructing me if she could find the time. To my complete amazement, she not only agreed but asked if I wanted to start right then! Like I said, Sri Lanka is a country of extremes – for long periods, nothing seems to happen and then in one afternoon, everything can change.

And so, just like that, we began. I spent about an hour going over what I hoped to achieve and writing down a long shopping list of what I'd need to make it all work. The private hospital where I lodged agreed to give me a room upstairs to work, with access to a courtyard for practical open air work. As long as I was sticking to basic stuff, I had no restrictions, but if I wanted to work on anything more, shall we say toxic, then I had access to their production facility outside of the city. This was a good arrangement as it allowed a lot of flexibility in my programme. The only downside was the rather harrowing journey to the outskirts of Colombo as and when more difficult operations were required. All that set aside, it was a great trade-off; I had got to my practical experience and found a place where I wouldn't get disturbed.

My new tutor was a woman of few words, and to be honest, her English was limited, but practical was practical. My instruction became a case of "I do, you watch," then "You do and I watch." She was busy every day and could not commit to seeing me regularly, so she suggested a compromise. Each time we met, she would give instructions on what to do next, often with

5 Shalya Tantra

a practical demonstration first. Sometimes, particularly at the weekend, she would take time to almost complete a preparation and have me repeat it. Due to the time taken to prepare each material, days in some cases, we would arrange to meet when I had reached a particular stage of the work, and from here she would inspect and assess. If things had gone well, we could move to the next stage; if things had gone awry, I'd have to scrap the batch and start again. If I wanted to double-check anything I could call in the evenings. In hindsight, it seems like a rather odd tutorial set-up, but in our case, it worked rather well.

Concurrently, twice a week, for four hours, I'd also take alchemy theory lessons with the original tutor. This was a great chance to fill the gaps, and there were many, allowing me to raise questions about any current practical projects. All in all, it was a good combination and worked well. Studies, now totally filling my day, helped my mind to focus, as in the extreme heat the body's automatic tendency is to drop to a low gear and coast. This is one of the biggest nightmares to overcome as everything starts to become an effort – just going to the local shop to buy bread or milk can seem like an assault on a mountain with dust blown in your face and everything that moves trying to assault you.

3/ Practical

Every day I felt the pace begin to quicken, and although I was not being chased timewise the days just slipped away. Daylight and darkness in Sri Lanka are roughly of equal length due to the island's positioning, about 8° north of the equator. By 6:00pm the light starts to fade, forcing you to retreat to the fluorescent flickering of neon and the nightly forays of the local mosquito gangs, ever eager for blood. Sunset brought large numbers of fruit bats flying overhead and a light evening breeze on the rooftop of the hospital. After dark, it was back to my room to hide under my mosquito net with a rather unreliable laptop, some CDs, and screen-savers from another life.

My accommodation was just off a busy main road in the centre of the city; the building itself was originally residential but had been converted into a Āyurvedic hospital, clinic and pharmacy. The privately owned establishment was owned and run by a team of doctors, each specialising in their various areas of expertise. These ranged from GP to ENT, gynaecology to Āyurvedic surgery. One unique aspect of the hospital was its manufacturing facility and direct access to freshly prepared medicine. From their dispensary they were able to issue their own medicine directly to patients, as well as trade over the counter with the general public. This obvious bonus was partly why I returned here after my ten-month absence. The hospital was also of special interest because of its independence from government suppliers and its reputation for producing high-quality medicines. Access to their production facility would prove to be of great advantage in the coming months ahead.

On more than one occasion, I got pretty sick acclimatising to life in Sri Lanka and was extremely glad to have a dispensing pharmacy twenty feet from my room! Tucked away on the top floor, I would spend each day working away, preparing materials. The top floor was relatively quiet and not often frequented by staff, but it was also very hot and stuffy. Starting at sun-up, armed with only a pestle and mortar and copious amounts of drinking water, I would stand for many an hour, sweating away the days – hammering away on my personal science project. The daily breaks in the monotony were lunch and tea times (tea being an essential) until at last darkness descended, and the lack of light finally stopped the play.

4/ Journey to the factory

It was now 5:45am, and after collecting my thoughts, and more importantly my materials, I left the safety of my workroom and made for the local bus stop. I'd made this journey many times, but knowing the routine didn't make it any easier. Picking your way on the dimly-lit pavements, avoiding the pot-holes, sleeping people and other unknown obstacles, often made walking on the road more desirable. Many street traders don't move their wares come night-fall, preferring to cover them with a tarpaulin and curl up inside the makeshift nests until morning light.

Waiting by the nearest bus stop, I strained to see any signs of my ride. This was a rather dilapidated 909 Tata bus, the number 177 heading to Rajagiriya, Malabe and Kaduwela. Slowly emerging out of the darkness, its dim headlights

Right, ferryman crossing the Kelani Gaṅgā

appeared, and what had been only minutes before been a deserted bus stop suddenly swarmed with locals who appeared from nowhere. Even clambering aboard speedily I found all the seats were taken, so I took a few moments to arrange my backpack and assorted boxes in an effort to take up less space. After ten minutes of standing and waiting, the atmosphere on the bus began to settle, but this was just the eye of the storm as the best was yet to come. Just when his captive audience was suitably lulled, the driver leapt down from his driving position to begin running up and down the pavement howling at the top of his voice for more passengers to make their way to his already fully packed bus! As incredible as it sounds, he would actually try to pack em, stack em and rack em, till not one more human body could be squeezed into the micro-spaces that remained.

After much groaning and squeezing and the collective weight of the vehicle doubling, the bus finally lumbered and lurched toward its next port of call. Incidentally, it's always a given that the one person who now finds themselves squished in the centre aisle wants to get off at the very next stop (usually about half a mile away), and the moment the bus moves off, they will start to fight their way toward the nearest exit point. I tried not to think about these excursions to Dompe, my final destination, in any great detail as it became quite depressing and, needless to say, it always seemed to take an eternity travelling Kotte Road until the change of bus at Kaduwela. This junction point, some seven miles out, at last saw a reduction in passenger numbers, partly due to a lower population density, and partly because we would be moving inexorably away from the city toward an environment that allowed you to breathe again!

Leaping aboard the number 143 at Kaduwela heralded the final part of the journey, just a few miles along Avissawella Road, winding around a country lane at hair-raising speeds, waiting for the driver to cry "Ranala Junction!" I had travelled this road enough now to spot one or two familiar landmarks, a shop or bus shelter, perhaps even the odd house. Getting off at the right stop was critical – even if these country roads were quieter, they still carried heavy traffic, and none had pavements.

After 7:00am it is light enough to locate a familiar local store and an old dirt track by the roadside that leads down to the Kelani Gaṅgā River and the final leg of the journey – a short hop over the river rapids, and I was there! The distance from Rajagiriya to Ranala Junction is probably no more than fourteen miles in total. By European standards, this would be a laughable distance, but in actuality it was a fairly gruelling trip, taking over an hour and a half and, if wrongly timed, almost double that.

This last part of the journey was generally the least troublesome part of the excursion and involved taking a ferry ride across the river. The ferry itself was a large floating pontoon with a canoe attached to one side. One lone ferryman who feverishly paddled its occupants across the 250-foot expanse of the fast-flowing river provided its propulsion – not an easy feat by any stretch of the imagination.

He was starting to get used to me arriving with boxes and bundles in the early morning, taking up extra space, so I always tried to pay a little extra to balance his lost trade due to my consuming his square footage on the pontoon. The ferryman liked to make each crossing count, so if I arrived alone, he'd usually wait on the other bank until enough passengers showed up. If no passengers arrived he'd eventually paddle over to take me across as a sole passenger, although this was rare. On this particular morning, my timing was excellent and the river seemed calm; I crossed within minutes of arrival and without any hassle. This was not always the case; if the spring rains were

Above, author's production facility in Sri Lanka

heavy, the banks of the Kelani would quickly burst, with the river effectively becoming one-third wider.

I jumped from the pontoon, paid my passage and walked the final half mile to my destination, a hidden factory set amongst dense foliage, well back from the main road. The last 200 yards was a steep driveway snaking up into the property. Finally, I dropped my boxes and backpack on the ground before taking a long drink and reminding myself why I was doing this and how it would someday be worth it!

5/ Setting up shop

By 8:00am, the factory workers had begun to appear. The foreman had already long opened up for them; his own property, where he lived with his wife and daughter and two cows, was built on a small piece of land right next to the main factory. When he saw me he waved and motioned toward another pile of stuff I had left a few days before. I had already started to amass equipment, stockpiling essentials on the factory grounds. After the journey to get there, you can understand why it was wise to travel light and, at the same time, make the most of each trip.

Though early, the temperature was building and time was wasting. It's always wise to move fast and avoid procrastination in Sri Lanka – time there has a mind of its own; hours will seem to go slowly, then suddenly gain speed, whisking away daylight only to unleash the sunset mosquitoes and their subsequent frustration.

Walking over to the main complex, I began setting up my workstation. The largest piece of equipment was a four-ring gas burner that was a two-man lift. I'd already positioned this a few days before, so it only remained to remove its protective tarpaulin and connect up one of three 15-kilo Lanka Gas bottles. I had been reliably informed that the whole process would take twelve to fourteen hours of constant heating.

Lastly, I scouted around for the large earthen pot I'd purchased a week before. These monsters are easily available in the Sri Lanka marketplace and used for all manner of things. Highly durable, sturdy and made from a locally-sourced terracotta-coloured clay, these vessels could be anywhere from 18 to 22 inches in diameter and weighed a ton!

Right, kūpī drying

Once the pot was positioned on the gas burner, I filled it with fine sand. The infill would distribute the heat more evenly. This so-called sand bath is technically referred to as *Vālukā Yantra*. At the centre and buried to its neck was positioned a Vat 69[6] bottle, heavily embalmed in a mud and cloth coated wrapper. This bottle had been prepared over a week ago, wrapped in seven successive layers and dried between each application of cloth and clay. Finally, the outer layer had been smoothed with one final application of fine wet clay. A bottle prepared in this manner is called *kūpī*, a word simply meaning 'bottle'.

6 Its coloured green glass and wide shoulders delivers a higher yield of finished material.

6/ Contents

I've already explained my prior selection of two remedies, Rasa Sindoora and Makara Dwaja, so now seems an opportune moment to explain why I was heating a glass bottle, what was going into it and how I had arrived at this point.

Both preparations are essentially mercuric sulphide, one differentiated by including a thin gold sheet, first forming an amalgam with its mercury content. After stringent purification of the mercury content, a thin sheet of 24-carat gold is triturated with the mercury until a gold amalgam is formed. When this step is complete, both batches of purified mercury are combined with purified sulphur and again triturated until a fine black powder is formed. This black sulphide of mercury is called *Kajjali* (see Part III for more details).

Above, sand bath and bottle

All these processes are time-consuming and labour-intensive, and as I had opted to produce Kajjali the old-fashioned way, i.e. by hand, each operation was a long succession of steps, each requiring a minimum of nine hours. In total, the time taken was approaching forty hours of manual labour. After the preparation was complete, the resultant powders were collected and placed into the bottom of the two glass bottles, which were then temporarily plugged until needed. The fruits of my labour were then buried neck-deep in a sand bath, and their contents were heated to temperatures of around 450–500°c for twelve to fourteen hours. After that, it was really just a case of sitting back and waiting, as once the heating had begun, there was not much one could do except watch. Eventually, after cooling, the contents could be retrieved and examined, but once committed to the heating stage, there was nothing to do except cross one's fingers. During this twelve-hour period there were a number of signs to look out for – marker points that confirmed things were on schedule. These would be a welcome sight; however, failure to spot these markers might indicate trouble brewing and the possibility that all of your work could literally be going up in smoke – a terrifying thought! I nervously ran over one last checklist, lit the gas burner and stood back, preparing for a long day.

7/ Final result

A buried kūpī should be thermally regulated in a sand bath, shielded from the higher temperatures at the base of the pot. From here, the heat is dissipated and percolates up through the sand content. It is hoped that this thermal regulation will eliminate hot spots which might stress and shatter the glass. As the internal temperature continues to climb, Kajjali will liquefy and burn, venting sulphur dioxide (SO_2). A strong odour of sulphur accompanied by yellow fumes will be seen at the mouth of the bottle, indicating that an internal temperature of around 250°c has been achieved. The mercury content of the Kajjali will vaporise at 357°c, being partially evacuated from the heated bottle as mercury vapour (HgO) along with sulphur dioxide. Needless to say, being in close proximity to this exiting gas is pretty hazardous.

Due to the constrictive nature of the bottle neck and its cooler temperature, vaporised material will condense and form stable mercury crystals, carbon and a number of other trace elements (see Part III). It is these silvery black crystals that are ultimately harvested and ground to produce the finished remedy. Over the twelve to fourteen hours of heating, much of the Kajjali will be vaporised, leaving a blackened crust at the base of the bottle. Ironically this residue contains the original gold[7] in the sample of Makara Dwaja. Sadly this residue is usually discarded after the operation, although there are some who reclaim the precious metal. This practice is not recommended, however, due to the essence of gold having already been captured, digested and assimilated by the mercury during the amalgam process. Reclamation at this point would recover a lifeless form of the gold, unable to be used in further operations; however, I suppose the gold would retain its intrinsic value.

During the heating process, the temperature of the sand bath will continually rise, eventually igniting the remaining sulphur deposits at the entrance to the bottle. This will display itself as a bluish flame or smoke, which then fades to an orange coloured flame. This is a sign of critical importance, signifying that the required level of heating has been achieved. Once seen, heating should be discontinued, and the entrance to the bottle should be covered with a piece of ceramic material. Most commentaries on kūpī and mercuric sulphide talk

7 The melting point of gold is 1064°c.

about peacock-blue emissions of smoke, signalling an end to the heating period – that was the theory, anyhow.

As I stood nervously looking at the kūpī I started to wonder if the bottle would endure the heat and remain intact. It's always possible that the heat-regulating sand, cloth and clay embalming and the painstaking work are not enough, and your silence is broken by a sudden 'chink' of cracking glass – announcing game over. It was now 8:30am, and the thought of hanging around for the day just looking at a gas burner and buried bottle neck was a little disconcerting.

Because of her work commitments my tutor could not be present for this part of the preparation; however, she felt all would be well and promised to call later in the day. We'd spent a few hours beforehand going over the whole scenario, the dos and don'ts. The factory had a portable pyrometer handy so I periodically checked the temperature of the sand bath. Finding this a little lower than expected, I arranged some additional sheets of corrugated iron to act as a windbreak, helping the burner to intensify its heat. These afterthoughts worked well, as temperatures started to increase.

The strategy for kūpī heating is often advised to be in three stages. The initial stage (low heat) is designed to bring the whole apparatus to an even temperature, protecting sensitive parts from thermal shock. The second stage (liquefaction) allows the contents of the kūpī to liquefy and boil, eventually burning off the sulphur. The third and final stage (vaporisation) brings the contents to temperatures above 450°c, finally igniting any remaining sulphur residue around the top lip of the bottle. When heating is complete, all equipment is left to cool for twenty-four hours.

Around noon the first visual sign of sulphur fumes began to appear from the submerged bottle. I'd smelled it sometime beforehand but could not detect any yellow vapour. By 1:00pm, sulphur was belching forth from the mouth of the bottle. From my tutor's notes, I knew that these heavy vapours would eventually block at its neck, clogging the exit point. For processing to continue, this obstruction must be cleared, which is best done using a long iron rod carefully lowered into the bottle until reaching its base. The rod is then gently rotated, stirring the contents of the bottle whilst simultaneously clearing the neck of any blockages. The iron rod is then slowly retracted, giving any liquid time to drip back into the bottle before its removal. As soon as this operation was complete, the emissions of sulphur returned. Around this time, the first gas bottle emptied, and I hurriedly swapped bottles before any noticeable drop in temperature could occur.

Occasionally I'd glance up from my work, distracted by voices or a familiar melody broadcast from a small crackling transistor radio on site. Although this was a relatively small factory, obscured by copious amounts of foliage, they produced a good variety of herbo-mineral-based medicines, having a good reputation in Colombo. During my stay I'd had the opportunity more than once to speak to a number of Āyurvedic practitioners who had their own medicinal resources; however, many still bought products from this particular company because of their excellent reputation.

One thing I liked about this environment was its abundance of raw materials; it was never more than fifteen feet in any direction to find wild herbs. Five minutes north landed you in an abandoned coconut plantation, overgrown but very fertile. I remember a few times running low on herbs when processing, only to have the foreman drop a bunch of fresh ones on my lap, not to mention a fresh supply of milk, dung (and cow urine) if needed. His cows were an invaluable resource indeed!

It was now 4:30pm and, again, the bottle neck clogged, requiring an additional stir, but soon after all sulphur activity just seemed to cease, together with the expiration of the second gas bottle. At this point, the heating had to be intensified, increasing gas consumption. Peeking carefully into the exposed bottle neck and using my Maglite on focused beam, I peered down into the opening. This, of course, is highly unadvised, but curiosity got the better of me. At its base, illuminated by torchlight, I could see the red boiling sulphur and mercury confined in its glassy prison – quite an amazing sight; it resembled blood.

At 5:00pm the factory began to empty and quieten down. The doors to the outbuildings were being locked up for another day. I'd long missed my quiet ride back to the city, the mid-afternoon buses being relatively empty. I knew this would be an all-day and early-evening job and even now started to prepare myself for an overpacked late bus ride back. I should mention here that I'd checked in with my tutor and gone over the events of the day; however, she seemed very relaxed about the whole thing and told me to keep going, saying she had not anticipated any problems and that all would be well.

At 7:30pm, after a day of staring at the bottle neck, it finally dawned on my conscious mind that the kūpī had just erupted in a plume of blue smoke.

It took a few seconds for me to react, but leaping forward, I killed the gas supply and watched the blue smoke wither to a blue flame and finally a deep orange flame before being swiftly receding back into the bottle neck. Hunting around, I located the small piece of ceramic tile I had saved and placed it carefully over the entrance to the bottle. Eleven hours, give or take. A last-minute pyrometer check showed the temperature of the sand to be well over 450°c, and the third and final full gas bottle was about to give up the ghost. It was late, it was dark, and I was exhausted! Scrambling around in the half-light, I packed up as much as possible and headed for the factory office. By the time I reached the building, the foreman was just waiting for me to finish. I waved and motioned toward my stuff and gave the thumbs up.

Racing down the drive, trying to fasten my backpack, I started to wonder what time the ferryman usually quit. I'd been told about 8:30pm – but you never know. It would be a complete nightmare to get stranded on this side of the river this late! Luckily by the time I reached the crossing the ferry was still running. Now the fun could really begin – just a case of backtracking my original steps from the morning. As usual it was a pretty stressful return journey, made more difficult by darkness. By the time I found my way back to base camp it was well after 10:00pm.

Above, recovery of mercuric sulphide from bottle

8/ Return

The next day I awoke with a kind of dread and excitement about what might lie ahead. Now it was time to head back and appraise the previous day's efforts. A few hours later, I was back at the factory and it was zero hour. I looked down at the equipment, cold and lifeless after the previous day. This was it – now or never! I cautiously retrieved the bottle from the sand bath and examined its exterior – no cracks! Considering its prolonged exposure to high heat, the exterior was relatively unscathed – a few scorches in places, but overall it was intact.

Using a metal spatula, I slowly teased off the remnants of the cloth and mud lagging. Although the outer layers were now only clay, the innermost section retained some support from residual cloth. Using a pruning knife, I painstakingly peeled away and discarded the last of this cloth, checking all the time for fracture lines. A shattering of glass now would be a disaster.

Freed from its wrappings, the green bottle showed a darkened rim around its neck. If all had gone according to plan this should be the place where the finished crystals would collect. The fact that this area had darkened was a positive first sign. Peering down into the neck of the bottle, I could detect a faint shimmering about the curved walls.

The task of cleaving the bottle proved to be an interesting experience; following my tutor's instruction on a sample bottle, I tied a piece of paraffin-soaked twine about its waist and ignited it. After a moment of burning, I wrapped a water-soaked cloth about it, extinguishing the flames. Following this action, there was a short 'click', and the deed was done. The bottle broke perfectly about its circumference. Once the bottle was parted, it became easy to inspect the inner chamber. The now upturned bottle displayed a dark silvery-red cluster of crystals just below the curvature of its neck. It seemed all the hard work had paid off; however, these crystals still needed to be dislodged and ground before I would know if all had been successful. Although the crystals seemed to resemble the descriptions given to me, I decided to pack up and return to the city to finish the final analysis.

Having arrived back at my accommodation, I unpacked the bottle neck and set to work removing its precious cargo. It was now mid-afternoon, and already another day had almost gone. Using a flat blade, I proceeded to dislodge the crystals from the inner glass wall. Having them drop down into the neck was a convenient way to collect the pieces in a small heap. I'd already stood the rim of the bottle neck on a sheet of white paper, so all I had to do was lift the bottle, and the contents would just drop out.

After prising the last few crystals from their resting position, the remains of the bottle were lifted and the last of its contents were tapped onto the sheet of paper. Not a bad haul – the total weight of the cache was 25g, a fairly respectable amount considering the low dosage required.[8] All that now remained

8 30–40mg

was to grind the material. As each crystal disintegrated in the mortar, it quickly reduced to a fine red powder. Working slowly, I collected the precious material and transferred it into an airtight glass jar, securing its lid. One down and one to go. I now turned my attention to the second batch of Kajjali I'd prepared. I would now repeat the procedure to produce a similar quantity of Rasa Sindoora.

Later that evening, my tutor arrived and examined the contents of the small glass jar, turning the sample over in her hands whilst holding it up to the light. Opening the lid, she put a little of the powder onto her index finger and circulated it between her finger and thumb, studying the powder closely. She then replaced the lid and smiled. "Good," she said.

I asked if she was happy with its colour and texture, not quite sure if I'd actually managed to pull it off. Again, she just smiled, tilted her head in the affirmative, and said "Good." Rummaging through her bag, she finally produced some paperwork wrapped in a heavy cardboard folder. Finding a grouped section of pages, she placed it on the table. "These," she said, "are my personal files, other preparations for you to do. You can do them, if you're quick."

Above, the author preparing Makara Dwaja

Grabbing a pen and paper, I scanned down the list of items, quizzing her about each one and what work might be involved. Obtaining some of these new materials might be a problem, but not an insurmountable one, although I also had to consider any extra equipment I might need. It seemed like there was maybe too much here to accomplish in the time I had left, but after my success I felt a lot more confident about tackling new projects.

All told I was able to purify and prepare just over forty materials during my stay in Sri Lanka, many of these being base ingredients for more complex formulations. Most of the work was to be the conversion of base materials into Bhasma[9] or Pisti.[10] These, however, were not their final destination, merely a stepping stone toward other more complex formulations – often involving a number of other ingredients, including minerals, metals, gemstones and herbs. This is essentially Rasa Shāstra: its complexity seems to grow and grow, but its principles are relatively straightforward. The key to its success is potentiating by purification and reduction. Its ultimate combination with organic and non-organic elements makes this system of medicine unique and ageless, as well as an untapped medicinal resource!

9 *Bhasma* = Alchemical ash (see Part III)

10 *Pisti* = Finely ground powder (see Part III)

Part I

Far left,
serpents

Setting the scene

To set the scene, we introduce a Puranic[11] tale called *Sagar Manthan*, the literal translation of which means 'churning the sea of milk'. This fantastic tale is set in an age of gods and magic and studded with priceless gems from which one can derive a thousand meanings. Indirectly, many elements from this story find their way into various sections of this book, its imagery and concepts hopefully coming to mind as the reader progresses.

This particular story is a great place to explore the concepts of alchemy, poison and astrology, encoding much information relevant to these subjects in its rich symbolism. Stories have always been an efficient way to commit complex information to memory, and this one is no different. In this skilfully woven tale, we are introduced to the origins of immortality, Āyurveda, astrology, herbs, purification techniques, auspicious gemstones and the origins of poison and its ultimate transmutation into *Amṛīta* (divine nectar).

One more detailed account in the *Vishnu Purana*[12] describes events preceding the churning as a period of constant warring.

Sagar Manthan (Churning the Milky Ocean)

The Devas (gods) engaged in endless battles with the Asuras (demons), each seeking victory over the other. Despite their long lives, they were not immune to the effects of time or ageing.

In their desperation, the gods sought an audience with Lord Vishnu, known

11 *Puranas*, meaning 'in ancient times', are a collection of 36 religious texts written in Sanskrit and in story form cataloguing the history of the universe (cosmology), genealogies of gods, demigods and kings, and the cycle of world ages. There is no agreement as to their age; the Puranas were an oral tradition long before being committed to writings.

12 One of eighteen *Mahapuranas*, a subdivision of the 36 religious texts. Vishnu Purana manuscripts survive into the modern era.

Previous pages, Churning the Milky Ocean. (reproduced by kind permission of Alex Florshultz)

as the 'preserver of the universe,' to inquire about their mortality. They wondered if it were possible to attain immortality. Finding Lord Vishnu reclining on a giant serpent, patiently awaiting the end of the current world age, they posed their question.

The gods expressed their concerns to Lord Vishnu, who attentively listened before suggesting they call a truce with the demons, their enemies. He advised them to set aside their differences and join forces to churn the ocean of milk collectively. Only through unity could they unlock the treasures of the ocean, including Amṛīta, the nectar of immortality.

Accepting Lord Vishnu's counsel, the gods negotiated a truce with the demons, presenting their plans and outlining the benefits of the joint endeavour. Knowing the demons' greed would compel them to seize the chance of gaining immortality, the gods strategised to prevent their drive to drink every single drop of Amṛīta, given the chance. Nevertheless, they needed the demons for now, so a truce was declared.

After much deliberation, it was decided that churning the ocean would require a massive object like a mountain. A gigantic serpent, Vasuki, would coil around the mountain, with both sides pulling on its head and tail. Lord Vishnu incarnated as a giant turtle named Kurma, supporting the mountain on his back, allowing it to rotate effortlessly and create mighty waves.

Meanwhile, on land, an argument ensued over who would pull on Vasuki's head and who would handle his tail. The venomous fangs of the snake were unbearable for the gods, who believed the demons could tolerate poison better and thus should take the head. The contrarian demons, however, wanted to pull the tail simply because the gods wanted it. Ultimately, it was agreed that the gods would take the serpent's tail while the demons would handle its venomous head, secretly planning to collect any dripping poison for future use against the gods.

When everything was prepared, the gods and demons commenced the task of churning the great ocean. As their efforts grew more vigorous, the waves rose higher. The thought of obtaining the rewards stirred their determination, leading to even greater intensity in their actions. Finally, the water foamed and turned a milky white. Just then, from the depths, an inky blackness floated up to the surface. This was *Hālāhala*, a terrible poison that had emerged as a consequence of their agitating the ocean.

In an act of selflessness, Lord Shiva rushed forward and consumed the poison, holding it in his throat. Despite its great toxicity, he was able to transmute it (being a master of the alchemical arts) and rendered it harmless. After he did so, his throat turned blue, like that of a peacock, and remained so. The poison, now transformed into nectar, made him immortal, however, some remaining Hālāhala remained toxic, and this he vomited. Upon solidification, this strange substance became like a rock and was known afterwards as *Sasyaka*.[13]

After this episode passed, a light appeared in the water, and the treasures of the ocean were revealed. The first gift to emerge was Kamadhenu, the mother of all cows and the source of earthly nourishment. Then appeared Kalpataru, the wish-fulfilling tree that granted all desires. Following closely was the Kaustubha Jewel, a magical gemstone radiating a brilliant glow brighter than the Sun. Each treasure was swiftly claimed, but as they scrambled to secure their prizes, another prize rose from the waves. First the Moon gracefully appeared, followed by the radiant Goddess Lakshmi, the embodiment of fortune, accompanied by her sister Alakshmi, the embodiment of misfortune. These were closely followed by a troupe of celestial nymphs known as *Apsaras*, then a figure who emerged holding a vessel brimming with the precious Amṛīta, the nectar of immortality. This divine figure was Dhanvantari, the god of healing in Āyurveda.

Above, Sasyaka, also known as peacock ore

With their coveted prize now in sight, the gods and demons ceased their churning and triumphantly seized the Amṛīta from the sea, bringing it onto the land. To commemorate their arduous labour and success, they decided to hold a grand banquet where each of the assembled beings could partake of the immortal nectar.

While the preparations for the celebration were underway, some of the gods gathered to discuss the situation, realising the need for a diversion to distract the demons from the Amṛīta. Having ascended from the depths, Lord Vishnu assumed his human form and stood among the assembly. It was then that the gods expressed their concerns regarding the distribution of the Amṛīta.

13 Copper Sulphate (blue vitriol) appears in the Maha Rasa section of alchemical materials.

The gods explained that if the demons were to consume the Amṛīta, it would lead to an eternal cycle of warfare, with no end to their battles. Since both sides would be immortal, they would constantly regenerate and continue to engage in conflict. Vishnu attentively listened to their worries and confidently approached the celebrating demons, declaring, "Carry on with your celebration as planned – leave the demons to me."

As the festivities commenced, each of the gods savoured the sweetness of the Amṛīta, drinking from the vessel and receiving the bestowed gift of immortality. The Asuras, their eyes blazing with anger, watched with envy and resentment as the gods selfishly chose to consume the sacred nectar first.

Rāhu and Ketu. Rāhu (far left) and Ketu (left)

The demons rose and began to shout, ready to confront the now-intoxicated gods. Conflict seemed inevitable as the demons drew their weapons, preparing to attack.

However, at that very moment an enchanting light, accompanied by a cool breeze and intoxicating scent, swept across the gathering. Before the demons there stood a mesmerising vision of unparalleled beauty – Mohini, the enchantress. As she gracefully moved among the demons, ethereal music played in their ears, and their eyes became transfixed on her heavenly sight. Mohini bewitched the demons with her captivating gestures and sweet melodies, erasing their quarrel with the gods from their minds since, completely entranced by this divine vision, they found themselves unable to divert their attention from the enchantress. Meanwhile, unbeknownst to them, the gods quietly consumed the remaining Amṛīta, nearly emptying the vessel.

However a lone serpent named Rāhu, known for his astuteness and quick thinking, observed the unfolding events from a distance and saw through Lord Vishnu's disguise. In a clever move, Rāhu transformed himself into a god and surreptitiously joined the entourage. Holding the nearly empty vessel of Amṛīta, he noticed a large droplet stubbornly clinging to the bottom. Despite its sweetness and stickiness, Rāhu managed to extract the final drop of nectar, licking it with his long serpent tongue.

Witnessing this horrifying act, the radiant Sun (Sūrya) and the recently born Moon (Chandra) sounded the alarm, alerting the gods to the presence of the impostor in their midst. In response, Vishnu swiftly launched his razor-sharp cakrika (edged disc) through the air, swiftly decapitating Rāhu and separating his head from his body. As the head and body tumbled to the ground, the demon reverted to his serpentine form, exposing the deception to all assembled.

To the horror of onlookers, both parts remained alive and fully conscious as Rāhu's tail writhed in desperation, futilely searching for its severed head. The sip of nectar that passed through the serpent's lips had bestowed upon it eternal life. The head retained the name Rāhu, while the tail came to be known as Ketu, and together they joined the ranks of the immortals.

The gods made futile attempts to capture the fragments of the former demon, but as they reached out for the severed halves their forms became shadow-like, powerless to grasp them. Helpless, they watched as both the head and tail ascended to opposite ends of the heavens, positioning themselves amidst the stars. Miraculously, the blood that flowed from Rāhu's wounds fertilised the ground, giving rise to the growth of garlic and red onions.

Henceforth, Rāhu and Ketu harboured eternal enmity towards those two celestial luminaries that had denounced them, the Sun and the Moon, and during eclipses, in the guise of the lunar nodes, they both devoured the light of these luminaries. While either node is capable of causing an eclipse, Ketu (the southern node) is considered more detrimental during a solar eclipse, whereas Rāhu (the northern node) is believed to induce more disturbance during a lunar eclipse.

Fortunately for the Sun and the Moon, the demonic nodes no longer possess physical substance and can only temporarily engulf or obscure their light. The eighteen-plus year orbit of the lunar nodes brings them in close proximity to all the planets, causing great foreboding and distress during each transit. It is said that during these celestial encounters, Rāhu illuminates certain aspects of future lives, while Ketu offers insights into past lives. Although both nodes are considered malefic in nature, they offer salvation from the cycle of reincarnation, granting Moksha – liberation from suffering.

Section 1

Overview of Āyurveda

1.1/ Ancient technology

> 'Sūrya (the Sun) evaporates the waters of Bhumi (Earth) for eight months in each year. This water evaporates then rains down for four months, nourishing the soils, producing different kinds of cereals for the nourishment of all. Evaporated water also nourishes Chandra (the Moon).
>
> Chandra himself does not consume that water; instead, he returns this water to the clouds. During the winter season, this water is released by Chandra to fall upon the earth as dew and the snows. Sūrya draws its water from Akasha Gaṅgā (Milky Way), also causing it to rain upon the earth. This water is sacred; its mere touch will destroy all disease and wash away sin. Rains that fall during Chandra's transit of the Nakshatras Krittika, Rohini and Ardra come straight from the waters of Akasha Gaṅgā.'
>
> *Garuda Purana*

Much of what we now classify as pseudo-science or ancient superstition actually originated as forms of technology in the past. There existed individuals known as necromancers, rainmakers, oracles, geomancers, abhicāra,[14] alchemists, medicine-men, shamans, and astrologers who possessed a profound connection to the world around them and were constantly vigilant for its signs. Their hope was to establish a bond of trust with nature, enabling them to gain insights into future events. While some of these disciplines have endured into our present century, others have been lost or forgotten. Alchemy was one discipline that has survived, although modern interpretations of this art have

14 *Abhicāra* = the use of magic to afflict others with misfortune.

15 Date based on the passing of Siddhartha Gautama Buddha (480 BCE) and the subsequent appearance of Nāgārjuna 400 years hence (Tibetan Buddhism sources). Modern researchers of Nāgārjuna place him somewhere around the first to third centuries CE.

Far left, Buddhist Ārchārya Nāgārjuna cir. 160 CE[15]

become significantly distorted. Its origins likely emerged from the amalgamation of various practices, including early metallurgy, herbalism, astrology, and geomancy.

This book aims to reintroduce a form of ancient technology that has endured as a living tradition in India, persistently striving to improve the quality of life and, in some cases, even extend it. Known as *Rasa Shāstra* or Indian alchemy, it was originally a unique and distinct discipline before becoming more closely aligned with Āyurveda.[16] Its integration into the indigenous system of medicine in India has added a new dimension to the healing arts, primarily through the introduction of herbo-metallic-mineral formulations.

These remedies were not only available to the sick and dying but also to those robust and healthy individuals who sought to prevent illness or enhance longevity. Over time, Rasa Shāstra ventured into the realm of elixirs that aimed to indefinitely render the human body impervious to the effects of ageing.

When exploring early Indian alchemical literature, one frequently encounters the name *Nāgārjuna*. In the eyes of my esteemed Sri Lankan tutors, Nāgārjuna held a position of reverence not only as an important figure in early Buddhism but also as an accomplished alchemist and philosopher.

Depictions of this enlightened being often portray him deep in meditation, seated upon the coiled body of a massive, multi-headed serpent. The image presented on the previous page is an adaptation of a Tibetan original, infused with a touch of artistic interpretation, of course.

In this portrayal, Nāgārjuna is seated and attired in robes, surrounded by various alchemical *yantra* (devices), many of which we will discuss in Section 6 (Preparation of Medicines). The ethereal figure to his left represents one of the Nāga beings. These extraordinary creatures, possessing a blend of human and serpent attributes, were renowned for their profound knowledge of the occult and fiercely guarded its secrets. Out of their admiration for the aforementioned alchemist, they bestowed upon him the name Nāgārjuna.

Whatever else Nāgārjuna may or may not have been, he is recognised as being both a brilliant philosopher and the author of the *Mādhyamakakārikā*,[17] and has

16 Attempts to put a historical date on Āyurveda places its written history around 500 BCE. India's own oral tradition asserts Āyurveda to be a staggering 10,000 years old.

17 Foundational text of the Mādhyamaka school of Mahāyāna Buddhist philosophy.

become so revered over time that all sorts of hagiographical legends have inevitably been attributed to his life story. Added to this, a number of other historical 'Nāgārjunas' have naturally gotten themselves woven into the same legend.

Nāgārjuna's multifaceted legacy spans various roles and traditions, from that of early philosopher,[18] to the first Patriarch in the lineage of Shingon Mikkyō, and even as a redactor of *Susrutha Saṃhitā*[19] or a later Tantric practitioner of Rasasiddha.[20] Many commonalities persist in association with him across Indian, Tibetan, and Japanese versions; these include his communion with the Nāgā, his advanced knowledge of metallurgy, his possession of thaumaturgical (saintly) powers, his expertise in the use of medicinal elixirs, and the pursuit of extreme longevity. While there may be few direct depictions showcasing his prowess as a physician, there are enough instances to suggest that this was yet another skill to add to his already impressive list of accomplishments.[21]

Within the realm of Rasa Shāstra, one notable classical work attributed directly to Nāgārjuna is called *Rasendramaṅgala.*[22]

The following provides an overview of the key points discussed in Parts 1–4 of this work:

1 The treatment of mercury (Pārada);
2 Mercurials, along with other minerals;
3 Metals, plants (herbal) and animal products;
4 Incineration and the melting of various materials;
5 The fixing or holding of mercury;
6 The manufacture of pills;
7 The treatment of disease caused through the vitiation of the three Dosha;
8 The preparation of fine powders (Bhasma);
9 Detoxification of the body through purgation, enema and *Netra Tarpana.*[23]

18 Nāgārjuna I (160 BCE) founder of the Mādhyamika school of Mahāyāna Buddhism.

19 Nāgārjuna II (seventh century).

20 Nāgārjuna III (ninth century). *Rasasiddha* = one highly skilled in the art of alchemy. The author of this work is almost certainly referred to by Al-Bīrūnī in his book *India* (1030 CE).

21 See *A History of Indian Medical Literature, Vol. 1* (1999) by G. Jan Meulenbeld.

22 *Rasendramaṅgala*, cir. 1300 CE means 'good luck associated with mercury'.

23 Rasāyana of the eyes, through direct application of collyrium or bathed in medicated oil.

1.2/ An early encounter with Āyurveda

I am often asked how I became involved with a system of medicine that incorporates remedies of an alchemical nature. I suppose the best answer to this question is that it found me, literally dropping into my lap back in the mid-1990s. A friend had just returned from a trip to India and threw a well-thumbed paperback at me as he chatted about his adventures. "Here," he said, "thought you'd find this right up your street! I picked it up in a bookshop in India and read in on my travels across the country."

As it turned out, this was my first encounter with Āyurveda and the first of many synchronicities that ultimately led me to be specialised in the study of Rasa Shāstra. The book looked complex at first glance, with lots of strange long words, Sanskrit and terrible graphics, however I persevered with it and eventually started to digest some of what it proposed. Having completed it, I set off looking for more material, interested to see if I could find something along the same lines but written a little closer to home. I started to search for Āyurvedic literature that focused more on its practical applications. What I eventually found set me on a collision course with the alchemy of India.

While wandering around a local bookstore I was amazed to find what, at the time, was a recent publication called *Practical Āyurveda* by Vaidya Ātreya Smith.[24] Its dark shiny cover seemed to just jump off the bookshelf and, having already ploughed through one book on Āyurveda and its concepts, I was quite excited about this new find. Thumbing through its pages in the store I quickly realised this was exactly what I'd been looking for; easy to read, easy to digest information and, best of all, it had only a modest amount of Sanskrit[25] terms (Sanskrit was still an alien language to me in those days, but it was becoming more familiar with time).

While later reading the section on herbal remedies, I came across a reference

24 Smith, Vaidya A. (1999) *Practical Āyurveda: Secrets for Physical, Sexual & Spiritual Health.* Available at: https://www.atreya.com/āyurveda/-Books-in-english-.html

25 Pre-Classical Sanskrit (Vedic Sanskrit or Devanagari) is an Indo-Aryan language dating back to around 1500BCE. Ancient Sanskrit texts are thought to be some of the world's oldest written material.

to a mineral-based medicine called *Shilajit*,[26] an exudate oozing from the Himalayas during the hot summer months. Further reading divulged this material to have been in use for several thousand years and also to be a remedy of highly prized medicinal value. For some reason, it was this commentary that set the needle spinning, catching my imagination. I became extremely curious and set about trying to obtain some of this exudate.

Above, Dhanvantari, the healing god of Āyurveda

The hunt was on, and after a few initial disappointments and dead ends I finally found a company importing the material, not fifty miles from my own front door. A few more enquiries via a phone call to the supplier made the useful discovery that one of the main business partners was not only a medical herbalist but also had some extensive training in Āyurveda, having recently studied in India. This just kept getting better – I could not believe my luck! Over the next eighteen months, I attended several of the foundation courses he offered and found that I really had a bug for Āyurveda and so pursued the next logical step – buying a computer! It was now 1999; I took the plunge and invested in a laptop. For me, the internet had finally arrived. I swiftly began the not-so-slow process of becoming an internet junkie, staring at a flat screen for hours on end and wrecking my eyesight in the process. Back then, connection speeds were painfully slow and every search seemed to take a lifetime, but at last, this new medium paid off, broadening my access to a growing online Āyurvedic community, eager to share ideas, texts and, most importantly, downloadable information! Rasa Shāstra was a little slower to emerge on the internet, but eventually, some material started to filter through. One early contributor to

26 For more information about Shilajit see Section 8 (Minerals), in Part III: Materials, formula and processing.

the digital revolution in Āyurveda was Dr Partap Chauhan,[27] an Āyurvedic physician of some note who had a special passion for Rasa medicines, as well as a strong inclination to actually use them in his practice. He was based in Faridabad and his speciality was the treatment of chronic conditions. My introduction to his work via the internet was to be yet another synchronicity, as I would eventually visit Dr Chauhan when I was in India and spend a few weeks at his clinic and alchemical pharmacy. At that time, however, it would be several years before we even spoke.

By now, I'd built up a decent library of Āyurvedic literature and enrolled on the four-year part-time college course that would bring me, ultimately, to Sri Lanka. As previously touched upon, I ended up spending about ten months there, returning to the UK in the autumn of 2005, laden with books, photographs, extensive notes and a case full of my alchemical samples (forty-two in total). Over the course of the next twelve months I did my best to catalogue everything I was able to do in my time abroad.

When I reviewed my little expedition, it occurred to me that there was enough material to piece together a kind of photo journal of preparations, taking the viewer through each individual process, step by step. When I actually got around to reviewing the material, I realised I had, in fact, more than I could ever use.

All this material eventually manifested as a series of slideshows entitled *Rasa Shāstra – The Art of Vedic Alchemy, volumes 1–4*.[28]

The revised edition of the book you now hold in your hands can be seen as the culmination of seventeen years of research, practical experimentation and additional studies of comparative techniques from around the globe.

---

27 Director of Jiva Institute, Faridabad, Delhi. Dr Chauhan is Jiva's principal Āyurvedic physician and a recognised specialist in the use of Rasa Shāstra medicine.

28 See https://www.neterapublishing.com/epublications.html

1.3/ Rasa Shāstra

अश्माचमे मृत्तिकाचमे गिरयश्चमे पर्वताश्चमे
सिकताश्चमे वनस्पतयश्चमे हिरण्ंयचमे अयश्चमे
सीसंचमे त्रपुश्चमे श्यामंचमे लोहंचमे
अग्निश्चम आपश्चमे वीरुधश्चम ओषधयश्चमे

Left, Shiva Prayer Sri Rudram Chamakam, Sanskrit verse

> 'I am rock, stone and ant hill, five types of soil and sand. I am the mountains and hills, the kingdom of trees, plants, flowers and vines. I am gold, silver, iron, copper, tin and lead, their alloys and their Bhasma. I am glass, I am fire and rain. Healing herbs grow tall when I nurture them, I am Shiva.'
>
> *Sri Rudram Chamakam*

Today, Rasa Shāstra is very much a form of medical alchemy, which uses its fusion of metals, minerals, gemstones, animal products and herbal ingredients to achieve highly medicinal compounds formulated both to cure chronic disease, and for rejuvenation and life-extension. Earlier, and on a more esoteric level, this same science provided a practical guide to the art of gold-making and immortality elixirs. Ultimately it was the medicinal side of this work that prospered, producing an array of highly potent formulas that aimed to relieve humanity from pain and suffering.

In Sanskrit, the word *Rasa* describes many things, including essence, juice, joy, taste, youth, nectar, elixir, flavour, drink, draught, fluid, seasoning, syrup, resin, fluid, love, desire, semen, sap, pleasure and aroma. Within the context of Rasa Shāstra, it is largely a synonym for the metal mercury, also known as quicksilver.

The word *Shāstra* means authority, knowledge, tradition, treatise, science, study, compendium, manual, rule, scripture, tool, method, discipline, body of teaching, excellence and reliable council. The term *Rasa Shāstra*, therefore, might be translated as 'science of mercury'. Legend says that this science was bequeathed to mankind by Lord Shiva himself, he being the deity most directly connected to the alchemical arts. All of the 108 names of Lord Shiva serve as synonyms for mercury, and all may be used interchangeably.

101 names of Lord Shiva
As previously mentioned, all of these names are also synonyms for mercury
Aashutosh | Aja | Akshayaguna | Anagha | Anantadrishti | Augadh | Avyayaprabhu | Bhairav Bhalanetra | Bholenath | Bhooteshwara | Bhudeva | Bhutapala | Chandrapal | Chandraprakash Dayalu | Devadeva | Dhanadeepa | Dhyanadeep | Dhyutidhara | Digambara | Durjaneeya Durjaya | Gaṅgādhara | Girijapati | Gunagrahin | Gurudeva | Hara | Jagadisha | Jaradhishamana Jatin | Kailas | Kailashadhipati | Kailashnath | Kamalakshana | Kantha | Kapalin | Khatvangin Kundalin | Lalataksha | Lingadhyaksha | Lingaraja | Lokankara | Lokapal | Mahabuddhi Mahadeva | Mahakala | Mahamaya | Mahamrityunjaya | Mahanidhi | Mahashaktimaya Mahayogi | Mahesha | Maheshwara | Nāgabhushana | Nataraja | Nilakantha | Nityasundara Nrityapriya | Omkara | Palanhaar | Parameshwara | Paramjyoti | Pashupati | Pinakin | Pranava Priyabhakta | Priyadarshana | Pushkara | Pushpalochana | Ravilochana | Rudra | Rudraksha Sadashiva | Sanatana | Sarvacharya | Sarvashiva | SarVātapana | Sarvayoni | Sarveshwara Shambhu | Shankara | Shiva | Shoolin | Shrikantha | Shrutiprakasha | Shuddhavigraha Skandaguru | Someshwara | Sukhada | Suprita | Suragana | Sureshwara | Swayambhu Tejaswani | Trilochana | Trilokpati | Tripurari | Trishoolin | Umapati | Vachaspati | VajraHastā Varada | Vedakarta | Veerabhadra | Vishalaksha | Vishveshwara | Vrishavahana

Above, the Hindu god Shiva, progenitor of Rasa Shāstra

Rasa Shāstra encompasses two inroads or pathways, firstly *Lohasiddhi*,[29] referring to the transmutation of lower metals into noble metals such as gold and silver, and the second pathway, *Dehasiddhi*, referring to the act of physical transformation, that is, to make the human body imperishable/immortal. The underlying aim of both practices was neither superficial nor materialistic but to establish a harmonious, prosperous and peaceful society through the use of these alchemical techniques.

Rasa Shāstra is very much a living tradition in both India and Sri Lanka, but these days it focuses on bodily rejuvenation, increased fertility and the reduction of senility in older people. This is achieved through the use of both organic and non-organic materials and is referred to as *Rasāyana*, a word meaning 'Mercury's pathway'. Many of these ingredients in their raw state are not normally associated with having medicinal value; however, through a sophisticated system of purification, reduction and refinement,

29 *Loha* here refers to metal in general; Loha can also refer directly to iron.

materials such as cinnabar, arsenic, iron pyrite, copper, zinc, and tin can be suitably processed to yield great medicinal benefits.

Āyurveda also uses the term *Rasāyana* to describe the actions of certain herbs and/or the practice of imbuing formula with regenerative/life-extending capabilities, all of which aim to impart the body with youthfulness. When reviewing the extensive range of Āyurvedic formulations, it is interesting to see just how many formulae have a Rasāyana energetic ascribed to them.

1.4/ Health and longevity in Āyurveda

> 'There is nothing in the world that does not have therapeutic value in the appropriate quantity, condition and situation.'
>
> *Caraka Saṃhitā*

Learning the basic principles of Āyurveda is not so difficult; however, learning to use that knowledge diagnostically can take a lifetime. Āyurveda is a life science, and the best way to understand it is to become a student of life.

Āyurveda recognises the physical constitution to be composed of three active principles that are constantly subject to imbalance. Imbalances can manifest internally as mental suffering, or they can arise externally through physical suffering brought about by an improper lifestyle or environmentally aggravating factors. Āyurveda uses the term *Dosha*, meaning 'that which spoils', to describe the three forces that regulate and maintain the body. When imbalanced, Dosha leads to diminished health and, if left unattended, manifests as various diseases. When the body's ability to repel disease has been compromised, as in the case of lowered immunity, then the end result can be death. In counterbalance to this, when the Doshas are balanced, we are said to be in a state of health and vitality, both of which support a strong immune system.

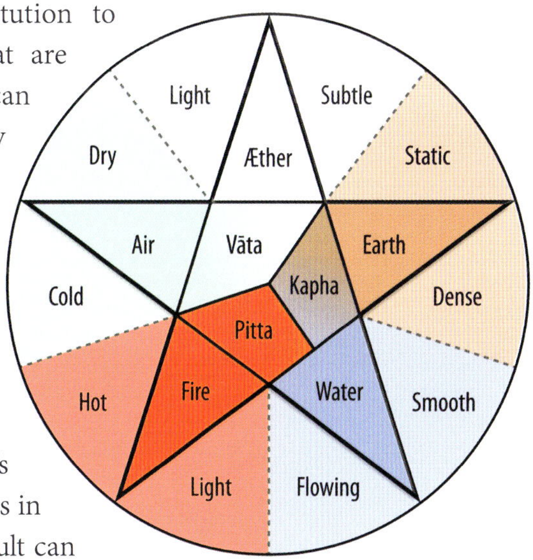

Above, Interaction of the three Doshas, five elements and their qualities in Āyurveda

Āyurveda recognises three types of Dosha: *Vāta*, *Pitta* and *Kapha*, each comprised of an element(s). The ancients recognised five great elements, called *Pañcha Mahābhūta* (see table opposite). Qualities of each element were seen to predominate in different substances, which helped define their attributes. Heavy, dull, dense and fragrant were all attributes of the earth element, while a concentration of this element brings stability, mass and endurance. Moist, slimy, cold and tasty substances tend to have a high concentration of the water element. This element brings cohesion, softness, compactness and contentment. The fire element brings heat, subtlety, sharpness and vision; it produces metabolism, radiance, colour and illumination. The air element brings movement, dryness and roughness; this force provides movement but is also restless and erodes the other elements. Finally, the subtlest of the five elements, æther, brings lightness, diffusion, sound and porosity. Æther provides the space in which all other elements abide.

The three Doshas regulate both the physical and psychological factors in the body, from the minute structuring of the cell to the complexity of mental functioning. Each Dosha is uniquely proportioned, giving rise to the multiplicity and variations of life forms. Their individual signature is called *Prakriti*, more commonly understood as one's 'constitution'. It is this constitution that determines physical appearance, behaviour, preferences and emotions. When the Doshas are aligned, or in a state of equilibrium, the body maintains health and vitality, but, as the name Dosha implies, its very nature tends toward imbalance and impermanence. It is this fundamental philosophy that required Āyurveda to develop a healing strategy based on avoidance rather than cure. This also happens to be the most logical course of action and one that seems to have been mostly forgotten in modern times, as avoiding the trap of an improper lifestyle appears to have been replaced by a 'fall into the trap' mentality.

The table opposite sets out the main attributes of the five elements and their impact on the five bodily senses. It is well worth committing these to memory, as the elemental attributes of the various materials explored in this book will help you understand their significations in relation to the body, its organs and the five senses.

Attributes of Pañcha Mahābhūta			
Sanskrit	**English**	**Attributes**	**Physical senses**
1 Akash	Æther	Spacious, light, refined, smooth, clear, soft, subtle and auditory. It forms an interactive medium for the remaining elements	Ears/hearing
2 Vāyu	Air	Light, cold, rough, mobile, subtle, restless, hard, diffuse and tactile	Skin/touch
3 Tejas	Fire	Hot, sharp, light, mobile, subtle, clear, transformative, assimilating and perceptive	Eyes/sight
4 Jala	Water	Cold, wet, heavy, smooth, dull, cloudy, liquid, and flavoursome	Tongue/taste
5 Prithvi	Earth	Heavy, dry, dull, hard, non-slimy, dense, gross, static, rough and fragrant	Nose/smell

Pañcha Mahābhūta	Dosha	Primary location
1 Akash and Vāyu (æther and air)	Vāta	Colon
2 Tejas (fire)	Pitta	Small intestine
3 Jala and Prithvi (water and earth)	Kapha	Stomach

Āyurveda places significant emphasis on understanding the Doshas and their interactions. Various works detail the attributes of each and how they relate to physiology and structure. The above table provides a brief overview of common associations attributed to the three Doshas, however, as with all diagnostic methods, it is important to not get too lost in the attributes of an individual constitution. Āyurveda states that all three Doshas are always present, and it is their ratio and interplay that give rise to multiple variations. It is the accumulation and/or imbalance of the Doshas that can quickly lead to improper bodily functioning, eventually resulting in one or more pathological

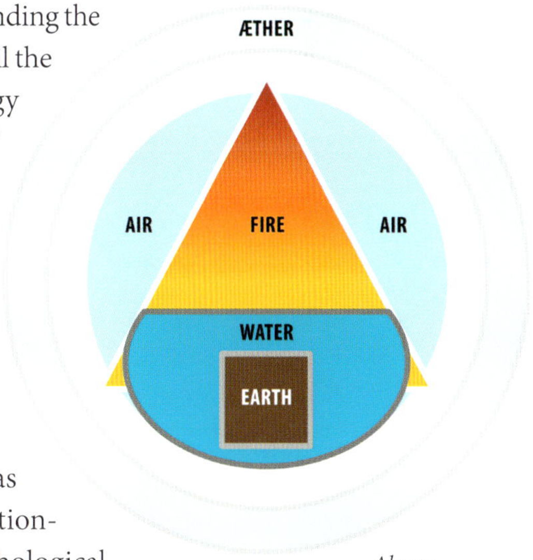

Above, Āyurvedic five elements

conditions. It was these preliminary observations of how the Doshas interacted that were studied to avoid disease. Those observing noted how the accumulation of specific Dosha(s), would increase their attributes in the body, and they carefully documented and addressed these signs and symptoms, hoping to stave off the eventual manifestation of disease. This aspect of Āyurveda can be a great revelation for those used to the Western model of health and disease, since the whole idea that preliminary stages of imbalance in the body's functionality will in any way have later ramifications has become an alien concept.

1.5/ Ṣaḍkiryakalas (six stages of disease)

Āyurveda's unique concept of health and disease sees the tiered system of encumbrance slowly deteriorating on a sliding scale from healthy Dosha functionality to a firmly entrenched disease. This process has been greatly simplified in the table opposite to help convey its concepts.

As can be seen, maintaining the body's optimal health was considered paramount, since fostering an environment of apathy could literally be an open invitation to death in the ancient world. Despite being simplistic in its analogy of the body's functionality, the Āyurvedic view of pathology is fascinating and very easily understood. It is amazing how a few simple corrections in lifestyle and diet can have massive ramifications on general health. By the simple regulation of eating and sleeping habits, adjusting what is eaten according to season, reducing one's food intake and limiting exposure to unwholesome impressions, disturbances in the sites of initial accumulation can be quickly abated. Continually practicing these techniques will often relieve the intensity of a disease, allowing the body to start its slow but sure repair processes.

These simple first steps are often a good way to establish a rapport with Āyurveda, as their implementation quickly improves health while empowering the patient. For my part, I cannot recall ever receiving advice from a healthcare professional that focused on an adjustment in diet or lifestyle peculiar to the symptoms I had presented. Modern allopathic modelling, it seems, favours a scenario where the individual is an aside from the disease; with its eventual manifestation being seen along the lines of a random lottery, where unlucky ones succumb while the lucky remain unaffected.

Doshic cycle and disease process	
1 **Sanchaya (increase)**	Dosha begins to increase, due to improper lifestyle and/or environmental factors. Each Dosha is most likely to accumulate in its primary organs: Vāta – colon; Pitta – small intestine, and Kapha – stomach.
2 **Prakopa (aggravation)**	As they continue to increase, the Dosha(s) become aggravated and start to act upon one another, leading to a suppression of their respective functions.
3 **Prasara (diffusion)**	Now intensified, the Doshas can no longer be contained and begin to move from their initial site of aggravation. Vāta Dosha moves quickly and with irregularity, while Pitta Dosha spreads outward like a fire, heating and burning. Kapha Dosha floods, saturates and blocks the tissues.
4 **Sthana Samsraya (relocation)**	Having now migrated, the Doshas take up residence in new locations and once again begin the processes of pathological disturbance, being fed by their initial site of imbalance, much like an army supply chain. Vāta Dosha, having a strong affinity toward the bones, will tend to concentrate its destructive activities in these locations, often favouring the joints. Pitta Dosha, having a strong affinity toward blood and bile, will tend to concentrate its destructive activities in these tissues, often favouring the mid digestive tract. Kapha Dosha, having a strong affinity toward mucous membranes, fatty tissues and dampness will tend to concentrate its destructive activities in the lungs (respiratory tract) and upper digestive tract (stomach).
5 **Vyakti (symptomatic)**	This fifth stage now sees the emergence of a specific disease with a known pathological pathway. Now well established with a constant source of disturbed Dosha, the disease fully integrates with its surrounding tissues. By this fifth stage the immune system is underactive or encumbered, it poses little threat to the progression of the disease. As the site of disease is slowly destroyed, the still-accumulating Dosha looks for additional fertile ground, ever seeking to expand its territory.
6 **Bheda (progression)**	This sixth and final stage sees the interaction of diseases; the pathways of other vitiated Doshas combine to create new diseases and multiple sites of manifestation. These secondary diseases are like the children of the original imbalances, now taking on their own unique characteristics, adapting to their environments and again diversifying into their own variations of the original disease. This latter stage is what would commonly be called 'complications.' As far as modern medicine is concerned, only stages 5 and 6 would be considered disease states.

> 'A flying bird is unable to avoid producing its own shadow, whichever be the direction in which it is flying. Similarly, no disease is produced without the involvement of Vāta, Pitta and Kapha. Hence the Doshas form the root cause of all diseases.'
>
> *Kāyachikitsa*

Knowing your predominant Dosha(s) will help determine how best to maintain a state of balance in the body. A very simplistic example of this could be that a person with a high degree of the fire element in their constitution will often display symptoms of accentuated heat, anger and feelings of competitiveness when internal or environmental factors cause their inner fire to burn with greater intensity. This might also be exacerbated by living in a hot climate, smoking, drinking alcohol, consuming hot, oily, and spicy foods, or being surrounded by competitive, argumentative individuals.

Even today, despite the fact that we tend to live in climate-controlled homes, you will often find your thoughts becoming more restless when the sky clears and the wind picks up, as *Akash* (emptiness) and *Vāyu* (wind) both increase the air element in one's constitution. With increased spaciousness and movement comes a surge of thoughts, ideas and impressions that stimulate the mind. The end result of this is restlessness. If your constitution also happens to be dominated by the air element, this effect will be felt doubly so, and so on and so forth.

The following tables outline a basic overview of each Dosha and its more commonly displayed attributes.

Attributes of Vāta	
Physique/build	Excessively tall or short, thin with less muscular development. Eating habits tend to be irregular with variable levels of digestion. Elimination also tends to be variable. A light sleeper, but dreams a lot, with frequent bouts of insomnia, worry or restlessness, particularly toward the early hours.
Skin	Cool, dry and rough
Hair	Dark, curly and dry
Eyes	Agitated, small, dry-looking

Temperament	Nervous, fluctuating moods, quick but short memory, highly creative
Energy	Sporadic bursts, over-active then exhausted, poor stamina
Activity	Vāta is most active in the autumn and between the hours of 2:00am–6:00am, then 2:00pm–6:00pm.
General	Vāta relates to the elements air and æther. It is associated with coldness, dryness, ageing, exhaustion and motion. Vāta resides in the colon, but it presides over the mind, nervous system, kidneys and motion of the physical body. Vāta has some affinity with any hollow organ, as well as the bones. It rules the last third of our lifespan, when we tend to lose weight and lessen our grip on material attachments.

Attributes of Pitta	
Physique/build	Medium build with good muscular development. Needs to eat regularly, with quick digestion. Elimination tends to be regular and somewhat loose. Generally a sound sleeper, with violent or vivid dreams.
Skin	Warm and oily yet prone to reddening, freckles or bruises/blemishes
Hair	Light-coloured, straight, fine, slightly oily
Eyes	Sharp, lustrous, prone to redness or inflammation/irritation
Temperament	Goal-oriented, organised, sharp memory and easily irritated
Energy	Moderate stamina, enjoys sports or other competitive physical activities
Activity	Pitta is most active in the summer and between the hours of 10:00am–2:00pm, then 10:00pm–2:00am.
General	Pitta relates to the fire element. It is associated with sight (perception), digestion, oiliness and discolouration. Pitta resides in the small intestine, but presides over the liver, gallbladder and blood. Pitta is the body's internal heat regulator, it has the power to absorb and digest nutrients. Pitta is strongly related to eyesight, as well as our ability to separate truth from falsehood. Pitta rules the middle portion of the lifespan, that period where we assert our will strongly as we strive to achieve and gain recognition for our efforts.

Attributes of Kapha	
Physique/build	Large, firm build, good muscular development. Steady appetite, slow digestion, regular elimination, and large stools. A deep sleeper, seldom experiences or remembers dreams.
Skin	Cool, smooth, moist
Hair	Thick and wavy
Eyes	Large, attractive and moist
Temperament	Calm, tranquil and compassionate, they also have long memories, i.e. they seldom forget injustice.
Energy	Steady with good endurance, yet prone to lethargy
Activity	Kapha is most active in the spring and between the hours of 6:00am–10:00am, then 6:00pm–10:00pm.
General	Kapha relates to the elements water and earth. It is associated with lubrication, protection and endurance. Kapha resides in the stomach, but presides over the lungs and mucous membranes. Like a mother, it protects our bodies from harm, lining the stomach with a protective membrane to contain strong stomach acids. Kapha predominates in the first third of the lifespan, during our formative years when we must grow the most, build mass and attain strength. Kapha represents cohesion, the sense of being grounded; it will slowly move us forward through what appear to be almost endless years.

'Rishis who were formerly residents in communities or nomadic between communities resorted to the diet, lifestyles and medicines of the ignorant villagers. As a result of these poor practices they became interested in the accumulation of wealth and so became lazy and thus no longer maintained their health and were unable to attend to their meditative practices. Realising their mistakes they returned to the Himalayas, back to the gods, the Gaṅgā, holy places and celestial drugs, protected by Lord Indra.'[30]

Caraka Saṃhitā

30 Indra is both King of the gods and an important figure in the lineage of Āyurveda.

The quote on the left can be found in the third volume of *Caraka Saṃhitā*.[31] This volume focuses almost exclusively on rejuvenation practices and paints an idyllic picture of life during the golden age, an age of Rishis, or enlightened individuals. It is inriguing to note how Caraka (the author) was only too aware of the almost instant effect of dietary regimes and environmental factors on daily health. This same volume also considers the impact of the unseen realm, described as infinitesimal or ultra-subtle forces that penetrate our defences and manifest disease within the body. These forces are collectively referred to as *Bhūta*, or what we might term 'spirits'. Caraka describes these supernatural organisms as parasites, that attach themselves to those individuals who offer fertile ground upon which to feed. The presence of Bhūta was thought to be particularly prominent in those cases of insanity, psychosis and depression (known as *unmada*). Some modern commentaries on *Caraka Saṃhitā* also liken these descriptions to an ancient awareness of bacteria and viruses.

The Eight Branches of Āyurveda details the full spectrum of this medical science, each subsection specialising in its own form of personal wellness:

Kāyachikitsa forms a large part of the Āyurvedic corpus, largely covering what we could recognise as internal medicine, diagnosis, diet and daily regimes.

Shalakya Tantra specialises in diseases of the head, eyes, ears, nose and throat. It is a fusion between internal medicine, surgery and ENT.

Rasāyana, previously spoken of, promotes rejuvenative practices through the use of longevity-promoting medicines. Many Rasa Shāstra formulations are to be found in this Āyurvedic category.

Kaumara Bhritya, also known as *Bala Roga*, specialises in the nursing of the newborn, childhood development, child psychology and dietary regimes to correct illness in the young.

Bhūtavidyā, known also as *Bhūtavijja*, was actually closer to exorcism or demonology, and focuses mainly on 'the removal, or the warding off' of

31 Caraka, author of the *Caraka Saṃhitā*, is believed to have composed this sometime between 100 BCE to 200 CE, while others prefer an earlier date of around 500–800 BCE.

malevolent ghosts/spirits. Modern interpretations of this Āyurvedic branch focused more on Bhūtavidyā as a method to heal the mind, a kind of twenty-first century Āyurvedic psychiatry, if you will.

Vājīkarana, also known as aphrodisiac therapy, aims to promote vitality and virility in those wishing to conceive strong progeny. The name *Vājīkarana* means 'to give one the strength of the horse'.

Agada Tantra,[32] known also as *Vishachikitsa*, specialises in the recovery and removal of toxins, via antidote. Poison comes from many sources: animals, insects, plants, metals and minerals. This branch of toxicology was developed to combat both endogenous and exogenous poisons, whether introduced by accident or aken purposely. Agada Tantra also discusses the purification of poisons, used in Rasa Shāstra for the further purification of materials such as Pārada (mercury).

Last, and not least, is **Shalya Tantra**, Āyurvedic surgery. This is perhaps one of the most interesting aspects of Āyurveda and one pioneered by the Susrutha School of Medicine. Susrutha's *Susrutha Saṃhitā* is regarded very highly, and he is held in the same esteem as Caraka. He is considered by many to be the father of surgery and, by extension, modern surgical techniques. This branch of Āyurveda tackles a number of life-threatening conditions such as complications during childbirth, broken or fractured bones, removal of tumours, teeth, wound management, anal-rectal diseases, urinary disorders, eye diseases and cosmetic surgery. A discussion of *Susrutha Saṃhitā* is unfortunately beyond the scope of this book; however, we will discuss some of his blood-letting techniques using leeches in the section on Raktamokshana.

Note: It is considered advantageous to read both *Caraka* and *Susrutha* when making a study of Āyurveda. Whether you wish to pursue surgery or internal medicine, both books are considered authoritative and full of wisdom. The more astute readers will note that I have used quotes from both works throughout this book.

32 *Agada* means disease or discomfort brought about by poison, *Tantra* means a doctrine or method of counteracting poison.

1.6/ Ojas and Sapta Dhātu

> 'Ojas is cooling, firm and contributes to the formation and growth of flesh. It maintains integrity, is mobile, soft and shiny and is possessed of the most efficacious nature. It should be regarded as the most important element, the seat of vitality. The whole body with its limbs and members is permeated by Ojas. Loss or diminution in its natural quantity leads to its eventual emaciation and ultimate dissolution.'
>
> *Susrutha Saṃhitā*

One of the key concepts of Āyurveda is how the body functions in regard to assimilating foods and its subsequent ability to maintain integrity, repair itself, and build tissue. This concept visualised a complex system of channels, each with interlocking dependencies, supporting and being supported by the collective whole. It was known as *Sapta Dhātu,* or seven tissues. The seven tissues in question were *Rasa, Rakta, Māṃsa, Medas, Asthi, Majjā* and *Shukra*, all culminating in the formation of *Ojas*,[33] or vital essence. In this section, we consider each of these seven tissues in turn, but first we will consider their ultimate culmination, Ojas.

As it states in the *Caraka Saṃhitā*:

> 'If Ojas is destroyed, the human body will perish. As bees collect their nectar from the fruits and flowers, so Ojas maintains the human being by virtue of its properties and action.'

Ojas as a substance is of two types – the superior or *Para-Ojas* resides in the area around the human heart, while the inferior *Apara-Ojas* freely circulates in amongst the tissues. The superior portion is fixed in quantity, being a total of eight drops,[34] its colouration is primarily white with a slight reddish-yellow

---

33 Sustainer of life, Ojas is also considered to be *Prana* (breath of life), suspended in a liquid medium.

34 A drop is a measurement calculated to be equal to the droplet falling from one's thumb after submerging it into water and then allowing it to drip.

tint. Para-Ojas is said to be Somātmaka[35] – that is, its nature is like nectar and pure water; it is said to be somewhat oily, cool and stable, unadulterated, and soft.

Its counterpart, Apara-Ojas, was reckoned to be half of one añjali,[36] or roughly a cupped handful. This inferior material is likened to ghee in appearance, having a taste like honey and a smell like fried paddy rice. This form of Ojas was thought to be primarily a distilled form of Rakta and Kapha, whereas Apara-Ojas fluctuates in quantity according to lifestyle and diet. Para-Ojas cannot be increased, and the loss of one single drop would see the termination of life. Direct injury to the heart or the direct action of poison are two possible examples of the sudden destruction of Para-Ojas.

Conversely, Ojas could be preserved and promoted by foods and herbs that promoted the strength of the heart and circulatory system, along with the avoidance of mental antagonism, anguish and nonviolence. Finally, it was recommended that people reside in tranquil settings to support Ojas. Quite simply, Ojas was, and still is, paramount for the maintenance of health; its presence is essential for life. Ojas and life are co-dependent.

Sapta Dhātu

The Sapta Dhātu (seven tissues) are, in effect, the unique filtration and transportation system that feeds Apara-Ojas. They perform their own alchemy by converting ingested foods into living tissue. Like an endless loop in the body, these seven tissues provide a framework upon which the overall strength of the body relies – both saturation, as well as deficiency can be problematic to any individual tissue at any level. It is thought that this cyclic transformation process, from the initial consumption of food, to the final rarefied Ojas, is completed during one synodic month.[37]

35 *Somā* is another name for the Moon, so, in this context, *Somātmaka* means 'to have a moon-like nature'.

36 One *añjali* = two palas or eight tolās (approximately 96ml). The actual amount of Ojas present in each body was assessed by an individual's hand; hence the amount of Apara-Ojas would vary in quantity relative to the size of one's hand.

37 A complete cycle of moon phases, averaging 29.530 mean solar days, or 29 days, 12 hours, 44 minutes and 3 seconds.

Three models of Dhātu formation co-exist and function in unison to provide the body with its nutritional requirements. Though each model could essentially explain the mechanism behind the body's functionality, it is as though all three models run concurrently with one another.

The three mechanisms are:

1 **Kṣīradadi Nyāya** (Dhātu transformation). In this first explanation, each of the seven tissues cook and transform nutrients into a medium that can be accepted and digested by a subsequent Dhātu. The analogy sometimes used is the transformation of milk into curd, then butter, and finally ghee.

2 **Kedārīkulyā Nyāya** (Dhātu transmission). In this explanation each Dhātu operates like a tiered irrigation system, filling and overflowing, with the expanding nutrients filtering down through each tissue, nourishing each subsequent tissue upon direct contact.

3 **Khalekapota Nyāya** (Dhātu selectivity). In this final process, nutrients pass freely between tissues, with each tissue systematically harvesting essential content relative to its respective Dhātu. The analogy often cited is that of grains and seeds being feasted upon by different birds. After taking their fill, each bird will then retreat to its own nesting ground.

Emaciation of Dhātu leads to the emaciation of the body, as each tissue is slowly starved, affecting both the preceding and adjacent tissues. This loss of tissue can be attributed to a multitude of factors, including excessive fasting, over-anxiety, the habitual intake of drying, over-processed foods, over-exposure to one of the six tastes (see Section 1.7), excessive intake of alcohol, prolonged exposure to the elements, loss of bodily oils, excessive sexual intercourse, blood loss, the natural effects of ageing, and finally, extraneous sources such as the presence of ghosts or malevolent spirits (Bhūta).

The following table gives a detailed account of each Dhātu, its ruling Dosha, origin, waste and signs of vitiation. It's worth noting that, for ease of memorisation, Ojas is given as an eighth tissue, although in truth, Apara-Ojas is really the culmination of the previous seven tissues.

Sapta Dhātu	
1 **Rasa (plasma)**	Underlying Dosha: Kapha Origin: heart and ten vessels Mala/waste: Kapha Vitiation of Rasa Dhātu: physical weakness after only minor exertions, chest pains with heart palpitations, a listless mind and hypersensitivity to loud sounds.
2 **Rakta (haemoglobin)**	Underlying Dosha: Pitta Origin: liver and spleen Mala/waste: Pitta Vitiation of Rakta Dhātu: physical dryness of the skin, resulting in cracks and lesions. The skin will lose its glow and lustre, appearing pale and transparent.
3 **Māṁsa (muscle tissue)**	Underlying Dosha: Kapha Origin: ligaments and skin Mala/waste: earwax, tears, nasal mucus, saliva and skin oil Vitiation of Māṁsa Dhātu: diminution (diminishing) leads to thinning of the tissues, especially around the buttocks, belly and neck.
4 **Medas (fat/adipose tissue)**	Underlying Dosha: Kapha Origin: kidneys and greater omentum Mala/waste: sweat Vitiation of Medas Dhātu: diminution leads to emaciation of the abdomen, weariness and dryness of the eyes, cracking of joints and general debility.
5 **Asthi (bone/adipose)**	Underlying Dosha: Vāta Origin: adipose tissue and buttocks Mala/waste: body hair, teeth and fingernails Vitiation of Asthi Dhātu: looseness of joints, improperly formed nails, teeth falling out and loss of body hair.
6 **Majjā (marrow)**	Underlying Dosha: Kapha Origin: bones and joints Mala/waste: none Vitiation of Majjā Dhātu: weakening and lightening of bones, the increase of Vāta diseases (in the hollow spaces), loss of lustre to the eyes (muddying of the sclera).

7	**Shukra (reproductive fluids)**	Underlying Dosha: Kapha Origin: testicles and vulva Mala/waste: lubrication of the eyes, skin and stool Vitiation of Shukra Dhātu: general debility, impotency, mouth dryness, pain upon exertion.
8	**Apara-Ojas (immunity/life-force) culmination of Sapta Dhātu**	Ojas is increased by the consumption of sattvic food, and decreased by the intake of rajasic/tamasic foods. Diminution of Ojas: Ojas is the culmination of all seven Dhātu functioning in equilibrium, rather than a Dhātu in itself. Diminution causes weakness, fear and anxiety, pallor, dryness and emaciation, loss of contentment, and decrease in the effectiveness of the sense organs.

1.7/ The six actions of taste

> 'Lord Punarvasu said, "There are only six types of taste viz., *Madhur* (sweet), *Amla* (sour), *Lavaṇa* (saline), *Kaṭu* (pungent), *Tikta* (bitter) and *Kaṣāya* (astringent)." The source material for the manifestation of all these tastes is *Jalamahābhūta*.'
>
> *Caraka Saṃhitā*

Understanding the actions of taste is a key factor in determining the energetics of foods, herbs, minerals and metals. Āyurveda uses a system of six tastes (see illustration overleaf) that helps categorise the general actions of taste. The basic premise is that a balanced mixture of these tastes should be followed during one's normal dietary regime, as over-consumption of foods predominating in one single taste can lead to ill health through the vitiation of Dhātu and the reduction of Ojas. Often, both herbs and minerals will be described in terms of their predominating taste(s), which helps to better indicate their underlying action upon the bodily tissues.

The six tastes are sweet, sour, salty, pungent, bitter and astringent; all require the presence of *Jalamahābhūta* (water) for *Rasa* (taste) to be apparent. Once sensed upon the tongue, the body's metabolism is stimulated accordingly, allowing foods to be quickly reduced to their respective elements and effects.

The following table gives a detailed account of taste, its elemental composition, its effect on the body and, finally, its planetary ruler.

Note: The inclusion of a planetary association is useful to an astrologer who determines taste preferences according to a client's horoscope, as the dominant planet/s tend to predetermine taste preferences.

The six tastes[38]

Taste	Elements	Effect	Planet
Madhur (sweet)	Earth and water	Heavy, cold, builds tissues	Jupiter
Amla (sour)	Fire and earth	Heating, promotes digestion, builds tissue	Venus
Lavaṇa (salty)	Water and fire	Heating, promotes digestion, improves taste, retains fluid	Moon
Katu (pungent)	Air and fire	Heating, drying, promotes digestion, burns toxins up	Mars
Tikta (bitter)	Air and æther	Light, cold, cleansing, reduces tissues	Sun
Kaṣāya (astringent)	Earth and æther	Drying, cold, alkalising, heals tissues	Saturn

In summary, tastes composed of air and fire have a light, upward motion, whereas the composition of water and earth increases weight and moves energy downward. Combinations of other tastes show a mix of qualities according to their composition of elements.

The six tastes

1 The sweet taste

The sweet taste promotes growth, strength, moisture and longevity. It helps build and maintain all seven tissues. Sweet improves the lustre and texture of the skin, it is soothing to mucous membranes, gives lubrication to the tissues and supports the immune system. Of all the six tastes, sweet is paramount as it basically builds and strengthens. In excess, the sweet taste increases Kapha Dosha and āma[39]

38 Tastes 1–3 are tonifying while 4–6 are emaciating in action. Planet Mercury represents Shad-Rasa, all six tastes in equal quantity.

39 *Āma* is the term used to describe undigested waste, usually the by-product of improperly digested food.

causing obesity, lethargy, low Agni[40] (reduced digestive capacity), parasites, excess mucus and catarrh, and vomiting.

Foods that are prevalent in the sweet taste include milk, honey, fruit sugars, grains, nuts, herbs (such as cinnamon) and vegetable starches. Minerals and gemstones that typically predominate include gold, mica, antimony, lead, iron oxide and alum, jade and cat's eye.

2 The sour taste

The sour taste stimulates the palate and appetite, promotes strength and reduces Vāta. Sourness refreshes the sense organs, nourishes the heart, increases bodily secretions and aids digestion through its overall moistening effect. In excess, it aggravates Pitta Dosha and Rakta (blood), causing oedema/swellings, itching, burning sensations in the chest, vertigo, ulcerations and the suppuration of wounds, turbid urine and the weakening of muscle tissue.

Foods abundant in the sour taste include cheese, yoghurt, sour fruits such as pomegranate and tamarind, fermented wines, pickled vegetables, and tomatoes. The metals *Varta Loha* (five alloy) and ferrous sulphate predominate in the sour taste.

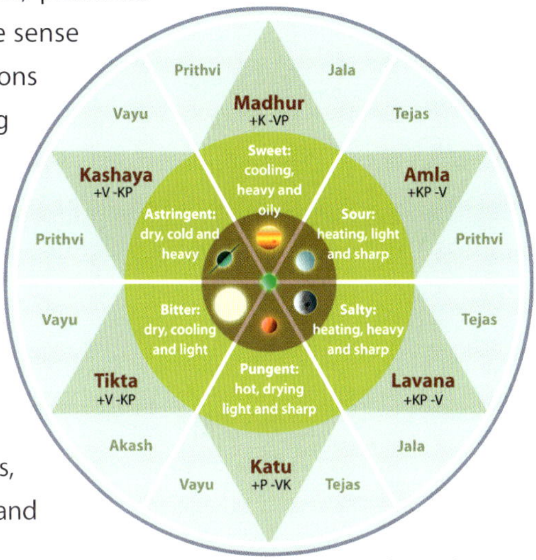

Above, the Āyurvedic six tastes and their associations

3 The salty taste

The salty taste helps digestion through its agglutinative effects, reduces Vāta, reduces accumulation and obstruction (it has a mildly laxative effect), hydrates and softens the tissues and liquefies Kapha. It neutralises all other tastes, improves circulation and reduces stiffness of the limbs. In excess, it aggravates Pitta and Rakta (blood), exacerbates skin diseases, causes early greying of hair, inflammation and stiffness of the joints, morbid thirst, and tooth loss; it wrinkles the skin, causes alopecia (patchy balding) and general ageing. The salty taste increases the desire to eat more and consume, both physically and materially.

40 In the context of Āyurveda, *Agni* is the word to describe your digestive capacity in the form of your allotted inner-fire, residing within the small intestine.

Foods and minerals that predominate in the salty taste include: rock, sea and table salts, processed foods (monosodium glutamate), sea foods and sea vegetables (seaweeds).

4 The pungent taste

The pungent taste promotes digestion and keeps the palate clean. It facilitates waste elimination and burning toxins, reduces obesity, and removes agglutinative substances. It reduces Kapha, kills pathogens and bacteria and helps remove clotted blood. In excess, it aggravates Pitta and Rakta, causing dryness, emaciation, burning sensations, pain in the extremities, bodily tremors, muscle mass reduction, light-headedness and morbid thirst.

Foods that are dominant in the pungent taste include garlic, chilli, ginger, cayenne pepper, and cardamom. Milder pungent tastes might include coffee or ginger. Minerals with greater pungency include copper sulphate, iron pyrite, calamine, cowrie shells, sulphur, arsenic disulphate, arsenic trioxide and cinnabar. Visha and Upavisha plant materials in the pungent taste include aconite, milk hedge and crown flower.

5 The bitter taste

The bitter taste enhances the tastes of other foods, promotes digestion and firms the skin. It dries excess moisture, reduces Kapha and Pitta and increases the removal of mala (natural bodily wastes). The bitter taste is cooling, germicidal and antibacterial, and it reduces skin itching, fevers and burning sensations. Used in excess, it depletes all Dhātu (emaciation), especially Rasa, Rakta, Māṁsa, Medas and Shukra. The bitter taste creates coldness, lightness and dryness of the palate and aggravates diseases of a Vāta nature.

Foods prevalent in the bitter taste include barley, dark leafy vegetables, bitter gourd, and spices such as sage and turmeric. Herbal bitters include: gentian root, neem, coptis, and goldenseal. Bitter tasting metals include bismuth, iron, tin, zinc and bronze. Upavisha plant materials include flame lily, cannabis and croton seed.

6 The astringent taste

The astringent taste reduces bodily fluids and is drying, binding, cold, heavy and stiffening. It reduces Kapha and relieves Pitta diseases, especially those involving Rakta (such as bleeding disorders). In excess, it causes tissue dryness, constipation, abdominal distension, reduced blood circulation, pain in the sides of the chest, weakness of the heart and a darkening complexion. The astringent taste causes spasms, joint stiffness and vitiation of Vāta Dosha.

Foods that typically predominate in astringency include dried pulses, beans, tofu, sprouts, lettuce and alfalfa. Fruits such as apple, pomegranate and plantain are a good source of the astringent taste. Minerals that predominate in the astringent taste include turquoise and red coral. Metals include copper and iron (especially Mandura/rust of iron). Upavisha plant materials dominant in the astringent taste include Dhattura and opium.

Taste remains a significant factor in the energetic determination of plants, animal products, minerals, metals and gemstones. Understanding the actions of the six tastes offers excellent insight into the therapeutic action of the many possible alchemical ingredients.

1.8/ Ṣaḍupakarmas (six catagories of therapeutics)

'One who knows how to reduce, to nourish, to dry, to oleate, to fomentate and the astringent therapies, is the real physician.'

Caraka Saṃhitā

Caraka divides therapeutics in Āyurveda into six categories: *Laṅghana* (lightening), *Bṛṃhaṇa* (increasing/building), *Rūkṣaṇa* (drying/depleting), *Snehana* (oleation/lubrication), *Svedana* (steaming/fomentation) and *Stambhana* (fixing/retaining). Through applying these six principles or three sets of opposites, it was understood that bodily ailments were to be successfully treated. These principles were considered paramount, second only to the three Dosha. If administered correctly with appropriate dosage and in accordance with the season, these therapeutics could eradicate all curable diseases. In as much as Vāta-Pitta-Kapha was responsible for the genesis of disease, so were these six therapies to be considered their nemesis.

1 Laṅghana therapies

Laṅghana therapies involved digestive stimulants, intense physical exercise, fasting, exposure to the elements and the use of certain medicated enemas. Traditionally, treatment to lighten the the body was more effective during the autumn months and on those with a stronger constitution who were afflicted with diseases of excess Pitta, Kapha and mala.[41] Lightening therapies are especially useful for the treatment of skin disease, obesity and diabetes. The energetics of favourable drugs included light, subtle, heating, rough, hard and non-sticky.

2 Bṛṃhaṇa therapies

Bṛṃhaṇa therapies included heavy diets rich with fresh meats, oils/ghee, dairy and sweet foods, warm bathing, nourishing enemas,[42] sleep and comfortable environments (sheltered from the elements). Traditionally, tissue nourishment was more effective during the summer months and was aimed at those suffering from diseases of high Vāta, including emaciation, over-exertion, piles, overindulgence (alcohol, sex, etc.) and the effects of old age. Favourable drug energetics included heavy, gross, cold, smooth, soft and sticky.

3 Rūkṣaṇa therapies

Rūkṣaṇa therapies were fasting, irregular meals, fomentation,[43] sleep reduction and the intake of diets rich in the tastes of pungent, astringent and bitter (oil cake, honey, etc.). Drying therapies targeted diseases that sought to obstruct the channels with Dosha and mala, such as gout, aching joints, cramps and urinary disorders. Traditionally, the drying of the body is more effective during the winter months and with those of a stronger constitution. The energetics of favourable drugs included light, dry, clear, heating, rough, hard and non-sticky. The use of astringent substances (earth and air elements) facilitates drying but does not cause too much lightness.

---

41 Excess bodily wastes including: sweat, mucus, urine, faeces etc.

42 Anuvāsana (Sneha) Basti uses warmed sesame seed oil.

43 Piṅḍa Svedana

4 Snehana therapies

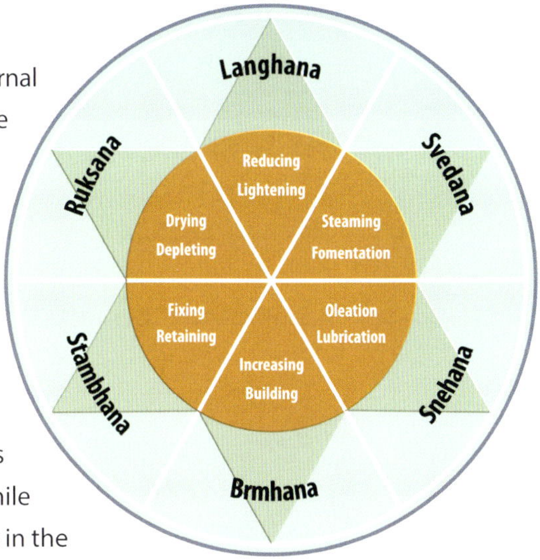

Above, Ṣaḍupakarmas (therapeutics)

Snehana therapies included the internal and external application of oily substances from both vegetable and animal sources. These were *Taila* (oils), *vasā* and *Majjā* (animal fats and marrow; ghee[44]). Oleation therapies target diseases of all Dosha but favour the reduction of Vāta, since Vāta causes dryness of the body, anxiety, fatigue, infertility, addiction (alcoholism) and low immunity.

Traditionally, internal oleation using ghee is most effective during the autumn months, while that of animal fats and marrow is more effective in the spring. The external application of Snehana therapy is usually combined with Svedana (steaming) and internal oils or fats are taken with suitable Anupāna[45] or within medicated formulas. All oleation therapies are contra-indicated during any extreme weather (excessive heat or coldness). Energetics of favourable drugs included heavy, moist, cooling, smooth, soft and sticky.

5 Svedana therapy

Svedana therapy involves the application of external heat to induce sweating. The fomentation of tissues by applied heat directly affects the Medas Dhātu (fat tissue), facilitating the removal of Dosha and mala by opening up the channels of elimination. Svedana therapy targets diseases of Vāta and Kapha, including spasms, abdominal distension, joint pains, constipation, sciatica and limb paralysis. Typically, these treatments come after internal and external

44 *Ghee* (clarified butter) is considered to be the most auspicious oily substance due to its Yogavahi properties, sweet taste, easy digestion and nutritive properties. It reduces both Vāta and Pitta, and promotes intelligence and good complexion. Ghee feeds Rasa/Shukra Dhātus and Ojas.

45 *Anupāna* = a unique 'delivery system' specific to Āyurveda that helps ensure a medicine is directed toward the targeted location for healing in the body. See Section 6:9/ Anupāna: Vehicle of delivery.

oleation, after digestion and in a closed environment (devoid of breezes). However, fomentation therapies are contra-indicated in close proximity to sensitive areas such as the eyes, heart and groin, which should be shielded by cloth, leaves, water and cool hands. Svedana is also contra-indicated with anything involving Pitta Dosha. Svedana therapy favours drugs with heating, weighty, penetrating, liquid, oily and stable energetics.

6 **Stambhana therapies**

Stambhana therapies seek to contract or retain the liquidity of the body, including sweat, urine, faeces and blood. These fixing and strengthening therapies have a direct action on patients afflicted by Pitta Dosha, diarrhoea, vomiting, poisoning or those exposed to excessive Svedana therapies. Stambhana seeks to restore strength to the body and provide relief from the aforementioned conditions. Signs of excessive fixation include constipation, stiffness, anxiety and chest pains. Drugs with cooling, dull, soft, subtle, smooth, dry and static and light energetics are favoured.

These six therapeutic actions similarly apply to the purification of many alchemical ingredients. During the practice of *Shodhana*, a term used to describe the purification or perfection of a material to deliver its optimum therapeutic outcome, many are regularly applied, singularly or in combination. One popular Shodhana of the element sulphur, for example, is to liquefy it using heat and then to nourish it by pouring it into milk, both of which help relieve the sulphur of its inherent toxicity. Similarly, the human body also benefits from Shodhana, which prepares the tissues to receive Rasāyana drugs. By pre-preparing the body, the medicine penetrates more deeply, allowing it to nourish the deepest tissues. Āyurveda uses the analogy of fertile soil, advising the human body to be treated as such – by one who desires to grow crops. Sowing seeds upon denatured, dry, stony or infested ground is unlikely to yield results, whereas soil cleared of pests, turned, fed and watered is likely to produce abundance.

The aforementioned technique is called *Pañcakarma*, or 'five therapies', and is perhaps one of Āyurveda's most important therapeutic strategies, possibly even its flagship. We will delve a little deeper into this in the next section.

1.9/ Pañcakarma (five purification therapies)

> 'Pañcakarma (five purification therapies) bestows happiness to both patients and healthy persons by promoting their strength and longevity, and also by curing their diseases.'
>
> *Caraka Saṃhitā*

Pañcakarma, derived from the Sanskrit words *pañcha* meaning five, and *karma* meaning action, is arguably one of the most powerful ways to eliminate deep-seated toxins from the body. It is a therapy that may be considered when attempting any return to long-term balanced health, provided the body is sufficiently strong enough to undergo the intensive therapeutic purgation of the five actions. The use of Pañcakarma (hereafter PK), should not be underestimated, as it forms an integral part of Āyurveda and Rasāyana therapy; indeed, its use is considered an important preparatory procedure before administering any rejuvenating medication.

As we shall see in later chapters, the Shodhana of minerals, metals, gemstones, and plant material forms the primary action with which to liberate their medicinal properties. According to classical texts it was through this application of elaborate purification procedures that harmful toxins were expelled

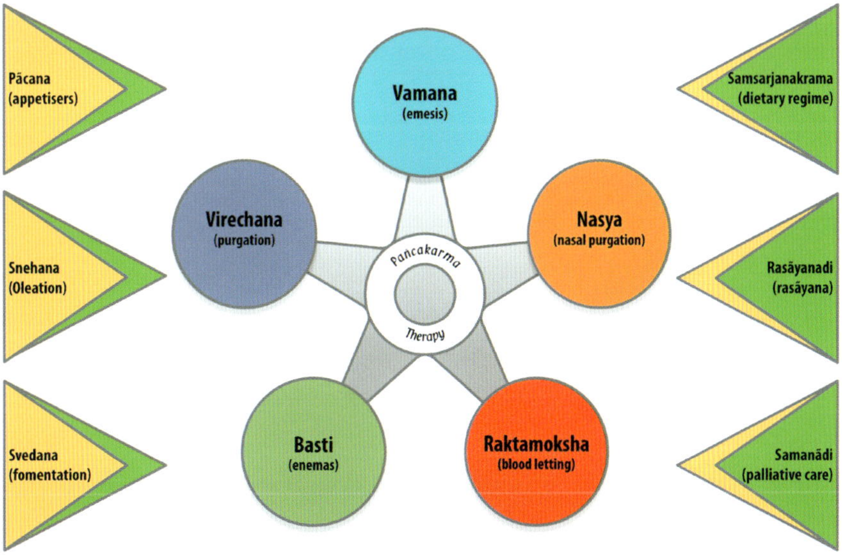

Left, Saṁśodhana Regimen: Purvakarma, Pañcakarma and Uttrakarma

from them, leaving the remaining material highly potentised. These exact same principles were being applied to the purification of the human body, in an effort to remove accumulated toxins, strengthen the tissues and extend life expectancy, however unlike the resilient structures of metals, gems and minerals, the Āyurvedic masters had to contend with the infirmities of the human body, a conscious biological organism complete with a digestive tract, vital organs, five senses and waste products, all of which made the process of purification extremely complicated.

Their eventual compromise was a purification system that worked in three distinct stages. The initial process prepared for purgation, the middle was the actual purging, and the third final stage was the recovery from purgation (see graphic on the previous page). This final stage, a sort of rebuilding or reprogramming of the body, was a combination of dietary recommendations and stringent daily routines and was, in a way, the most critical, as any reduced diligence at this stage could effectively undo all the previous hard work. You might think of this final stage as the 'home straight', that last part of the race where the end is in sight but the finish line has not been crossed. This really is the most dangerous time due to the temptation to coast or lose focus. Anyone reading this who has had this happen will be painfully aware of how badly wrong PK can go when you stray from the path; conversely, if done correctly, there really is nothing to describe how good you feel afterwards.

The uses of Saṁśodhana or rectifying techniques (such as PK) were also seen as a vital prerequisite for the reception of Rasa Shāstra medicines, which is why they have been given space here in the book. It is worth familiarising yourself with these practices as they also offer insight into the principles behind some of the techniques used in Indian alchemy.

Different schools, different styles

Some variations on the five main therapies exist. The classical PK of *Ātreya/ Caraka, Dhanvantari/Susrutha* adhere mainly to similar procedures, differing in the application of *Basti* (enema) and the supplementation of blood-letting therapies. *Keralīya*, a Southern Indian style of PK, details a system which seems to have reinterpreted earlier classical techniques and adapted these methods in a simplified form; these have been briefly covered at the end of this subsection.

Saṁśodhana regimen table (Ātreya and Dhanvantari schools)		
No.	**Sanskrit**	**English & Unani**
Pūrvakarma (preparatory treatment)		
1	Pācana	Digestive appetisers and stimulants
2	Snehana	Internal oleation with medicated ghees. External Abhyaṅga (oil massage)
3	Svedana	Tareeq (medicated fomentation via steaming and sweating)
Pañcakarma/Pradhanakarma (five therapies)		
1	Vamana	Qai (therapeutic vomiting) –K*
2	Virechana	Ishal (purgation) –P*
3	Nirūha Basti	Medicated enema –V*
4	Anuvāsana Basti	Oil enema –V*
5	Nasya (Śirovirechana)	Nasal purgation can be either Virechana (purgating) or Bṛṃhaṇa (nourishing)
Dhanvantari School variation		
3 & 4	Basti	Nirūha and Anuvāsana (as combined treatments) –V*
4	Raktamokshana	Blood-letting Fasad (incision), Hajamat (cupping) and Taleeq (leeching) –P*
Uttrakarma (post-therapy)		
1	Saṃsarjanakrama	Dietary regimen
2	Rasāyanadi	Rasāyana therapy
3	Śamanādi	Palliative therapies
–V* (reducing Vāta Dosha), –P* (reducing Pitta Dosha), –K* (reducing Kapha Dosha)		

The following pages are a breakdown of each of the three stages.

1 Pūrvakarma (preparatory treatment)

Pācana

Pācana helps promote and strengthen the digestive process, allowing the body to burn up toxins and properly convert ingested food, minimising its production of āma. This is usually achieved with the introduction of foods easily assimilated by each Doshic type along with various appetising spices, for example long pepper, black pepper, ginger, cinnamon, cardamom and fennel. To aid in the Shodhana of the body, a number of foods and beverages are to be avoided throughout the process of PK, including dairy (excluding ghee), fermented foods, stimulants such as alcohol, cocoa, caffeine, tea, coffee and recreational drugs, meats, seafoods (including shellfish), certain spices such as chilli and garlic, and the minimisation of salty foods.

Snehana

Snehana means to make smooth by the application of oils, both through diet and by external application. The best dietary oil advised is ghee, whereas certain plant-based oils (such as sesame and castor) nourish muscle, fat and marrow. Ghee[46] and, in some instances, olive oil are, in practice, preferred due to their ability to be digested. Snehana is usually performed over a seven-day period, slowly increasing the dosage in accordance with an individual's digestive tolerance. At the end of this period, the tissues will have become saturated, no longer requiring oleation. Continuation beyond this point invites the re-accumulation of toxins (āma).

Externally, medicated oils are applied liberally to the tissues using regular strokes called *Abhyaṅga*. Cured sesame oil[47] is the oil of choice for this procedure, since due to its slightly warming nature it penetrates deeply into tissue and bone. Sesame is slightly sweet, bitter and astringent in taste, pacifying Vāta without aggravating Kapha. The application is performed in a warm, calm environment, with periodic steaming sessions (Svedana) which help to fully activate the potential of externally applied oils.

---

46 Different medicated ghees are preferred to aid in the digestion of āma, for example *Dadimadi Grtha* (blood and heart disorders), *Vasā Grtha* (lungs), and *Pippali Grtha* (diabetes).

47 Cold-pressed sesame oil is heated on a very low flame (less than 100°c) for approximately two hours until cured.

Svedana

'Even dry pieces of wood bend freely by means of oleation and fomentation when duly applied. Then why living human beings should not also be benefited.'

Caraka Saṃhitā

Svedana (applied heat) was used to induce sweating, dilating the bodily channels and promoting the absorption of oil and the removal of toxins. This was, and is, commonly applied via steam-heated cloths, warmed stones, bolus, poultice or bath.[48]

Caraka Saṃhitā recommends a number of applications for Svedana; these include the use of subterranean cellars heated by fires, with sufficient exhaust points for smoke, and specialised water tubs filled with a mixture of milk, ghee and oils. It also describes a number of elaborate poultices prepared with items such as ghee, various types of meat, oils and horse gram, sand, stone, animal faeces and iron oxide. The above quote from this same work extols the benefits of Svedana, likening a sickened body to that of dried wood. With just the application of warmth and moisture, even dried wood can be coerced to 'bend' without breaking – in this case allowing encumbered channels within the body to freely transport their waste to be expelled.

In *Caraka Saṃhitā* there is an interesting reference to *Nāḍī Sveda* that appeals to physicians who are technologically inclined, advising the use of advanced fomentation equipment. These were actually akin to modern pressure cookers, where properly harnessed steam was administered through a tube and used in conjunction with a potent Vāta-reducing decoction that contained ingredients such as castor seeds, sesame seeds, rice, milk and blood. Where Pitta and Kapha treatments were required, alternate decoctions were given. Apparatus such as *Nāḍī Sveda Yantra* (steaming apparatus) are a common sight in modern Āyurvedic hospitals and are indeed literally sophisticated pressure cookers. They are particularly useful for outpatients that require treatment upon awkward or less accessible parts of the body. During my stay in Colombo, I witnessed several of these localised steaming treatments, used mainly in the treatment of stiff arthritic joints.

---

48 Modern interpretations of Svedana tend to favour the use of wooden steam cabinets or custom-made water-resistant body tents.

2 Pañcakarma/Pradhanakarma (five therapies)

Āyurvedic texts place some emphasis on the correct timing for the undertaking of purgation therapy, specifically depending on how much excess Dosha the patient desires to be purgated. This emphasis is linked to the distinguishable six Indian seasons: Spring, Summer, Monsoon, Windy, Pre-winter and Winter. When reproducing these treatments in the West we correlate each Dosha with the following seasons: Kapha in the Spring, Pitta in the Summer and Vāta during the Autumn.

The original texts recommended the following Indian lunar months: *Chaitra* (March/April) for the removal of excess Kapha, *Shrávana* (July/August), for the removal of excess Pitta, and finally *Mgrashirsha* (November/December), for the removal of excess Vāta.

Vamana

Vamana is the act of therapeutic vomiting. This action rids the body of excess Kapha, and secondarily Pitta, expelling it from the stomach. The ejected material is composed mainly of the water and earth elements. Following sufficient oleation and sweating (four to seven days), the bodily pathways should be suitably lubricated and dilated to facilitate the removal of larger amounts of material. Vamana was not only undertaken in the appropriate season but also in the morning hours, a period dominated by Kapha.

The patient is kept in a warm environment, where they are to consume large quantities of foods that provoke Kapha.[49] Having eaten, they retire early and sleep well before rising early to receive emetic drugs. No further food is consumed, and the patient is kept warm and comfortable until vomiting begins. Vomiting is continued while clear and copious amounts of vomit are seen; however, with the appearance of bile (Pitta), the treatment is discontinued. After Vamana, the patient recovers for two to three hours or until hunger returns, at this point watery rice gruel can be given in very small quantities.

Conditions that benefit from Vamana include cough, asthma, rhinitis, tuberculosis, diabetes, acute fever, skin disease, anaemia, diarrhoea, nausea, poisoning, tumours, piles, heart disease and excess phlegm/mucus.

49 Heavy dairy foods such as soft cheeses, yoghurt and sweetened milk/cream.

Virechana

Virechana is the second Pañcakarma, relating to purgation. This stage primarily aims to rid the body of excess Pitta Dosha accumulated in the small intestine, liver and gallbladder. Although Virechana is the primary means used to eliminate excess Pitta, this method facilitates a secondary reduction of Kapha Dosha. Virechana was undertaken in the appropriate season and later morning hour(s), a period dominated by Pitta. Virechana is preceded by a sufficient period of oleation and sweating, allowing the bodily channels of elimination to become sufficiently lubricated and dilated and facilitating a strong downward purgation. Prior to the commencement of Virechana, the stomach should be empty and the previous meal fully digested. There are a good number of purgative ingredients recommended for this procedure, some of the most popular including Eranda Sneha (castor oil), Trivruta (*Operculine turpethum*) and Jayapāla (*Croton tiglium*). Purgation is usually continued until mucus is seen in the evacuated stool. Generally, this procedure is considered the least complicated to undertake.

Conditions that benefit from Virechana include fever, haemorrhoids, diabetes, ulcers, skin diseases (such as spots and pigmentations), chronic itching, inflammation, constipation, poisoning, gout, chest pains, heart disease, eye disorders and anaemia. Those who have previously received Vamana also benefit from Virechana.

Nirūha Basti and Anuvāsana Basti (medicated enema)

> 'No therapeutic measures other than Basti cleanse the body as quickly and easily, causing both depletion and nourishment instantaneously and free from adverse effects.'
>
> *Caraka Saṃhitā*

Nirūha Basti (decoction) and Anuvāsana Basti (oil) are, respectively, the third and fourth stages of PK. Caraka says that while Vamana, Virechana and Nasya represent fifty percent of the five actions' potential, Basti constitutes the remaining fifty percent of its total medicinal power. Each type of Basti had a specific function to perform within the colon; the action of Nirūha is both purifying and purgating, whereas Anuvāsana produces a solely tonifying effect.

Basti (enema) is the primary means of reducing excess or vitiated Vāta. Traditionally, Basti (either the decoction or oil) was administered through the use of a specially prepared animal's bladder and nozzle. Using this type of apparatus, medicated materials could be easily introduced into a number of orifices, including the anus, urethra (penis and vagina) and certain types of wounds. Basti was undertaken in the appropriate season and in the early morning hours, which are dominated by Vāta.

Nirūha Basti

A number of different ingredients were recommended for Nirūha Basti and included such diverse ingredients as milk, rock salt, and honey, along with herbal decoctions such as Triphala,[50] Dashamula,[51] Vaccha,[52] Sunthi[53] and Shatapushpa.[54] The total contents in this type of Basti could be from 500ml–1 litre.[55] Favourable oils included in Nirūha Basti were castor and sesame.

Anuvāsana Basti

Anuvāsana Basti uses a range of medicated sesame oils in quantities of about 100–250ml. After administrating, the patient is encouraged to hold the oil for as long as possible, sitting quietly and breathing regularly. The longer the oil is held, the more excess Vāta it will attract and collect. After the evacuation of stools and other waste, the patient is advised to rest and take a warm bath before eating a light meal.

Conditions that benefit from Basti include sciatica, infertility, chest pain, tremors, joint pains, abdominal bloating, pelvic pain syndrome, emaciation/atrophy,

---

50 *Triphala* = A mixture of three medicines: Amla, Bibhītaki, and Harītakī, each of which form a synergistic and balanced partnership when combined. See Section 6.5/ Mediums used for Bhavana.

51 *Dasha* means ten, *mula* means root. The ten ingredients are: Bilva (*Aegle marmelos*), Agnimantha (*Premna integrifolia*), Shyonaka (*Oroxylum indicum*), Patala (*Stereospermym suaveolens*), Kashmari (*Gmelina arborea*), Bruhati (*Solanum indicum*), Kantakari (*Solanum xanthocarpum*), Shalaparni (*Desmodium gangeticum*), Prushniparni (*Uraria picta*) and Gokshura (*Tribulus terrestris*).

52 Sweet flag (*Acorus calamus*)

53 Ginger root usually cooked in milk or soaked in lime water prior to drying and powdering.

54 Fennel (*Anethum sowa*)

55 Caraka gives a number of different recipes and treatment protocols dependent upon constitution, season, and age or time of life.

and retention of flatus, urine, semen or stool. Basti also aids in the reduction of obesity, improves complexion (skin lustre) and strengthens Agni (digestive power). It also gives strength and longevity to the body by reducing the drying and ageing effects of Vāta Dosha.

Nasya (nasal medication)

> 'The skin, shoulders, neck, face and chest become thick, well developed and bright; the body parts and the sense organs become strong along with the disappearance of grey hairs by those persons who are habituated to nasal medication.'
>
> *Aṣṭāñga Hṛdayam*

Nasya, or Śirovirechana, is the fifth and final stage of PK, centring on the application of medicated oils into the nasal passage. *Nasya* means 'of benefit to the nose', and the nose was seen as a direct pathway to the brain. Subsequently, medicated oils and powders introduced via this passageway directly pacified all three Doshas, as well as having a positive direct effect on the head, neck and shoulders.

Nasya was of three kinds: *Virechana* (purgatory), *Bṛṃhaṇa* (nourishing) and *Śamana* (palliative). It could be used in the initial stages of Pūrvakarma as a part of the internal oleation process or introduced latterly, during Paścatakarma, to subdue Doshas or help rejuvenate the circulatory channels of the head. Recommended ingredients for Nasya included ghee, milk and medicated oils.[56] Other variations included liquid herbal extracts, dry powders (snuff) and medicated herbal cigarettes.

Nasya was generally performed in accordance with daily predominating Doshas, i.e. early morning for Kapha, late morning for Pitta or late afternoon for Vāta. Nasya treatments were advised to be given in specially constructed clean rooms, devoid of dust and with comfortable seating (reclining if possible), along with receptacles for discharged sputum. Following a light head and facial massage, warm towels and herbal bolas were applied about the neck and head to induce sweating and dilation of circulatory channels. Extracts, oils or liquids were then dropped individually into each nostril, the dosage and material dependent upon individual treatments.

---

56 Medicated oils best suited for Nasya include Anu Taila, Vaccha Taila and Bramhi Taila.

Following inhalation, the patient was then given another light massage and allowed to rest. Any burning sensation in the throat or unpleasant tastes in the mouth were alleviated by gargling with warm water.

Conditions that benefit from Nasya include migraine, blurred vision, rhinitis, ADD (attention deficit disorder), hay fever, tinnitus, diminished hearing, tonsillitis, toothache, stammering, facial pigmentations, cataracts, stiff neck, tetanus, goitre or hoarseness of the voice.

Raktamokshana (medical blood-letting)

> 'A person accustomed to blood-letting enjoys a kind of immunity from all types of skin diseases, sarcomata, aneurism, oedema and disorders brought about by the vitiation of blood, such as ovarian tumour, carbuncle and erysipelas.'[57]
>
> *Susrutha Saṃhitā*

Raktamokshana, or bloodletting via leeches, is less commonly seen outside of India; this is largely due to other more practical incision techniques as well as the fact that using leeches incurs a minor risk of infection.[58] That being said, the use of Raktamokshana in PK is an extremely interesting topic and has therefore been included in this subsection on Saṁśodhana regimens.

Raktamokshana might indeed better be termed 'the art of bloodletting', since it aims to relieve the body of toxins within the blood itself. It is a technique that was developed and adopted by the more surgically orientated Dhanvantari School of Pañcakarma.[59] In this school of thought, the practice of bloodletting was considered a primary therapy. This idea was similarly supported by the master of Āyurvedic surgical techniques, Susrutha, who proposed that the three Dosha

57 A form of cellulitis

58 The gut of the leech contains an endosymbiotic bacterium (*Aeromonas hydrophila*). Like most bacterial species *Aeromonas hydrophila* has been implicated as a pathogen under exceptional circumstances.

59 See: An early encounter with Āyurveda; the opening image shows Dhanvantari, the deity associated with the surgical school of Āyurvedic practice. In this image, his lower hand (left side) holds a leech, denoting the importance of blood-letting therapies. Readers may also recall the emergence of Dhanvantari during the story of Sagar Manthan, the churning of the milky ocean.

Vāta-Pitta-Kapha be accompanied by *Rakta*, or blood tissue. His classical and ancient work *Susrutha Saṃhitā* remains required reading for modern-day students of Āyurveda.

Despite this technique sounding somewhat medieval, perhaps even a little gruesome, it should be remembered that Raktamokshana was an effective means of removing vitiated blood from the body. Two primary methods for incising and exuding small amounts of blood from the body have been given; these were: *Sashastra*, to cut or puncture the skin with a sharp instrument, and *Ashastra*, to pierce the skin via the use of medicinal-grade leeches, known as *Jalukā*.[60]

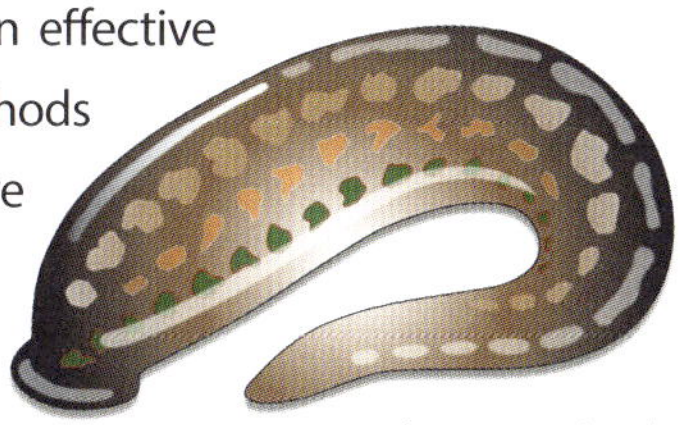

Above, medical leech

The following is a more detailed account of both practices:

Sashastra The wound, once opened, was relieved of blood through a number of different methods, these included *Sirā-vyadha* (venepuncture), *alābu*[61] (dried bottle gourd), *Śrnga*[62] (animal horn), or multiple incisions made by a small sharp instrument, a process known as *Pracchānakarma*.

Ashastra After cleansing the skin, a small amount of milk was applied to the desired locality, or the skin was pricked to induce a small amount of blood to be available. Freshly collected leeches were then positioned and allowed to feed until fully engorged or naturally detaching themselves. There was no specific time given for this, but in practice, leech feeding lasts about thirty to ninety minutes.[63] Upon

---

60 *Jala* means water and *oka* means to dwell in. *Susrutha Saṃhitā* goes into significant details on the type of leech best suited for the task of blood-letting, including physical appearances and suitable environments to collect from.

61 This procedure is synonymous with *Hajamat* (cupping), whereby small incisions were made into the skin and a dried hollow gourd was heated by flame, positioned near the wound. The subsequent vacuum in the gourd sucked the wound, facilitating bloodletting.

62 Usually the horn of a bull, open at both ends. Several incisions are made into the patient, covering the wounds with its broad end. The practitioner then sucks on the apex of the horn to extract blood, placing his thumb over the aperture while recovering his breath.

63 *Susrutha* recommends a wet cloth be kept over the leech while feeding. It is not clear if this was to make the leech feel less vulnerable or, being a water dweller, to keep it moist. Modern leeching practices have noted that leeches urinate excess water via their many kidneys, to allow greater quantities of blood to be retained in their crop.

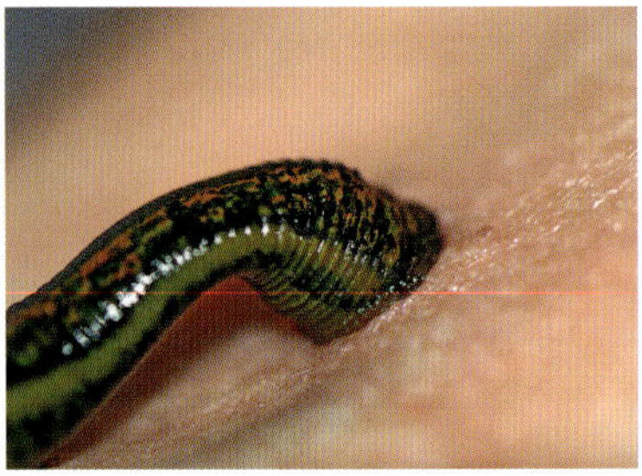
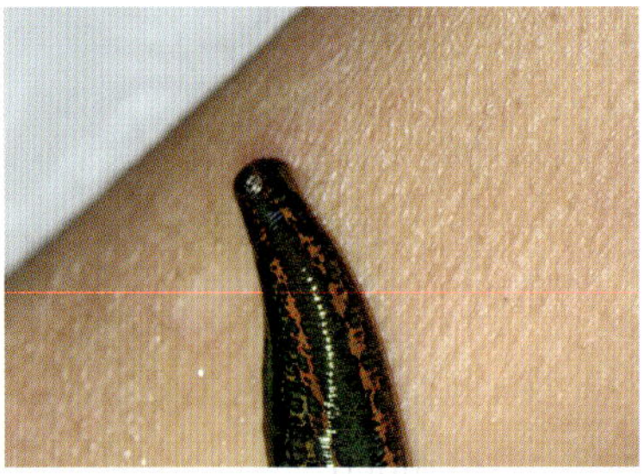

Above, from top right, Hirudo medicinalis *during the initial stages of feeding*

detaching, the wound was allowed to bleed for some time before being washed in cold water. It was then dried before honey was applied to the bite area. In some instances, alkalising powders would be applied to the wound. These powders, known as *Kasáya*, were themselves alchemically prepared, either by calcining barley ears or from slaking limstone. The resulting 'ash' from these preparations was mixed with powdered turmeric and applied to wounds prior to dressing; the natural antimicrobial qualities of turmeric, paired with the slightly caustic properties of the ash would help clean and knit the wound.

Note: *Susrutha Saṃhitā* devotes a rather lengthy section to the application of leeches; it also goes into some detail on the identification of the best qualities of leech. It also advises on the preparation and handling of leeches, and their general therapeutic actions.

In modern times, the identification and classification of leeches that are most favoured for medicinal purposes are *Hirudi medicinalis* or the European medicinal leech, *Hirudo granulosa*, the preferred Asian leech (used in India) and finally, *Hirudo nipponica*, another variety of Asian leech, favoured for internal consumption, usually dried and taken in the form of medicinal tea.

Raktamokshana benefits both traditional and modern conditions including skin disease, skin discolourations, particularly of a bluish colouration, boils/inflammation, migraines, the reduction of cysts and tumours, diabetes and glaucoma, blood pressure, gastritis, gout, arthritis, varicose veins and Pitta-type (very high) fevers.

I also want to draw the reader's attention to the saliva of the leech. Something like twenty-plus unique substances within the saliva have been identified and named to date, with many more still awaiting identification. As well as having

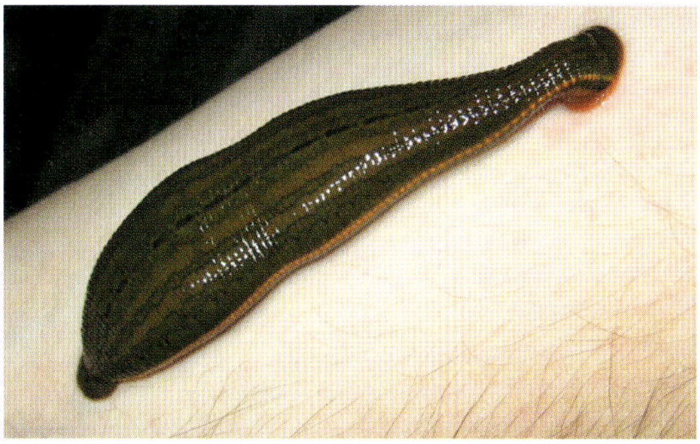

Above, Hirudo granulosa *during final stages of feeding*

vasodilator (anticoagulant), analgesic, and anti-inflammatory properties, there remain a number of unidentified substances simply labelled *Factor X*. These have demonstrated the ability to repair varicose veins or improve other venous diseases. It may well be that the saliva of the humble leech is as much a contributory factor in health benefits as the actual bloodletting process itself.

Although the West no longer practices medical leeching treatments such as those I have outlined, modern use of the leech continues in many instances, for example reconstructive surgery, where blood flow in newly repaired or reattached limbs, fingers or toes is facilitated by the highly efficient biological micro-pumping action of the leech. To date, the use of leeches in certain types of reconstructive surgery has been highly beneficial, as well as in cases of venous congestion, deep vein thrombosis and osteoarthritis (usually the knee and elbow). Other areas where leeches are used are hearing loss, tinnitus, shingles, abscesses, bruising (hematoma), rheumatic and inflammatory conditions, trapped nerves and muscle tension. All of these benefit from the application of the humble leech.

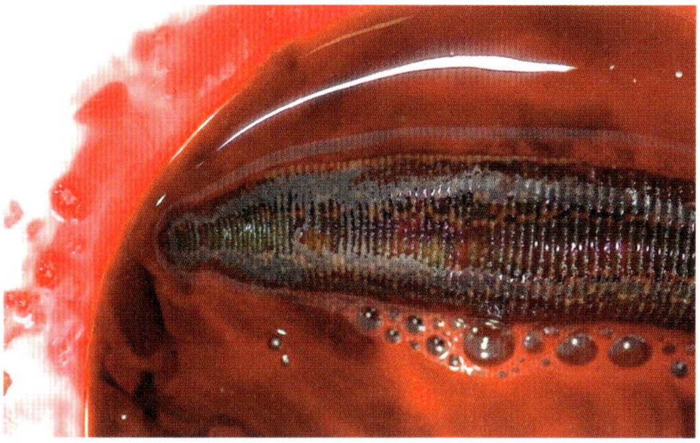

Above, Hirudo medicinalis *during the removal of blood*

Raktamokshana is an intriguing therapy, and during my stay in Sri Lanka I was able to observe it firsthand at a local Āyurvedic hospital and also at the clinic where I was staying. A visiting doctor, an ENT specialist, allowed me to participate in the procedure during one of his weekly clinics, fulfilling my curiosity to experience the entire process. The procedure took approximately an hour and was largely painless apart from the initial leech bite. Afterwards, the bite wound was a little tender and itchy but healed very quickly.

You may be curious about what becomes of the leech after its use. For the sake of hygiene and to prevent possible infections, they are usually incinerated. However, traditional practices still use a technique known as *Jalukā Shodhana*. This involves placing the leech in a copper vessel filled with eight pala (about

300ml) of mineral (not tap) water, mixed with around five grams of turmeric powder. The lid is sealed, the water is changed regularly, and the leech is stored until required for another Raktamokshana session. According to the texts, turmeric in the water causes the leech to salivate and so clean itself in preparation for the next feeding. Of course, it's one leech for one patient; your own designated leech follows you through the treatment process.

In the subsequent years, I have become even more interested in Raktamokshana, and with the help of a successful Āyurvedic surgeon I have been able to go a little deeper into the application and methodology of this therapy. It is worth noting that the United Kingdom, or more specifically Hendy, South Wales, is home to Biopharm Leeches[64] – innovators in the cultivation of medicinal leeches, who currently supply a large majority of the leeches used worldwide in modern medicine.

3 Paścatakarma/Uttrakarma

> 'Freshly hatched eggs are to be handled with tenderness as is an oil pot filled to its brim or cattle protected by their herder. Similarly the physician should carefully protect the patient from unwholesome diets and regimens.'
>
> *Caraka Saṃhitā*

Saṃsarjanakrama

This prime post-therapy phase addresses the patient's dietary regimen and the gradual re-introduction of simple foods after each of the five purgative therapies. *Saṃsarjanakrama* is nothing short of the rekindling of one's Agni (digestive fire) and the rebuilding of Ojas – the essence of vitality.

This phase of the procedure is highly precarious and one in which the patient should be carefully watched over and administered to. Incorrect restoration of Agni (and Ojas) can not only undo all the effects of the purgative therapies but also potentially trigger a whole string of new conditions in the recovering patient.

Following strong purgation, the body enters a state of exhaustion. Digestion is impaired, ligaments and joints feel loose and the body hollow. To this end, Caraka recommends a gradual reintroduction of nourishment, starting with *peyā* (watery gruel) and slowly building to meat soups. As one's digestive power returns, the pleasing tastes of sweet, sour and oily can be reintroduced. Later on,

64 See: https://www.biopharm-leeches.com/

sour and saline are added, followed by sweet, bitter and, finally, astringent and pungent tastes. The general rule here is that a 'light and less' kindles the digestive fire and hunger like a small flame is able to devour tinder. As the strength of Agni grows, more substantial fuels can be reintroduced into the diet, but as mentioned previously, this is a dangerous time for the patient who will crave all six tastes, and in large quantities. Giving in to these tastes would be a terrible mistake, as the sudden diversity of tastes and heaviness (quantity) of foods would stifle the digestive capacity as surely as a wet blanket will smother a fire.

Rasāyanadi

'A person undergoing rejuvenation therapy attains longevity, memory, intellect, freedom from disease, youth, lustre and complexion. One attains potentiality of the body and sense organs, the power of speech (vāk-Siddhi), respect and brilliance.'

Caraka Saṃhitā

It should be borne in mind that Āyurveda sprang from a culture deeply enmeshed in the lineage continuity and the potential for strong offspring. Individual kingdoms relied on the stability of their monarchs and heirs, so it is of little surprise that PK and rejuvenative therapy were predominantly reserved for royalty and the super-wealthy and that two Āyurvedic treatment branches were dedicated wholly to Rasāyana and aphrodisiacs. *Rasāyanadi* focuses on the application of these life-promoting medicines in an effort to regain lost youth and restore vigour. After undergoing the aforementioned therapies, the body waited like fertile land, to be newly seeded.

The introduction of Rasāyana drugs was best considered a protracted process, starting with low doses before slowly increasing both quantity and frequency. This practice would continue over several weeks and months until reaching the required dosage. At this point, the dosage would be reduced over a similar time frame, eventually arriving back at the original dose.

Caraka Saṃhitā Cikitsāsathānam gives us many longevity-promoting formulae,[65] and this section of the classical work is replete with diverse Rasāyana recipes, some promising a century of life while others offer a more generous 10,000 years of longevity. It is particularly relevant to this book due to its inclusion of

65 Chapters 1.4 and 2.4

elixirs and 'celestial drugs', some of which include metals, minerals and herbal components.

The Rasāyana section in *Cikitsāsathānam* primarily focuses on herbal materials such as Harītakī (*Terminalia chebula*), Āmalakī (*Emblica officinalis*), Bibhītaki (*Terminalia belerica*), Haridra (*Curcuma longa*), Pippali (*Piper longum*), Vaccha (*Acorus calamus*), Nāgabalā (*Sida veronicaefolia*), Bṛāhmi (*Bacopa monnieri*), Bhallātaka (*Semicarpus anacardium*) and Guggulu (*Commiphora mukul*), but it also includes the use of curd, pasted til (sesame seeds), milk, ghee, honey, oils and jaggery. This same section also mentions recipes that contain metals such as gold, silver, iron and copper, and minerals such as copper pyrite, copper sulphate, sulphur and Himalayan Bitumen.

Cyavana Prāśa

The Āyurvedic pharmacy is divided into a number of important categories, including Kwatha (herbal decoctions), Tailam (medicinal oils), Lepas (medicated balms) and Avaleha.[66] The latter are often a blend of pulped fruit, decoctions, honey, cane sugar and an assortment of herbal powders.

Although there is no absolute agreement on the exact recipe of Cyavana Prāśa Avalehsa, it requires about 35–40 ingredients, of which a high percentage are Āmalakī, ghee and jaggery. The finished paste is aromatic, sweet, sour, pungent, bitter and astringent, with a post-digestive sweetness. Considered to balance all three Doshas, this formula is typically recommended when undergoing Rasāyanadi therapy. It promotes strength, aids in respiratory weakness, alleviates cough, colds and symptoms of gout, improves digestion, builds blood in cases of anaemia, strengthens the heart, and, of course, is an excellent Rasāyana.

Rishi Cyavana's tale

Within the Āyurvedic pharmacy, Cyavana Prāśa is perhaps the most renowned Avaleha. Its name derives from the Rishi, Cyavana, who acquired the formula from the Ashwin Twins. Numerous stories and events surround its legendary emergence into the world, and below is just one example.

Rishi Cyavana had dedicated a significant portion of his life to the pursuit of enlightenment and seclusion. Immersed in deep Samadhī, his unwavering

66 *Avaleha* = medicated jams

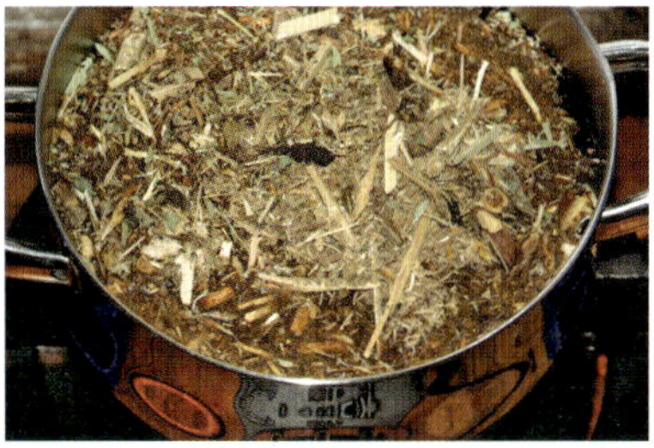

Above, from left to right, various stages of Cyavana Prāśa being prepared

devotion caught the attention of the gods, who, growing anxious about his extreme commitment, feared that it might eventually grant him rightful access to the heavenly realm. To prevent this outcome, they summoned Menakā, the most exquisite among the Apsaras (celestial nymphs), and gave her a task: to divert the elderly man from his meditative endeavours.

Traditionally known as 'those moving in water,' Apsaras frequented the rivers and lakes of the world, often assuming the form of exotic birds. Menakā transformed into her human form, and embarked on her mission to locate the aged Rishi. Her plan was simple: she stealthily cleared the undergrowth around where the meditating sage sat and encouraged fragrant flowers to bloom. Additionally, she arranged a comfortable resting place for him and provided him with simple yet flavourful sustenance. Day after day, without uttering a word of complaint, she faithfully repeated these tasks for months on end.

As time went by, Cyavana found himself gradually drawn away from his deep meditation, growing increasingly curious about his mysterious attendant. Eventually, Menakā revealed herself, unveiling her immense kindness and breathtaking beauty. Complicating matters further, their prolonged proximity had also captivated Menakā, leaving both of them hopelessly in love. However, a predicament arose – they were trapped in a love that seemed impossible to fulfil. Cyavana, being mortal and aged, contrasted with Menakā, who was immortal and eternally youthful.

Confronted with their dilemma, Cyavana made a decision. He temporarily set aside his spiritual practice and ventured deep into the forest to undergo six months of *Pañcakarma*, a purification and rejuvenation process. At its culmination, he would prepare a potent elixir that could reverse ageing, thereby transforming him into a young man once again. Recognising that such an elixir could only come from a divine source, Cyavana approached the horse-headed Ashwin Twins, divine beings renowned for their mastery over medicinal and regenerative powers. The twins had already proven their ability to retrieve souls from the afterlife using the powerful Somā drug, so restoring youth to an old man should be a relatively simple task.

Through skilful negotiation, partly helped by their acknowleging the austerity of Cyavana's life, the twins were persuaded to share their formula for restoring youth. After completing his Pañcakarma, Cyavana diligently prepared the recipe, resulting in the creation of *Cyavana Prāśa* – the elixir that would reverse his ageing process.

True to his word, Cyavana returned from his transformative journey, now radiating with rejuvenation. His once-grey hair and beard were now vibrant, his skin had regained its firmness and youthfulness, his teeth and eyes shone with brightness and his vision was sharp. Cyavana and Menakā established their abode in the forest and together they brought forth numerous children, their love and union flourishing.

The divine formula bestowed upon Cyavana was eventually enshrined in the ancient Āyurvedic classic, the *Caraka Saṃhitā*, ensuring its preservation for all time. While the text contains various rejuvenation recipes, none can rival the popularity and delight of Cyavana Prāśa.

It is worth noting that Cyavana Prāśa comprises an extensive range of herbal ingredients, many of which are initially prepared as decoctions before being blended with jaggery and the pulped Amla fruit, also known as Indian gooseberry. Amla is renowned for being a rich source of natural vitamin C, as well as containing multiple types of organic acids and tannoids. The mineral composition of this formula includes iron, magnesium, potassium, calcium, sodium, and zinc. Scientific studies have demonstrated that Cyavana Prāśa possesses protective antioxidants, capable of blocking potential carcinogens and aiding in the elimination of free radicals. Additionally, its benefits extend to lowering blood sugar and lipids, improving circulation, and providing energy and stamina to the body's tissues.

Śamanādi

Śamanādi promotes the long-term preservation of the patient through judicious lifestyle choices, eating habits and exercise. Āyurveda uses the term *Dinacharya*, meaning 'pathway through the day,' to help the individual decide how best to promote health, happiness and wellbeing whilst managing to avoid what most of us would call today a modern lifestyle.

The ancient Rishis noted and, accordingly, promoted or discouraged a number of factors that determined health or ill health. They observed that wellness was enhanced by consuming wholesome, fresh foods; reducing the intake of dried and saline foods; avoiding old, putrid or stale foods, taking in food regularly; abstaining from addictive substances; a moderate pursuit of pleasure; avoiding excessive physical strain and mental stress; waking and sleeping early; and not giving in to lower emotions, such as fear, grief, greed and jealousy.

Keralīya Pañcakarma (a popular variation on Āyurvedic Pañcakarma)

This 'contemporary' Pañcakarma therapy was developed in South India, specifically the southern state of Kerala, hence 'Keralīya Pañcakarma'. This alternate approach might be seen more as a physiotherapy-Pañcakarma, incorporating the essential classical principles of Āyurvedic PK with more simplified techniques. Whereas Āyurvedic PK is purificatory in its aims, Keralīya PK could be said to be a more palliative procedure. Many of the former techniques have a lightening energetic, whereas the latter are Bṛṃhaṇa therapies (nutritive and tonifying).

Keralīya PK is non seasonal and non time-dependent, and it places greater emphasis on Snehana and Svedana. Keralīya PK also targets more specific ailments and places emphasis on tonification and rejuvenation with regard to specific conditions, rather than the stronger purgation and general rejuvenation.

Keralīya PK has become somewhat iconic within the last decade. A search for 'Āyurveda' images on your web browser will return a disproportionately high amount of high-gloss Keralīya PK examples. This is not to undermine or undervalue its importance as a therapeutic option; however, it is important to keep in mind the clear distinction between traditional Pañcakarma and Keralīya Pañcakarma.

The following is a brief overview of some of the more popular therapies that are offered.

Śirodhara/Dhārā Karma

Śirodhara/Dhārā Karma uses a number of warmed liquid mediums,[67] which are poured upon the forehead whilst the patient lies in comfort. The administrator holds the suspended reservoir and skillfully guides and manipulates the vessel back and forth while refilling its contents. *Śirodhara* tends to be the more popular name for this therapy and has been found to be effective in the treatment of a number of

67 Warm milk, buttermilk, curd, ghee, herbal decoctions, coconut milk and water.

mental disorders, such as stress, memory loss, confusion and insomnia. Other conditions that benefit from this treatment include facial palsy/tics, rhinitis, sinusitis, fainting, coma, impaired hearing, tinnitus, cataracts and diminished vision.

Pizichhil/Kāya Seka

Pizichhil/Kāya Seka, also known as 'the royal treatment', employs a similar principle of gently pouring warm medicated oils[68] onto the body. This is done in conjunction with a light massage using long, even strokes. This therapy is usually performed by two masseurs, who work in unison. Pizichhil (its more popular name) aims to kindle the body's digestive power while improving and enlivening the senses. Pizichhil is thought to build Ojas while pacifiying Vāta Dosha. The application of warmed (and sometimes medicated) oil helps remove muscle tension and spasm while rejuvenating the skin, bone and muscle tissue.

Piṅḍa Sweda

Piṅḍa Sweda is quite literally fomentation with the use of a herbal bolas. This is considered to be the most potent of the five Keralīya therapies. First, the patient undergoes oleation and steaming before being treated with decocted herbs cooked with navara rice[69] and Kulatha.[70] This fusion is then tied into cloth bolas and gently tamped over the body to induce sweating, with special care being taken to foment the area about the joints and extremities. Sensitive body parts are covered by cloth during this procedure. After the fomentation is completed, the bolas is opened, and its contents are removed and then massaged into the tissues. This process is ended with a warm bath and a rest. The materials typically used in the medicated bolas include milk, paddy rice, Eranda leaves and a decoction of Bala (*Sida cordifolia*). Piṅḍa Sweda promotes the rejuvenation of the neuromuscular system; its use is also indicated in cases of osteoarthritis, osteoporosis, Parkinson's disease, multiple sclerosis, sciatica, muscular cramping and deteriorating joints.

---

68 Usually a mixture of ghee and sesame oil, medicated with Rasāyana herbs such as Shatāvari, Ashwagandhā and Guḍūchī.

69 Also know as śāli rice or red rice. There are many medicinal qualities already attributed to red rice, so its selection for use in an herbal bolas makes sense.

70 Also known as horse gram.

Anna Lepa

Anna Lepa involves the direct application of *Lepas* (herbal pastes or balms) onto the body. Quite often, this treatment is used in conjunction with Piṅḍa Sweda if Lepas are not sufficent in themselves to remove deep-seated toxins from the body. Typically, the applied paste tends to be one of three primary energetics: warming to pacify Vāta Dosha, cooling to pacify Pitta Dosha and, finally, astringent to pacify Kapha Dosha. Lepas can be highly effective when applied to arthritic joints, ulcers, boils, severe skin conditions and lesions or bleeding wounds. Lepas, like Piṅḍa Sweda Bolas, tend to be formulated with medicated grains cooked in herbal decoctions, milk and ghee. When the treatment is complete, the paste is removed, and the area bathed in lukewarm water.

Śiro Lepa

Śiro Lepa applies medicated oil directly to the hair and scalp area, after which Lepas can be pasted over the top. The whole of the treated area is then usually enclosed in banana leaves. The patient then receives whole-body oleation therapy and rests for approximately one hour before the leaves, paste and oil are removed. After cleaning, medicated oils are again applied to the scalp, after which the patient receives a warm medicated bath. Śiro Lepa is highly favoured for mental ailments, head and neck injuries, early greying or falling of hair, facial paralysis, stroke, fatigue and stress.

1.10/ Patients unsuited for Saṁśodhana

Caraka Saṃhitā includes some interesting commentaries on the contraindications of patients, meaning those least likely to benefit from treatment. Most of its recommendations are what we would term good practice or common sense, however, one or two are not really of concern to modern physicians. The practicioners of the past faced as much of a risk to their reputation by accepting a new patient as their patients did placing their life in the hands of the practitioner, and this should be remembered when reading the following ten contraindications. It was therefore of great importance to consider the prior treatment since an unsuccessful outcome might result in complications or re-manifestation of the original symptoms and therefore could not only be dangerous for the patient, but also cast doubt upon the healer's ability to heal and so end his livelihood.

Contraindications of the treatment of a patient:

1 One who considers his knowledge to be superior to that of the physician;
2 One who cannot organise or arrange his affairs prior to treatment;
3 One who has an aversion to kings and physicians and is himself despised by them;
4 One who is sceptical of mind;
5 One who is unwilling to carry out instructions;
6 One who is grief-stricken;
7 One without belief in god(s);
8 One who is rash, fierce, fickle, ungrateful or cowardly;
9 One who is hostile toward the physician;
10 One who is terminally ill and destined to die.

In conclusion, it should also be mentioned that these types of therapy were, and still are, highly labour-intensive, requiring a trained doctor, nurse and manual workers or orderlies to be present. Treatment also includes a certain quantity of medicated oils and ghee, herbal decoctions and Rasāyana drugs, all usually prepared on demand. If they were to undertake treatment and gain benefit, the individual needed to be able to support themselves financially for extended periods; hence PK was understood to be the treatment of kings, and not something generally within the reach of the common man. Saṁśodhana, in its entirety, was a time-consuming therapy; most importantly, its post-period was allocated to Paścatakarma. An insufficient recovery time being allocated during this final stage after the strongly purgating acts of P, could easily result in the body becoming imbalanced.

Caraka Saṃhitā's *Sutrasthanam* speaks of the The Essential Four Pillars of Treatment. These are portrayed as four foundational stones that should, like the four legs of the table, be laid at the outset to ensure successful treatments and outcomes and, as such, the removal of even one leg causes immediate instability and loss of functionality.

Very briefly, the Four Pillars of Treatment were considered to be:

Viadya/physician
Aushdhi/drug
Paricharak/attendant (or nurse)
Rogi/patient.

Viadya (the physician) should possess proficiency in both practical and theoretical knowledge. His hands should be dexterous and highly skilled. They should be pure in body and mind, not motivated by either fame or greed.

Aushdhi (the drug/s) should be abundant, fresh, rich in flavour and taste, and free from parasites (insects). All drugs should be purified and prepared in accordance with their requirements to behave in a therapeutic manner. The pharmacist or agent who prepares these 'drugs' should likewise possess proficiency in both practical and theoretical knowledge, such as knowing how to collect, store, purify and formulate.

Paricharak (the attendant) should be intelligent and alert, attentive to the needs of the patient, and always have their best interests at heart. They, too, should be pure in both body and mind.

Rogi (the patient) should be of strong mind, display stamina and be obedient to the physician. The patient should be able to descrsibe their condition and disease clearly to the physician.

In light of this, the aforementioned contraindications of the treatment of a patient might be thought of as risk-reduction regarding the patient.

While all four pillars are of equal importance, the patient is somewhat paramount as no cure will be possible if the patient is unwilling to cooperate or follow the advice of the physician, take the remedies when they are given, or accept help from the attendant.

Section 2

Metals and metal-working

2.1/ Early metallurgy

The manipulation of metals and the technologies that enabled alchemists to work with these materials are of particular interest to us in this book, since, as will be demonstrated, they are crucial to how a final medicinal product is purified and ultimately produced.

The manipulation and mastery of metals is one of the most remarkable advancements in human history, and its evolution and complexity are closely intertwined with the development of society. From the early smelting of copper ore to the alloying of it with tin and the resulting emergence of bronze, the progression of this art is awe-inspiring. Metal working must have seemed like pure magic to those unfamiliar with the craft. Here was a substance unlike any other, capable of withstanding extreme heat, shaped through being hammered and drawn, liquefied and cast, only to be returned to the flames, transformed and reborn anew.

If ever there was a process that could inspire the notion of birth, death, and rebirth, heaven and hell, it was surely found within the fires of the furnace. The manipulation of fire and the mastery of temperature eventually provided the means through which metal yielded to the boundless imagination of humanity.

Just as the flames of the hearth softens food, allowing the digestion of fibrous plants and indigestible meat, the flames of the forge and furnace have the power to produce charcoal, crack stone, and extract metal from various ores. The extraction of copper, smelted from malachite ore, is believed to have been a practice known as early as 7000 BCE in the ancient world. The process involved simple earthen pits filled with charcoal, which were used to heat ceramic crucibles containing higher silicates percentages. Crude bellows made from animal skins and bladders were repeatedly and steadily operated to blow a continuous stream of air through carved bone or stone nozzles, raising the temperature sufficiently to smelt the ore.

Far left, casting Kanwsdsya (bronze) ingots

If you ever have the opportunity to witness this ancient process, it is a truly remarkable experience. The simplicity of the procedure is awe-inspiring, and despite the backbreaking work involved the results are well worth the effort. Not so far from my home, there is a permanent living-history project aimed at recreating such ancient practices through reenactments. These practical experiments not only provide a glimpse into the past but also stimulate historical research by presenting unique challenges that require innovative solutions.

Recently, this establishment organised an open day, during which several visiting archaeologists set up shop to demonstrate the simplicity and effectiveness of these ancient technologies. Everything was handcrafted and appeared exceptionally crude, yet with nothing more than primitive mortars, a hole in the ground, turf, and animal-skin bellows, they were able to produce a substantial amount of pure copper globules in less than an hour. The site was quickly set up, a hole dug, charcoal ignited, and bellows brought into action. The rhythmic movements of the bellows operator seemed almost hypnotic, and before long, the temperature reached 1000°c. Pieces of malachite ($Cu_2CO_3(OH)_2$) were dropped into a handmade crucible composed of clay, sand, bran, horse dung and refractory.[71] I later learned that this specific crucible composition had been developed through extensive experimentation, resulting in a formula that could withstand the high temperatures without cracking.

As the process unfolded, small shiny metallic beads of copper began to form on the surface of the ore, eventually coalescing into a small pool of liquid metal that rolled about at the bottom of the crucible. Working swiftly, the molten copper was poured into a pre-prepared ingot mould carved from the soft interior of a cuttlefish bone. Upon cooling, the resulting ingot was bright and square-shaped, ready to be further refined. As mentioned before, the entire process took less than an hour, yet it was immensely successful and satisfying.

71 Previously fired pottery, reground and added to the mixture to create greater stability.

Bronze Age and beyond

Above, bronze ingot

The advent of the Bronze Age marked a significant milestone in metallurgy, as the crystal lattice of copper was strengthened through the alloying process.[72] It is believed that the introduction of arsenic, and later tin, was initially accidental, however it soon became evident that the incorporation of a softer metal resulted in a new material with enhanced properties that surpassed those of its individual components. Early experiments with bronze involved varying ratios, ranging from as low as 5% to as high as 25%, in search of the ideal alloy that offered both strength and flexibility while retaining the ability to reproduce intricate details when moulded. Ancient Chinese bronze maters were particularly devoted to perfecting this art and achieved an unparalleled level of quality in the ancient world.

Above, very simple forge being used to cast bronze

These ancient metallurgical examples are especially intriguing because we possess such limited knowledge regarding some of the production techniques that were employed, and even the latest research has revealed that much remains shrouded in mystery. We can only speculate upon how certain cultures managed to refine their metalworking skills to such an advanced level of sophistication. However, it is evident to anyone who has worked with metals and participated in metal casting that it is a meticulous and skilled craft. Contrary to what one might hear, pouring liquid metal is an art in itself.

Three notable examples of exceptional metalworking from ancient times

72 This is also true in the reverse, where miniscule amouts of copper (about 1%) are introduced into tin, during the process of tinning. The recovered tin, cast into tin ingots, no longer 'cries' when stressed as the small amount of copper sures up the crystalline property of the tin.

deserve mention: (1) Wootz/Damascus steel,[73] renowned for its use in crafting Saracen scimitars; (2) the seemingly corrosion-resistant forge-welded iron showcased in the monumental six-ton Ashokan pillar in Delhi, India; and (3) the Nihontō Katana,[74] better known as the Samurai sword. These exceptional works[75] continue to inspire and amaze due to their outstanding quality, ingenuity, and practicality, and serve as a testament to the artistry of metallurgy.

In the past, I ventured into the realm of blacksmithing, inspired by a one-day taster session at a local college. After a brief introduction to workplace safety, the first lesson focused on the art of lighting a forge. This seemingly simple task proved to be quite demanding, as many of us struggled to just keep the forge alight beyond ten minutes. It was during this experience that I gained a newfound appreciation for the skill required in blacksmithing.

I observed adept students skilfully twisting, stretching, and moulding heated metal as if it were plastic. I was particularly impressed when one student managed to create a flawless coiled spring from a simple metal tube. However, it was at this point that the distinction between armchair speculation and the

73 Swartz/Damascus steel is believed to have been manufactured from around 300 BCE in Southern India. Favoured by both Persian and Arab armouries, this material was revered for its armour-piercing abilities. One key process thought to have been used in Wootz manufacture were ingots heated in a closed crucible. For more information see Verhoeven, J.D., Pendray, A.H. and Dauksch, W.E. (1998) 'The key role of impurities in ancient Damascus steel blades.' Member Journal of the Minerals, Metals and Materials Society 50, 9, 58–64.

74 Manufactured from tamahagane or gem-steel, this is a fine iron sand that contains small amounts of carbon. This mixture gives the blade its hard, yet flexible, quality.

75 (1) The Antikythera Mechanism, recovered from the waters around the Greek island of Antikythera in 1901. For more information see www.antikythera-mechanism.gr or Marchant, J. (2008) Decoding the Heavens: Solving the Mystery of the World's First Computer. London: Windmill. (2) The blue-coloured tiles of Ishtar's Gate, Babylon (600 BCE). Chaldean ruler Nebuchadnezzar emblazoned the walls of his processional thoroughfare with glazed tiles produced by a combination of sodium, silicate, cobalt and copper oxides. When heated to precise temperatures in excess of 1000°c this glaze can be reproduced using modern methods; however, it is still difficult to explain how over 20,000 of these tiles were manufactured, with almost no colour variation.

reality of blacksmithing became apparent to me – it can be both a challenging and hazardous craft.

While my brief period of study did not allow me to master this art, it proved to be an invaluable experience. Witnessing the mesmerising sight of semi-molten iron being forge-welded, accompanied by the dazzling shower of sparks before being plunged into cold water, evoked mythological imagery of Vulcan's forge, that Roman god of fire fashioning lightning bolts for Jupiter, King of the gods.

2.2/ Loha (metal)

Categories of metal in Rasa Shāstra fall into one of three types: *Sudhā* (pure), *Puti* (impure) and *Misra* (alloy/mixed). Additionally, metals are also assigned a social caste, known as *Varna*. Gold and silver represent the highest or purest form of metal and so attain a kind of royal status. Iron, tin, copper and lead are considered impure metals, and so in terms of hierarchy occupy the various tiers of courtly favour, from priest to commander-in-chief to the lowly servant caste.

Above, Varta Loha, a curious alloy containing bronze, lead, copper, brass and iron

The remaining alloys, brass and bronze, were considered to be of mixed caste and so duly considered outsiders or untouchables within this hierarchy.

The final metal, mercury, was unique, since it was considered impure, but also had royal status. This metal was special due to its liquid form, which was nevertheless able to form an amalgam with five of the pure and impure metals, as well as the two alloys. This ability made mercury a magical metal, yet it was not without impurities. As we will see in later chapters, the process of removing these impurities required many protracted and complex procedures.

Metals, caste and planets

The association of metal with caste comes largely from Jyotish,[76] better known as Vedic astrology. This places great emphasis on the social hierarchy of planets, each being framed as one of nine important components within the celestial royal court. Once the role of each planet has been ascertained, it becomes easier to predict how that planet will act in any given situation. For example, the King, represented by the Sun, naturally outshines his courtiers. Moving slowly, he presides over statesmanship and power, yet being a 'hot' planet is given to bouts of temper or vindictiveness. The Moon, by contrast, moves quickly and surreptitiously behind the intrigues of the court, excerpting a subtle or reflective energetic. She, being naturally gregarious and nurturing, spends much of the time interacting with the other planets, listening to or consoling them. Hence the Moon exudes a motherly or caring demeanour.

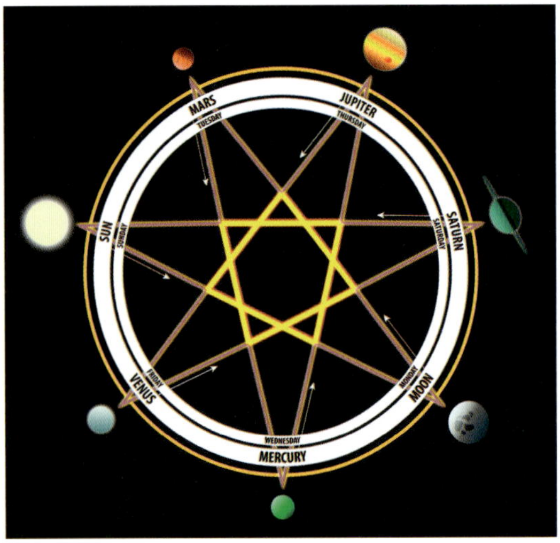

Above, the planetary relationships and their respective weekdays

Alchemically speaking, the connection between metal and planet grew steadily more influential over time. As astrology and alchemy intertwined, their separate meanings became synonymous with one another. For example, the Sun and gold both signify vital breath, life force, the physical heart and the immune system. The metal gold is therefore associated with resistance to disease and imbued with life-extending properties.

76 One translation of Jyotish might be 'the science of starlight'.

Planetary status and qualities of metals				
Metal (Loha)	**Planet (Graha)**	**Quality (Guna)**	**Caste (Varna)**	**Tissue (Dhātu)**
Gold (Swarna)	Sun (Sūrya)	Sudhā (pure)	King	Bone
Silver (Rajata)	Moon (Chandra)	Sudhā (pure)	Queen	Blood
Iron (Loha)	Mars (Kuja)	Puti (impure)	Commander-in-chief	Muscle/sinew
Mercury (Pārada)	Mercury (Budha)	Puti (impure)	Prince	Plasma
Tin (Vanga)	Jupiter (Brihaspati)	Puti (impure)	Priest/advisor	Adipose fat
Copper (Tamra)	Venus (Shukra)	Puti (impure)	Priest/advisor	Reproductive
Lead (Nāga)	Saturn (Shani)	Puti (impure)	Servants	Nervous tissue
Brass (Pittala)	Rāhu (north node)	Misra (alloy/mixed)	Militia/outcaste	–
Bronze (Kansya)	Ketu (south node)	Misra (alloy/mixed)	Militia/outcaste	–

Seven of the nine planets were also significators for specific bodily tissues (see Dhātu) in the table above:

The Sun presided over bone tissue as well as one's stature; gold also benefited this tissue.[77]

The Moon presided over bodily fluids, the stomach/digestion and particularly the blood. Alchemical silver similarly strengthened these same tissues.

Mercury (both the planet and the metal) was associated with royalty, firstly regent (direct offspring of the Sun and Moon) and secondly, its essential part in the extraction of gold and silver from their native ores. Mercury's ability to readily amalgam with either royal metal made it indispensable in the processing of both. After forming an amalgam, mercury releases its bond at 357°c and vaporises, quickly re-condensing, free of both metals. This ability earned Mercury the reputation of being a magician, healer and trickster.

77 Modern research into colloidal gold reveals it may reduce damage to the cartilage and bone, thereby reducing joint pain.

Iron relates to physical prowess, muscle and sinew; these attributes are aptly conveyed in the astrological portrait of the planet Mars. Impurities and chemical reactions cause iron to rust and corrode, yet rusted iron reheated in the forge becomes malleable and renewed. In the ancient world, iron was known to resist mercury, so iron flasks were used to transport the liquid metal without fear of damage or amalgam forming. This uneasy and strained relationship also endures between the two planets; Mercury and Mars being enemies, so to speak.

Tin relates to adipose tissue and general heaviness in the body; again, both attributes are aptly conveyed in the astrological portrait of Jupiter. Tin is deemed an impure metal, or *Puti*, a word meaning impure or sweating, which implies its odour upon liquification. Impurities and chemical reactions similarly cause tin to corrode or oxidise. Tin is unusual in that it has a very low melting point (232°c), yet a high boiling point (2586°c). Tin is largely an unreactive metal, however when liquefied and brought into contact with water it becomes volatile. These attributes made tin a metal of extremes. To the ancients, tin was noted to build tissue, yet act upon the subtle elements æther and air. When liquefied, tin looks like mercury; when cool, it looks like silver; it was perhaps these attributes that elevated it in the celestial court to priestly status.

Copper relates to the reproductive tissues and is therefore naturally associated with planet Venus. Alchemically prepared copper rejuvenates the organs and de-ages the body whilst adding elasticity to the tissues. Although considered an impure metal, copper, like tin, attains priestly status. It has a mid-range melting point of 1085°c, ironically close to the melting point of gold (1064°c). In several important Indian alchemical works, copper is almost always an essential ingredient in the manufacture of artificial gold, having both the right kind of colour and weight to mimic it. Astrologically, Venus never strays more than 48° from the Sun and so takes on some of its radiance, indeed in some instances, it is even referred to as Sūryaloha (Sun-like metal).

Lead relates to the nervous system, its health and its tissues. It is associated with the planet Saturn. Saturn is the most feared of planets in Vedic astrology, principally due to its ability to age and/or debilitate the body. Astrologically,

Saturn occupies the lowest position within the celestial court, whereas the Sun occupies the highest. In the Vedic planetary mythology, Saturn was the son of the Sun, but by birth circumstances was denied recognition by his father. Alchemically, lead frequently represents the lowest metal aspiring to transform into gold, hence the saying 'turning to lead into gold', to represent the idea of changing something base to something valuable, however all the allegories I have discussed so far tell us that both the planet and the metal have deep, hidden and highly interconnected potential. In other words, do not look down on this metal; it has more influence than you might think. One of the names for lead is *Sindoorakara*, a name meaning 'in tribute to redness'– a reminder that within the dullness of lead, minium, or red lead, is waiting to come forth.

The alloys, **bronze** and **brass** do not relate to any bodily tissue directly. Instead, their effects are understood through their composition. They are a fusion of two metals, bronze is a mixture of copper and tin, and brass is a mixture of copper and zinc. In both cases, the overwhelming ingredient is copper. Both alloys impact blood purity, the liver, skin diseases and the removal of parasites. Astrologically, both alloys are assigned to the lunar nodes known as Rāhu and Ketu. Of course, these nodes are essentially one being, Rāhu (the head) and Ketu (the tail). While both nodes attain planetary status, they have no physical presence. The nodes are mathematical points along the ecliptic[78] where the Moon and Sun intersect, occasionally leading to an eclipse.

2.3/ Reduction and conversion of metal

Ariloha, or 'metal's enemy' is one technique used to reduce and convert each of the nine planetary metals into Bhasma (alchemical ash).

The idea here is to melt and introduce two antagonistic metals; this fusion then weakens the structure of both. Just as a small percentage of tin or arsenic strengthens copper and produces bronze or arsenical bronze, an addition of unfriendly metals produces unstable alloys. As cohesion lessens,

78 In essence, the ecliptic is the Earth's orbital path around the Sun; hence it represents the apparent pathway of the Sun through the twelve constellations.

Right, an alloy of copper and iron produces an unstable alloy but does also produce a beautiful red/purple colouration

the combined metals are more easily processed and reduced to a powder.

Having combined unfriendly metals, the further addition of corrosive elements such as sulphur, orpiment or cinnabar further destabilise and reduce. *Rasendra Sāra Saṃgraha*[79] additionally notes that copper pyrite is useful in the reduction of silver, sulphur in the reduction of copper, realgar in the reduction of lead, orpiment in the reduction of tin and cinnabar in the reduction of iron.

The following table describes Ariloha metals.

Ariloha – natural antagonism	
Planet and ruling metal	**Ariloha (antagonistic metal)**
Sun (gold)	Saturn (lead)
Moon (silver)	Jupiter (tin)
Mars (iron)	Venus (copper), Ketu (bronze), and Rāhu (brass)
Mercury (Pārada)	Mars (iron ore)
Jupiter (tin)	Moon (silver)
Venus (copper)	Mars (iron)
Saturn (lead)	Sun (gold)
Alloy	
Rāhu (north node) Brass Cu+Zn	Mars (iron)
Ketu (south node) Bronze Cu+Sn	Mars (iron)
Varta Loha: Cu, Cu+Sn, Cu+Zn, Pb, Fe	Venus (copper), Ketu (bronze), Rāhu (brass), Saturn (lead), and Mars (iron)
Tri-Loha: Au+Ag+Cu	Sun (gold), Moon (silver), and Venus (copper)

79 Gopālakṛṣṇa's Rasendrasārasaṃgraha's 'The Collection of the Essence of Mercury', 16–18th century text

Āvāpa, Nirvāpa and Ḍhālana

Aside from Ariloha, a number of metal reduction methods have been given in Rasa Shāstra texts. These methods are called *Āvāpa*, *Nirvāpa* and *Ḍhālana*. All are routinely applied to metals in most modern pharmacies. All require some level of heating, and all include the use of liquids, oils and, on occasion, dried herbs. Each approach is, for the most part, quite simplistic and highly effective, rendering even the most resilient metals into either small granules or powder.

Left, lime leaves (Citrus hystrix) *calcined with zinc*

Far left, Apamarga (Achyranthes aspera) *calcined with tin (early stage); left,* Apamarga *calcined with tin (late stage)*

Let's take a look at each individually to see what is involved and how it is implemented.

Āvāpa – adding

The word *Āvāpa* simply means 'to add'. In effect, this means liquefying a metal, usually metals with lower melting points like tin, lead and zinc, before combining them with herbal powders/granules or whole dried plant materials. This processing method comes after a prior pre-purification technique, such as Ḍhālana.

Having liquefied the metal in question, a small amount of dried herbs are added and stirred in.

*Right, bamboo manna (*Bambusa arundinacea*); far right, bamboo manna calcined with liquid copper*

The four most preferred herbal ingredients used are:

1. Poppy (*Papaver somniferum*)
2. Lime leaves (*Citrus hystrix*)
3. Apamarga (prickly chaff flower, *Achyranthes aspera*)
4. Tabasheer (bamboo manna, *Bambusa arundinacea*)

Of particular interest is Tabasheer, which is added to high-temperature metals such as copper. Its high silica content enables it to resist high temperatures. Tabasheer is collected from resin that naturally occurs as a whitish/blue substance at the nodal points of the female bamboo tree. As the herb slowly incinerates and is reduced to ash, more is added. Having added the required amount, heating is discontinued, and vigorous stirring begins. As the temperature drops, the metal cools but is prevented from aggregating into larger pieces due to the presence of the calcined herb. Eventually, the ash and metal mix down until only the ash remains apparent.

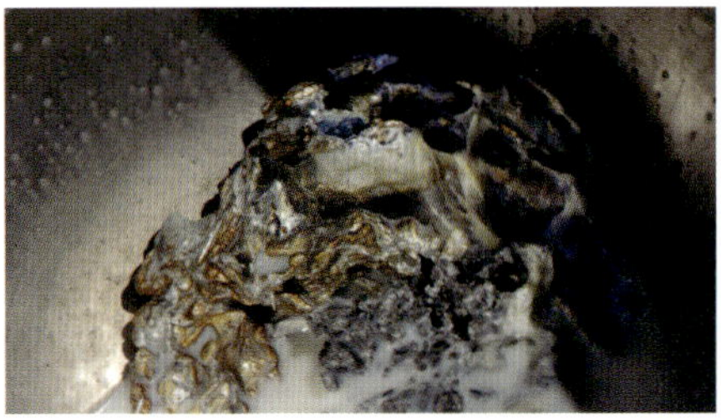

Right, zinc is liquefied and poured into milk

Typically, this ash is referred to as tin ash or lead ash etc. The final part of processing involves washing the ash. This involves collecting the ash and mixing it with water in a container. The ash is agitated by hand, and the water is decanted off, slowly removing the loose carbon and very fine ash particles.

The remaining powder is then triturated with various sour liquids, such as orange or lemon juice, or decoctions such as Triphala or lotus seed. In some cases, milk or aloe gel is recommended. Any of these substances is ground with the various ashes to produce a thick paste which can then be formed into cakrika (small rounded discs). These discs when dry are then sealed in larger crucibles and calcined at a very high temperature, resulting in Bhasmas.

Nirvāpa – without adding

Nirvāpa simply means 'without adding'. In effect, this means that heating or liquefying a metal in itself is the process; however, to fully complete the procedure, the metal must be quenched (see Ḍhālana).

Ḍhālana – quenching in liquid

Ḍhālana means 'to quench'. This method can be used for both high and low-temperature metals. This process involves either heating metal sheets or fully liquefying metals before pouring them into various prescribed liquids.

Quenching liquid metal is a highly dangerous process and requires some foreknowledge of how metals will react when introduced into liquids. For instance, zinc, a normally passive metal, can be poured into milk with little reaction, whereas tin and lead can produce explosive results when subjected to quenching. Copper, usually a passive metal, can, under some circumstances, 'explode' when introduced into a cooling liquid. For this reason, Ḍhālana should only be performed by an experienced hand. These safety concerns aside, the two most common methods of Ḍhālana are single or five-liquid methods.

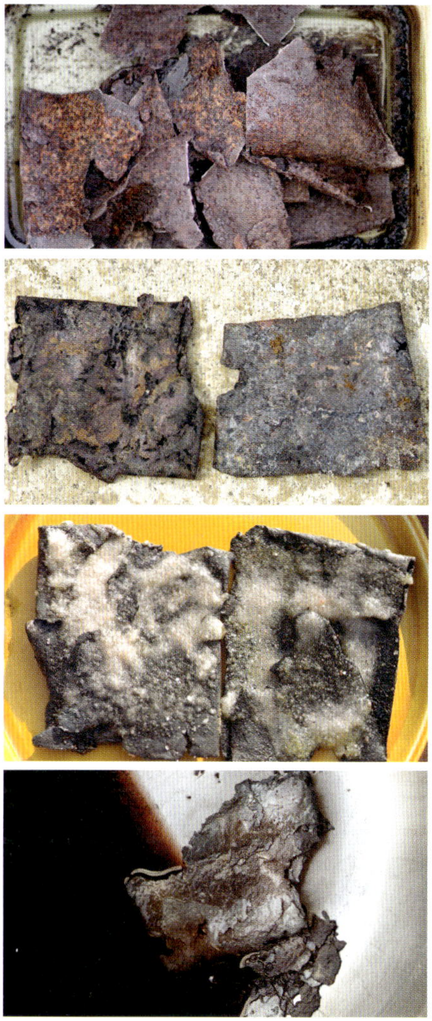

Above, rusted iron sheet is coated in salt and heated until red hot. It is then quenched in cow urine.

In most cases, Rasa Shāstra recommends that metal undergoes purification using five quenching mediums; these have been detailed in the table on the next page. The qualities of each medium ensure that the metal is thoroughly purified.

While alternate quenching methods are available in various alchemical texts, the following procedure is generally considered the best and most satisfying for the metals copper, gold, silver, zinc, tin and lead. Iron is a special case and has been dealt with at the end of this section. The two alloys, bronze and brass, are treated as copper and processed similarly. As an aside, the two remaining alloys, known as Varta Loha and Tri-Loha, are both suited to the five-liquid process, but their preparation is far less common, as is their

Five-liquid process			
1	**Taila**	Sesame seed oil	7x
2	**Takra**	Curd or buttermilk	7x
3	**Gomutra**	Cows' urine	7x
4	**Kāñjī**	Vinegar	7x
5	**Kulatha**	Horse gram decoction	7x
		Total	**35**

use as a medicinal ash. I don't think I ever saw either offered as a remedy, but they none the less remain interesting.

The rapid change of temperature incurred during this process stresses and weakens the metal's structure. From the initial smoothness of oil-quenching metal in sesame to the harsh and somewhat discolouring effect of cows' urine, each medium makes its mark upon the metal.

The five-liquid process is most practical for processing tin, lead and sometimes zinc, as liquefaction occurs at a low temperature and requires less energy and time to complete the process. Although I have included zinc here, this metal is in practice more frequently purified in either milk or milk infused with the juice of turmeric. In other instances, I have seen a decoction of Nirgundhi (*Vitex negundo*) and turmeric juice recommended. In both instances, melting and quenching are repeated twenty-one times, so even using these alternate methods, the processing still remains a somewhat protracted process.

Copper

Copper is not an easy metal to process and requires considerable energy to reduce to its liquid form, it is therefore more commonly purified in its sheeted form. Thin sheets of copper are washed, and a paste of lime juice and rock salt is liberally applied. The sheets are then folded and allowed to stand until the formation of verdigris covers its surface. The loose pigment is then removed by brushing before each sheet is heated over a strong flame, roasting it until red hot. At this point, the sheet is quenched in Kāñjī (vinegar). Once retrieved, the sheet is washed and lightly abraded and dried before once again applying pasted rock salt and lime juice and repeating the whole process. Indeed, this entire process is repeated seven times in total.

At this point, the copper sheets are considered purified and ready for further processing. Further details of this process are detailed in the section titled Use of metal-based medicines, where I describe how copper sheet is reduced to ash using sulphur and cinnabar as digesting agents.

Copper (alternate method)

Continuing with the copper theme, it is worth noting that metal extracted directly from a copper ore, such as chalcopyrite, bornite, malachite etc., or recovered from sheets, wires or pipes, can also be purified using the Ḍhālana method. By heating ore in a closed crucible to around 1000°c, copper can be poured off. Wires and sheets can take a little longer to melt. Once liquid, copper is poured into buttermilk and allowed to cool. The metal is then washed before melting and repeating the process. In total, this method is repeated eight times.

Another popular method of metal purification is simply to boil pieces of metal that have been infused with the so-called *Pancalavaṇa*, or five salts, in cows' urine. This method, in practice, is quite popular since it requires little effort to prepare and process metal. Having added the five salts to the urine, the liquid is brought to the boil, and the metal is left to boil for three hours. Of course, this purification method does not reduce metal to a granular form or ash, but it is considered to purify prior to its reduction via other methods, such as Āvāpa or Ḍhālana.

Above, various stages of copper purification using the buttermilk method

Pancalavaṇa process	
Saindhava Lavaṇa	Rock salt
Padelon Lavaṇa	Black salt (naturally occurring)
Samudra Lavaṇa	Sea salt
Parasara Lavaṇa	Sal ammoniac
Sambar Lavaṇa	Romaka salt (a mix of sodium chloride, sodium carbonate, and sodium bicarbonate)

Iron (a special case)

Purifying and processing iron involves a different method, primarily because of its higher melting point (1538°c).[80] Iron is an abundant metal, so there are many sources for processing. Rusted wrought iron is considered the best, particularly when weathered and old, in fact the older the better. The old iron is usually boiled in Triphala decoction, continually heated until the liquid evaporates. At this point, fresh decoction is added and the evaporation is repeated. This process is repeated seven times in total. Upon completion, iron pieces are washed and dried, then coated in pasted rock salt and lime juice. These are allowed to dry, heated until red hot, then quenched in cows' urine. This process is repeated seven times in total. Cows' urine is considered to have a special affinity with iron, which has the power to cleanse and scrape iron clean while potentising it many times. In many Rasa Shāstra pharmacies, it is common practice to soak pieces of iron in cow urine for a number of seasons. Over time, even the most corroded iron will become blackened and restructured as if regenerating.

80 This varies according to purity; munda, so called pig-iron, has the greater carbon content, and due to these impurities it melts at a lower temperature. For more information about iron see Appendix: Catalogue of materials and their use.

2.4/ Metallic immune booster

An interesting herbo-metal-mineral formulation, still favoured in India, serves as a potent immune booster for children and elderly people. Paediatric physicians often recommend this Avaleha (medicinal jam) to stave off infant diseases, particularly where the immune system is compromised. There are a few reasons why this formula finds favour with its patients; firstly, it is potent and easily assimilated, and secondly, it tastes rather good and requires little or no coaxing for regular consumption. The formula is comprised of: *Swarna Bhasma* (calcined gold); amber Pisti; fresh stems of Vaccha (*Acorus calamus)* or sweet flag; powdered almonds; honey, and a little ghee (clarified butter). Lastly, a very thin gold thread, literally 24-carat gold, is included during its preparation. The method of preparation is simple and interesting. Beginning with a short section of fresh Vaccha stem, the golden thread is passed through its centre longitudinally and trimmed at each end. One end of the stem is then ground in a circular motion on a course grinding stone. The stone's rough surface both macerates and mixes plant stem and gold into a gelatinous substance with colloidal-sized pieces of gold suspended in it. The gel is then collected and combined with the Swarna Bhasma, ground amber (copal) and almond powder, with small amounts of these being used. Finally, the ghee, along with a drop or two of honey, is added to sweeten the mixture. To administer, one simply dips a finger into the finished paste and dangles the nectar above the baby's curious fingers and hungry mouth.

Section 3

Visha (toxins)

3.1/ Origins of Visha

> 'Because poison has its origins in water, it becomes sticky like jaggery on contact with water. This allows for its rapid spreading during the rainy seasons. However, the appearance of the star Agastya (Canopus)[81] signals the end of the rainy season, counteracting the effects of poison. Therefore the effects of Visha (poison) become milder after the rains have subsided.'
>
> *Caraka Saṃhitā*

The use of *Visha* (poison) is perhaps one of the most interesting and unique aspects of Āyurvedic and alchemical thought, both of which acknowledge toxins as both potent medicines and mediums through which other poisons can be purified. Much of the following information in this section has been gleaned from Agada Tantra, a specialist branch of Āyurveda that deals with toxicology.

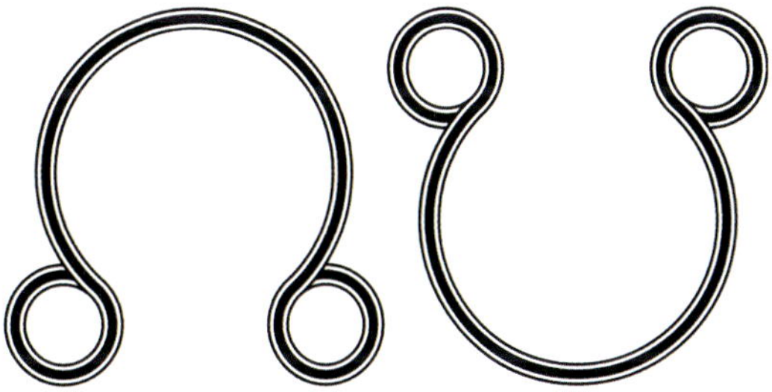

Above, 'cobra eyes' as seen on the back of a cobra's hood

Purified poison can have a miraculous medicinal effect upon diseases, while certain poisons play a role in purifying toxic material, for example, *Navavisha*,[82] or nine poisons, is

81 Canopus (also known as *Agastya* in Vedic astrology) is the brightest star in the southern constellation of Carina, being visible to 37° north of the equatorial plane.

82 These are: Sarpa-Visha (cobra venom), Vrischikavisha (scorpion poison), Haritāla (yellow arsenic), Vatsanābha (*Aconitum nepellus*), Dhattura (*Datura metel*), Snuhi (*Euphorbia ligularis*), Gunja (*Abrus precatorius*), Langali (*Gloriosa superba*) and Kuchala (*Nux vomica*). Other souces refer to nine Mahāvisha, these are given as: Hālāhala, Kālakūṭa, Śruṅgaka, Pradīpana, Saurāṣṭrika, Brahmaputra, Hāridraka, Saktuka and Vatsanābha. Of these only Vatsanābha (aconite) is easily identifiable.

Far left, Fu She (dried viper snake).

used to significant effect when employed in both the purification and fixation of mercury. Mercury is inherently toxic, and introducing other strong poisons actually coaxes the resulting combined toxicity to unite and separate out. This like-attracts-like action is a valuable technique in taming the toxicity of the liquid metal. Through this process, Mercury is ultimately subdued, forcing it to release its poison and become like nectar. We will take a more detailed look at mercurial toxicity later on in the book.[83]

As noted in the introductory passages of *Caraka Saṃhitā*, the ancients were well acquainted with the effects of poison, since they resided in a world where exposure to poison was an ever-present danger. Visha was understood to originate from two[84] primary sources, mobile and immobile. Today, we recognise these distinctions within the framework of inorganic and organic poisons. Mobile Visha was to be found in various types of insects, spiders, leeches, snakes, fish, reptiles and mammals,[85] and these poisons could be introduced into the human body through stings, bites and stratches. Immobile Visha was derived from toxic plants, minerals and metals. These included elements such as copper, zinc, bismuth, various forms of mercury including cinnabar, and compounds of arsenic and antimony.

Mobile and immobile poisons were observed to have two directional movements within the body, as well as physical symptoms. Poisons derived from animal sources were noted to move downward, whereas those derived from plants were noted to move upward.[86] Consequently, each could be used in micro-doses to counteract the other's action.[87]

---

83 See 7.2/ Origins of mercury and its impurities.

84 Susrutha mentions a third category of Visha called *Gara*, or artificial poisons, being specifically engineered to kill; taken over longer periods these poisons attracted less suspicion and became tools of the unscrupulous.

85 The full list of mobile poisons includes: snakes, insects, spiders, scorpions, lizards, leeches, fish, frogs, locusts, chameleons, dogs, tigers, lions, jackals, hyenas and mongooses.

86 Gaṅgādhara's edition of *Caraka Saṃhitā* reverses this direction.

87 The commentary on *Caraka* notes this mutual contradictory property of Visha is only specific to *Prabhava* (its specific action). It is noted that some Visha are contradictory in action, i.e., Kuchala counteracts the effects of Vatsanābha.

Toxic substances displayed the following qualities:

1 *Vyavayi*/fast acting
2 *Vikasi*/pervasive, not requiring digestion
3 *Tīkṣṇa*/sharp
4 *Usna*/heating

Immobile poisons produced a range of responses from the body, including high fever, choking, vomiting of white froth, fingertip sensitivity, shortness of breath, fainting spells and anorexia over the long-term. These were considered slow poisons, their effects taking many hours, days or sometimes even months to manifest fully. *Susrutha Saṃhitā* makes some mention of mineral-based poisons, and comments on their propensity to strike at one's heart (the seat of Ojas), causing burning sensations and fainting spells. The overall effect of Visha was noted to be in direct opposition to that of Ojas. See the table on the following page for comparative energetics.

Many, if not all, of the immobile substances (plants and minerals) were purposely developed into medicines of one sort or another through the practice of Shodhana/purification (see Section 5.4 in Part II), wherein their toxicological effects were transmuted into Amṛīta (nectar).

In contrast, mobile sources of Visha produced inflammation, oedema, fatigue, fever, drowsiness and, eventually, coma. Therefore although these substances were also harvested for medicinal use, their use fell into decline in Āyurveda. In contemporary terms, some rural doctors in southern India, who are usually trained in the Siddha Tradition, do continue to favour the use of cobra venom. Classical Chinese medicine is also a tradition that still actively promotes the use of mobile poisons, and a number of formulae rely on their inclusion; these include ingredients such as *Quan Xie* (scorpion), *Wu Gong* (centipede) and *Fu She* (viper). In each case, specific seasons to harvest and prepare are given, for instance, a viper's organs are removed and the body dried, while centipedes are best caught in the spring and dried along slivers of wood. Scorpion is administered to clear the liver channels and eliminate toxins, and helps remove endogenous liver wind, a precursor to paralysis, convulsions and spasms. Centipedes have similar medicinal uses, dispelling convulsions, both chronic and acute. Viper has a warming nature and detoxifies the body, building both blood and energy.

Visha	Ojas
Laghu – light	Guru – heavy
Ruska – rough	Snigdha – smooth
Tīkṣṇa – sharp	Manda – dull
Usna – heating	Śita – cooling
Sūkṣma – fine	Bahulu – thickening, building
Viśada – non-slimy	Picchila – slimy
Vyavāyi – spreading	Sthira – stable
Anirdeśya – indefinable taste	Madura – sweet
Vikasi – causing looseness	Slaksna – connectivity
Āśukārī – fast-acting	Avikara – orderly
Avipāki – indigestible	Prasāda – nectar

In summary, Visha, when used correctly, can have miraculous effects, partly due to its pervasive assimilation properties and inherent potency. Visha's ability to penetrate and spread effectively has the potential to neutralise Ojas, yet, conversely, minute purified quantities of the same material have the opposite effect, boosting Ojas and stimulating the immune system. Once administered, it moves quickly and relatively unobstructed, able to penetrate the deepest tissues. With an appropriately selected Anupāna, its course can be directed and its effect even more pronounced.

Visha as an antidote

Agada Tantra recommends introducing a small amount of poison in the event of accidental poisoning or after direct exposure to toxins, for example after being bitten or stung.

The following examples are taken from classical works that advise on the counteracting of poisons. The first example relates to scorpion poison, and the second addresses both mobile and immobile poisoning, notably poison tubers and the bite of *Krishna Sarpa* (black cobra).

> 'A pill made from Hingu (*Asafoetida*) and Haritāla (yellow arsenic), along with the juice of Matulunga (*Citrus medica*), is applied on the bite and collyrium (eye-salve) is best to remove the poison of a scorpion.'
>
> *Aṣṭāñga Hṛdayam*

> 'Equal quantities of Sudhā Tankana (borax), Sudhā Tuttha (copper sulphate), Haridra (*Curcuma longa*), Jatikosa (*Myristica fragrans*) are ground together well with the juice of Devadali (*Luffa echinata*[75]) – equal to the amount of the drugs. This formulation is known as *Viṣavajrapāta Rasa* which can overcome all kinds of Visha Vikara (poisoning).'
>
> *Rasendra Sāra Saṅgraha*

This latter formula claims to raise one from their death bed upon introduction. Typically, up to 3g of this formula could be administered, along with an Anupāna of human urine to induce vomiting and consequent purgation.

Borax and turmeric were both known to have powerful anti-Visha effects, whether administered singularly or in unison. It was also observed that a combination of Kajjali (mercuric sulphide) and Abhraka (mica) improved resistance to toxins or enabled the tissues to purge any remaining toxic residue, which conventional herbal preparations were less likely to relieve.

Above, dried Quan Xie (scorpion),

General treatment of poisoning

The general process of removing low-concentrated poisons in the body begins with identifying the Dosha or Doshas most affected by its presence. Poisons aggravating Vāta Dosha are combated by the introduction of sweet, oily, sour and salty foods and the administration of oil enema or Anuvāsana Basti.

Poisons aggravating Pitta are combated by the use of mild purgation or Virechana – the application of cooling poultices and foods that predominate in bitter, astringent and sweet tastes. Other beneficial anti-Pitta therapies include the use of pearls ground with rosewater.

Finally, poisons aggravating Kapha respond well to emesis or Vamana, along with foods predominating in the astringent, bitter and pungent tastes.

Furthermore, a close examination of a Doshas 'home' was also undertaken to determine the depth of toxicity. For example, if the stomach (Kapha) was heavily implicated in poisoning, treatment would be light, and a result would quickly be achieved. If the small intestine (Pitta) was the most affected, treatment would be middling and curable with some due diligence. In cases of the large intestine (Vāta) being seriously poisoned, treatment would be protracted and difficult, requiring multiple remedies and a highly skilled physician.

Above, Wu Gong (centipede)

Selecting an appropriate antidote

Having already identified a number of suitable substances for processing alchemical materials in Rasa Shāstra, it is logical to then select an appropriate antidote, inasmuch as any substance used for purification purposes automatically has potential antidote properties attached to it. Tankana (borax), for example, was routinely used in the purification of Sasyaka (copper sulphate), and so in cases of poisoning by impure Sasyaka its administration was automatically recommended. Likewise, garlic (Himalayan single clove) is regularly used in the purification of Pārada (mercury) but is also highly efficient in the removal of impure Pārada from the body.

There were a number of different materials known to have excellent anti-Visha effects, which were used topically or internally, as required by Āyurvedic physicians. In the opposite table are twenty-two commonly used remedies to alleviate or remove the effects of poisoning, in addition to the common materials used in the purification of alchemical medicines.

General anti-Visha substances	
1 Borax*	12 Goat's or cow's milk
2 Rock salt	13 Arjuna (*Terminalia arjuna*)
3 Garlic (Himalayan single clove pearls)	14 Triphala decoction*
4 Betel leaves (*Piper betel*)	15 Turmeric
5 Kanksi*	16 Cow's urine (external wash only)
6 Lime water	17 Lemon juice
7 Sulphur*	18 Black pepper (*Piper nigrum*)
8 Ginger juice	19 Lime juice
9 Ghee*	20 Sarpakshi (*Rauvolfia serpentina*)
10 Bee honey	21 Aloe juice
11 Gaireeka (red ochre)	22 Gojihva (*Elephantopus scaber Linn*)
*** Materials requiring prior purification or preparation**	

The table below and on the following page shows twelve specific antidotes for poisonous plants and the protocols commonly used to alleviate or remove their effects. For individual information about these, see Appendix: Catalogue of materials and their use.

Specific antidotes for Visha and Upavisha plants	
Visha	**Antidote**
Vatsanābha (*Aconitum napellus/ferox*)	• Warm goat's milk with ghee • Kuchala (*Nux vomica*) • Curcuma with amaranthus (*Amaranthus polygonoides*) • Sarpakshi – Indian mongoose herb (*Ophiorrhiza mungos*) • Butter
Arka (*Calatropis gigantea/ Calatropis procera*)	• Water mixed with Gaireeka* (red ochre) • Tamarind (*Tamarindus indica*) leaf, macerated in water
Sehunda (*Euphorbia ligularis/neriifolia*)	• Water mixed with Gaireeka* (red ochre)

Dhattura (*Datura stramonium/ metel*)	• Saline water • Cow's milk sweetened with sugar or juice of eggplant (*Solanum melongena*)
Langali (*Gloriosa superba*)	• Goat's milk with ghee
Karaveera (*Nerium indicum*)	• Sugar candy mixed with buffalo milk • Ground Arka (crown flower) bark with milk or water
Gunja (*Abrus precatorius*)	• Honey, dates, grapes, tamarind, sour pomegranates and Āmalakī, ground together • Prickly amaranth (*Amaranthus spinosus*), mixed with sugar and taken with milk
Khasabeeja (*Papaver somniferum*)	• Rock salt, Pippali (*Piper longum*) and emetic nut (*Xeromphis spinosa*), ground, then taken as a drink with hot water • Tankana* (borax) and Sasyaka* (copper sulphate), taken with ghee • Brihati (*Solanum indicum*) – poison berry, mixed with milk
Vijaya (*Cannabis sattva/ indica*)	• Yoghurt and ginger juice • Juice of jackfruit leaves (*Artocarpus heterophyllus*) • Sandesaro (*Poinciana elata*) macerated in water
Jayapāla (*Croton tiglium*)	• Coriander seeds with sugar and yoghurt • Ghee and milk drink
Bhallātaka (*Semicarpus anacardium*)	• Swelling is relieved by butter and juice of prickly amaranth (*Amaranthus spinosus*). Applied externally • Milk with pasted sesame seeds, applied externally • Dēvadārū (*Cedrus deodara*), Sarṣapa (*Brassica campestris*) and mustā (*Cyperus rotundus*), applied externally • Warm milk sweetened with dark jaggery, taken internally as a drink
Kuchala (*Nux vomica*)	• Vatsanābha* (*Aconitum napellus*) • Warm goat's milk with honey, ghee, or camphor (*Cinnamomum camphora*) • Coffee
*** Materials requiring prior purification or preparation**	

3.2/ Signs of Visha

The following list outlines the general signs and symptoms associated with Visha, whether from a mobile or immobile source:

1 Topically, there is a discolouration of the skin about the area where Visha first entered;
2 A person begins to shiver;
3 There is inflammation in the bodily tissues;
4 Disfigurement or contortion of the whole body occurs;
5 Foam is seen issuing from the mouth;
6 Extreme contraction of the shoulders and arching of the back;
7 Loss of movement in all the limbs;
8 Final stages of death, including respiratory and cardiac failure, are observed.

In the event any of the above a physician is advised to determine the quantity of poison present and treat the patient using the first or second protocol. Protocol 1 involves treating patients where a small amount of poison has been ingested. Protocol 2 is required for the ingestion of larger quantities.

Protocol 1 The patient is first made to vomit; this is best achieved by drinking goat's milk. After the patient has vomited, he must again be given goat's milk until he can drink without the need to vomit. In other words, the patient is not to be given any additional milk when he no longer feels the need to vomit after drinking.

Protocol 2 The patient is again made to vomit by drinking goat's milk, salt water or water from washed fishes. He must drink until he vomits only bile. After this vomiting stage, he is given purgation drugs until mucus is seen in his stool. Finally, the patient is given ghee (clarified butter) to help strengthen the tissues and heal the internal organs. Ghee is considered to be one of the best substances for the removal of poison, and at the same time it helps repair the tissues from the damage sustained. Following these treatments, the patient will regularly drink honey water (weakly diluted honey) for an agreed period of time; this has the long-term effect of removing any residual traces of Visha still embedded deep in the tissues.

3.3/ Visha as medicine

The following two commentaries from *Susrutha Saṃhitā* outline treatment protocols for the use and avoidance of purified Visha. Medicines containing Visha were specifically prescribed for rejuvenation and the prevention of senility. The use of Visha-based medicines outside of these guidelines was ill-advised:

> 'Visha may be prescribed to one who consumes ghee with regularity and who maintains a salutary diet (containing milk/ghee). One who is willing to follow the directives of his physician is permitted to take Visha.'
>
> *Susrutha Saṃhitā*

> 'Persons disposed to anger, irritability or have an excess of Pitta Dosha, impotent or are members of a royal family. Those who are exhausted from excess hunger, thirst, physical strain, perspiration or are consumptive. The elderly (eighty years or more), pregnant, very young (less than nine years) or showing signs of dry patchy skin conditions are contra-indicated.'
>
> *Susrutha Saṃhitā*

3.4/ Visha and caste

The classic work on Rasa Shāstra, *Rasa Jala Nidhi*,[88] categorises poison by caste. To determine caste, the physician is instructed to introduce poison into fresh milk, as the purity of the latter will reveal the true nature of the former. We are advised that:

> 'Visha deposited into fresh milk will effect a discernable colour change in this medium. Milk remaining unchanged but causing the Visha to be transformed to a similar whitish shade denotes poison of Brahmin[89] caste. These signs are auspicious and should be taken to elevate its usage as a remover of disease and senile decay. Milk turning red upon its introduction is deemed to be suitable for

88 The ocean of Indian alchemy.

89 Brahmin: spiritual, teacher, lawgiver, considered the pinnacle of the Vedic caste system.

alchemical workings with Pārada (mercury) and is of Kshatriya[90] caste. The yellowing of milk or Visha indicates it to be a curer of kustha (skin diseases) and of Vaishya[91] caste. Finally, the blacking or darkening of both milk and Visha indicates poison to be of a Shudra[92] caste and consequently the bringer of death.'

Rasa Jala Nidhi

The text further suggests that each caste of Visha has its respective uses in the art of medicine. *Brahmanical Visha* can be used in the cure of the most serious diseases, whereas *Kshatriya Visha* is best administered to one who has inadvertently swallowed poison. *Vaishya Visha* is to be used in the treatment of minor diseases, whereas *Shudra Visha* is sometimes useful for the treatment of common snake bites.

Left, milk reacts with impure zinc, producing a silvery paste on the surface

Snake bite and caste

Caste classification also extended to the perpetrators of venomous attacks; so the snakes themselves could also be typed by caste.

The classical surgical text *Susrutha Saṃhitā* states:

'Brāhmaṇasarpa: have large thick scales of silver and gold. Their fragrance is sweet smelling their sheen pearlescent. They roam in clean places; possess marking like that of a sacred thread. Brāhmaṇasarpa are aromatic like flowers or Guggulu, they bite head-on (mid-section) and their venom aggravates Kapha Dosha.'

'Kṣhatriyasarpa: have the temperament of warriors, are proud, brave and have reddish eyes. They are easily provoked and irritated. They have glossy scales which depict logos of the Sun and Moon, cakrika, conch shells, spears and ploughs. They ambush or frequent hidden places; they have a sour aroma, bite the right side with venom that aggravates Pitta Dosha.'

90 Kshatriya: fighting elite, governing through fear of reprisal, the warrior caste.

91 Vaishya: trader, merchant caste

92 Shudra: servile or servant caste

'Vaiśyasarpa: have a rich black or reddish shine; they can also be ash-grey or coloured like a pigeon. Their bodies are covered with smoky, dotted or diamond patches. Their movement is ridged or may contort in unusual ways, they frequent all places. They have an aroma of ghee, sheep's milk or smell like a goat. They bite the left side; their venom aggravates Vāta, Pitta and Kapha.'

'Śūdrasarpa: have dark or dull lifeless scales, their skins are rough and the colouration of a buffalo, wheat or an elephant. Their bodies are covered with either dots or lines, they frequent dirty places. Śūdrasarpa have the aroma of blood or beer, they bite at your back and have venom that aggravates Vāta Dosha.'

Susrutha Saṃhitā observes that snake venom doubles in potency during the heat of the summer months and that the venom obtained from a youthful dark-coloured cobra was particularly prized for its medicinal effect. This same text makes an eighty-fold categorisation of all snakes, and for simplicity groups them into five distinguishable types. These species are: hooded and hoodless; painted with circular patches; hoodless and striped; non-venomous and slightly venomous.

Further more detailed sub-categories of snakes include painted hoods that depict logos of cakrika, ploughs, spears, umbrellas, elephant goads and swastikas. Snakes bearing these symbols often strike with speed, agility and aggression,[93] whereas full-bodied (large) snakes that are marked with partially coloured rings, spots or a bright fiery lustre strike with great force. Snakes with glossy, partly coloured scales and perpendicular lines running the length of their bodies possess characteristics of both the aforementioned types.

93 Many of these symbols appear in my book, *Hastā Rekha Shāstra, Indian Palmistry*. See Section 5: Selection of Apprentices.

Additional note: In India, scientific analysis of *Masi Kalpana* (calcined herbs and animal products) has conducted trials on the curative effects of *Krishna Sarpa Masi Bhasma* (calcined black cobra) as an antidote for snake bites. The preparation of 'snake ash' involves traditional puṭa; see Puṭa paka (temperature) for more information. After death, the black cobra's body (minus the head, tail and internal organs) is sealed into a clay crucible and fully incinerated. The classical Āyurvedic text *Astāñga Hṛdayam* also recommends this ash be mixed with Bibhītaki (*Terminalia bellirica*) oil and applied to the skin as a paste in cases of leucoderma (vitiligo).

3.5/ Snake venom

In India, snakes evoke both fear and reverence. In the countryside, they lurk in the grasses, posing a threat to unsuspecting travellers along the footpaths. Within temples, encounters with Nāgāfolk, semi-divine beings with a half-human, half-snake form, are common. Even in the celestial sphere, serpents hold significance, since constellations and individual stars are directly associated with them, for example, *Kaliya* (α Hydrae) is linked to the head of Rāhu, while the rest of the constellation, known as *Asleshā-bhava*, represents Rāhu's tail (Ketu) and body.

In the realm of yogic practice, the snake symbolises *Kundalini Shakti* – a dormant energy that lies deep within the human body. This primordial force, characterised as feminine, awakens and rises to unite with its masculine counterpart. Signs of this awakening include heightened perception, intelligence, enhanced psychic abilities, and robust physical health. Often likened to an inner fire, Kundalini Shakti once ignited unfurls and ascends the spine, evoking the image of a serpent in motion. Its upward and spiralling journey intersects various key energy centres before reaching the crown of the head, leading to a state of enlightenment.

Lord Shiva, revered in Rasa Shāstra and Yoga, is frequently depicted with a coiled serpent encircling his neck. In some portrayals, the snake's head occupies an elevated position, signifying its ability to ascend to the divine. Additionally, *Kala* (time) is represented as a coiled serpent, further linking the snake to longevity. Through the process of periodic skin shedding, snakes naturally embody the concept of regeneration, rejuvenation, and, ultimately, immortality.

Snake venom as medicine and elixir

Although snake venom is highly toxic, once purified it is believed to release a highly potent healing effect. Classical works such as *Caraka Saṃhitā*, *Susrutha Saṃhitā*, *Astāñga Hṛdayam*[94] and *Astāñga Sangraha* mention its benefits in treating gastro-entomological conditions like ascites (accumulation of abdominal fluid), as well as certain types of chronic fever, while later Unani[95] sources mention its use as both hepatic and aphrodisiac. From a more shamanic standpoint, purified cobra venom has been credited with consciousness-expanding properties and is used in ritualistic practices.

While modern Āyurvedic pharmacies utilise cobra venom, its use is not widespread, due mainly to unavailability. When it is used it is formulated with a sizeable herbal component. Common ingredients include Triphala,[96] chitraka,[97] turmeric,[98] liquorice,[99] ginger,[100] black pepper[101] and long pepper,[102] with the addition of rock salt, honey and Shilajit.[103] Some of the more common conditions treated with snake venom include chronic fever, diabetes, bleeding disorders, anaemia and intestinal parasites, along with a significant number of skin diseases. Formulated remedies are prescribed for both external and internal conditions.

Modern venom collection involves milking captive cobras over a two to three-week period. Extracted venom can be sun-dried until crystallised or triturated directly with various fresh juices such as ginger or cow urine, or herbal decoctions such as chitraka or liquorice.

---

94 *Aṣṭāñga Hṛdayam* = The Heart of Eight Branches. These are: surgery, gynecology and obstetrics, pediatrics, psychiatry, toxicology, rejuvenation and aphrodisiacs.

95 Greek-Arabic traditional medicine.

96 *Emblica officinalis, Terminalia bellirica* and *Terminalia chebula*

97 *Plumbago zeylanica*

98 *Curcuma longa*

99 *Glycyrrhiza glabra*

100 *Zingiber officinale*

101 *Piper nigrum*

102 *Piper longum*

103 Himalayan Bitumen

Having thoroughly triturated, the mixture is evaporated until paste-like and more juice or decoction is added. This process is usually repeated seven times. *Bhavana* (purification through impregnation) is similarly applied to a number of toxic plant materials, such as opium latex.

By its nature, poison naturally disperses oils and dehydrates Kapha Dosha. Its inherent heat aggravates Pitta, and its sharpness and quickness aggravate Vāta. Upon entering Rakta (blood), it invades the subtle channels of the body, eventually dissipating Ojas. However, after being subjected to Bhavana, then dried and crystallised, it can be rehydrated in milk (Kapha) and sweetened with honey. The toxicity of the venom can now be digested in the small intestine (Pitta) and finally eliminated through the colon (Vāta).

When digested in this manner, snake venom stimulates the immune system and coaxes any vitiated Dosha back to its normal abode. Kapha returns to the stomach, Pitta to the small intestine and Vāta to the colon. Consumed orally via the gastrointestinal tract, venom exposure to Doshas occurs in a fundamentally different manner, and so poses little threat.

Note: the use of snake venom for therapeutic purposes is a highly specialised field and should only be undertaken by experienced practitioners.

Part II

Workshop, equipment, method and apparatus

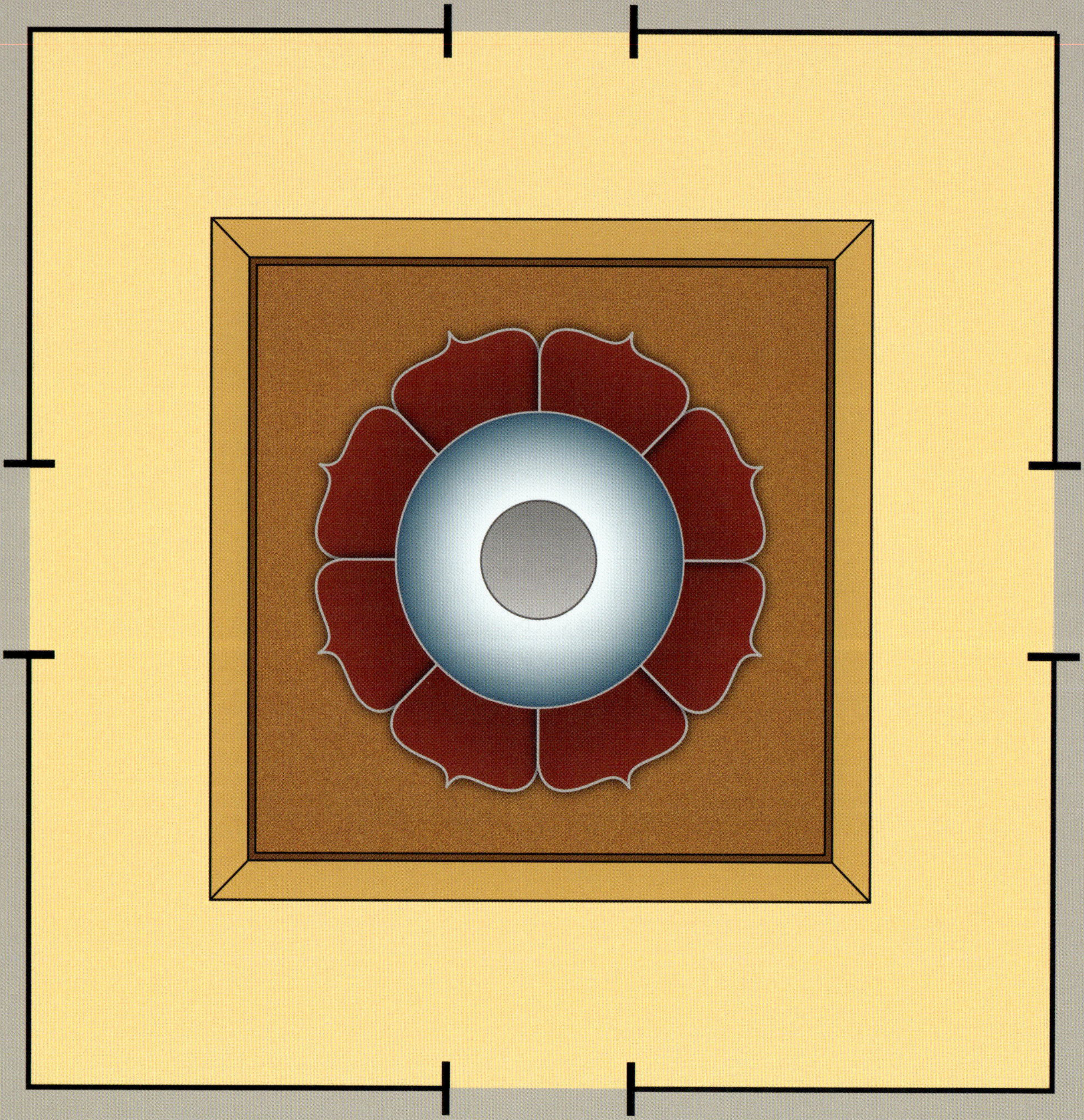

Section 4

Siting of Rasashala

4.1/ Siting of the workshop

What is a Rasashala? Quite simply, *Rasa* means mercury and *shala* means a protective covering. To put that another way, *Rasashala* means mercury in a protective place, in other words, the pharmacy where alchemical works are undertaken.

There are many recommendations made for the siting and construction of this all-important building, some very practical, others more challenging to accommodate. In this section we will take a closer look at some of the basic requirements for this most auspicious of buildings.

First and foremost, the location of an alchemical pharmacy should be situated away from residential areas. This was originally in order to protect the populace from dangerous fumes, and, more importantly, to keep this art secretive and away from prying eyes. The prospective site therefore needed to be secluded and preferably shielded by foliage, yet still be able to receive light breezes – the building could not be at the mercy of the elements. Access to fresh water, along with a supply of combustible materials, such as wood and dried cow dung, was essential. Equally important was access to sand, clay and stone. Of these essentials, cow dung was paramount, being the primary heat source. Once dried, this vital commodity would be used to heat a number of different apparatuses engaged in the process of baking (calcination), frying and boiling.

It was specified that the surrounding terrain be suitably fertile and boast a good selection of fresh herbal supplies. In this regard, *Susrutha Saṃhitā* has an interesting commentary, titled 'Discourse on the general features of ground recommended for the culture of medicinal plants or herbs'. It observes that plants mirror the features of the land on which they spring, physically, as well as energetically. Therefore, auspicious land, ripe for the bearing of medicinal-grade plants, should be firm, devoid of anthills, and away from religious rites, cremation sites or grounds used for execution. Soil should be heavy, rich, compact

Far left, Rasashala Yantra (from The Mystery of Kūkai by Tamotsu Sato)

Right, central floor mandala, drawn two cubits square, rendered using cinnabar paint

and devoid of sand and potash. It should be blackish, red or yellowish in colouration and close to water if possible. Once collected, herbs should be tied in cloth and placed inside earthen vessels or hollow tubes of wood. These should be stored in the north-eastern direction of the pharmacy, north-east being the traditional direction least likely to spoil perishable goods.

If all of this sounds like a tall order, even more considerations were advised, such as the pharmacy having almost uninterrupted sunlight yet still being able to provide some shade during the hottest part of the day and season. The pharmacy should also not encroach upon the habitats of wild animals or disturb their nesting, migratory and feeding routes. Buildings should be sited away from burial grounds, crossroads, marshy lands, or lands subject to regular flooding.

> 'Just as here in the reign of men the flame of the fire raises upward into the air and heavy objects fall when thrown to earth; so it also happens below, in the realm of the Asuras.'
>
> *Pañcha Siddhântikâ, Varāhamihira*

The diagrams shown here and overleaf present classical floor plans recommended for the construction of Rasashala. There are some variations in design, so the information given here is not to be taken as absolute.

When considering these graphics, it should be remembered that the art of Vāstu[104] would also influence siting, that is, a building's harmonics set following the eight cardinal points. Some points of interest also clearly derive from Jyotish (Vedic astrology). Both Jyotish and Vāstu are detailed sciences in their own

104 The art of laying out a construction in accordance with subtle energy principles and practical considerations, using direction, astrology, water, timing and propitiation.

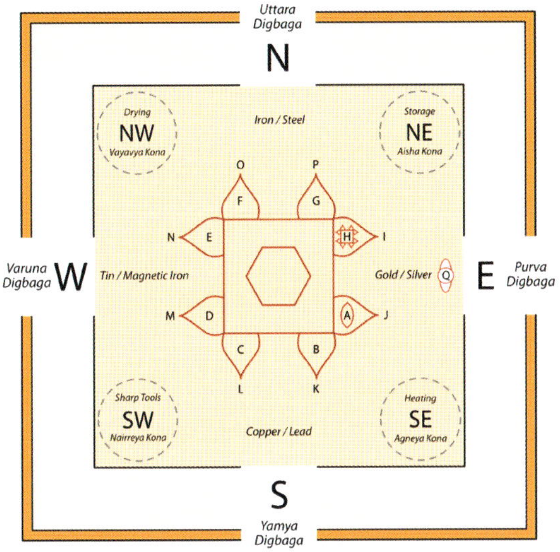

A	Parada - Mercury	**J**	Vimala - Iron pyrite
B	Hiraka - Diamond	**K**	Swarna Maksika - Copper pyrite
C	Ratna - Assorted gemstones	**L**	Chapala - Bismuth Sulphide
D	Abraka - Mica	**M**	Tuttha - Copper sulphate
E	Kanta Pashana - Iodestone	**N**	Anjana - Antimony sulphide
F	Tankana - Borax	**O**	Hingula - Cinnabar
G	Bhunaga - Earthworm Sattva	**P**	Sasyaka - Bornite
H	See numbered table below	**Q**	Shiva statue
I	Rasaka - Calamine		

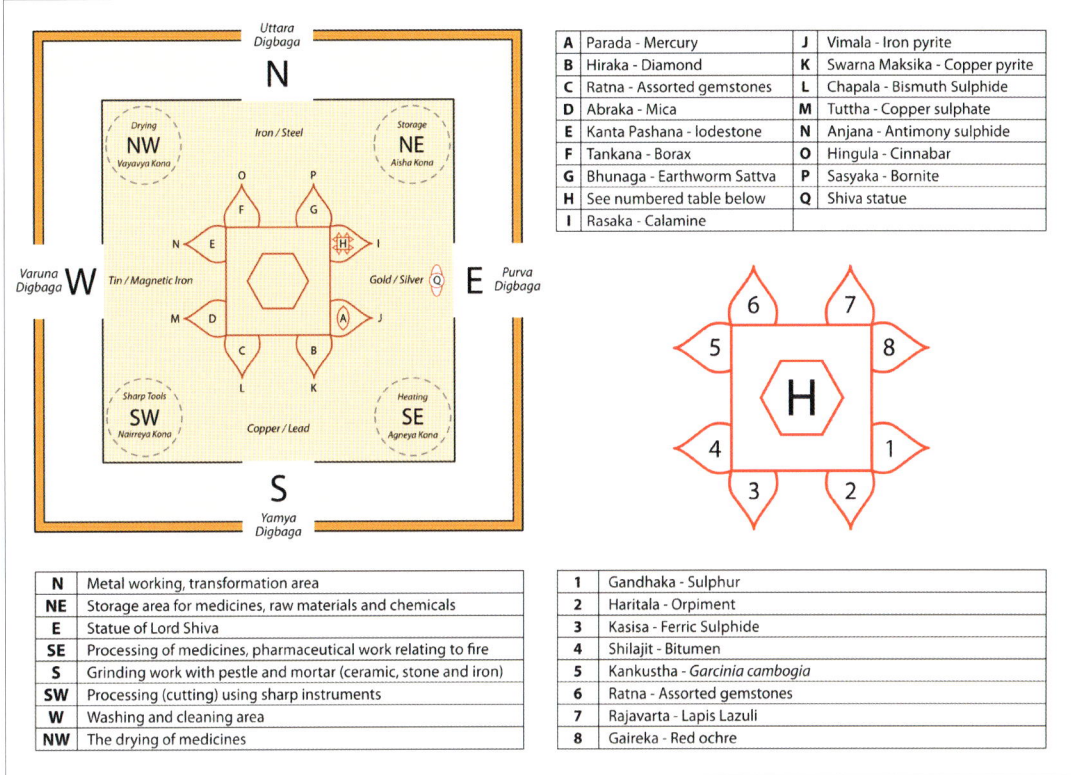

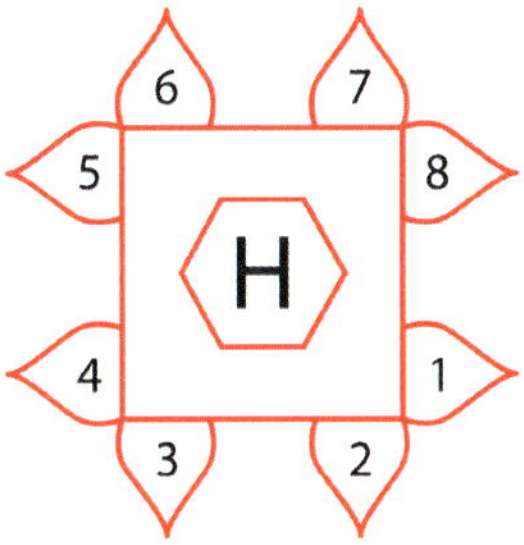

N	Metal working, transformation area
NE	Storage area for medicines, raw materials and chemicals
E	Statue of Lord Shiva
SE	Processing of medicines, pharmaceutical work relating to fire
S	Grinding work with pestle and mortar (ceramic, stone and iron)
SW	Processing (cutting) using sharp instruments
W	Washing and cleaning area
NW	The drying of medicines

1	Gandhaka - Sulphur
2	Haritala - Orpiment
3	Kasisa - Ferric Sulphide
4	Shilajit - Bitumen
5	Kankustha - *Garcinia cambogia*
6	Ratna - Assorted gemstones
7	Rajavarta - Lapis Lazuli
8	Gaireka - Red ochre

Above, classical floor plan for Rasashala: the centrally inscribed Yantra hosts offerings of fifteen individual Rasa materials along with a self-contained Yantra (H) featuring an offering of eight additional materials

right and beyond the scope of this book. However, it is worth noting that both would influence any construction of sacred spaces.

The general energetics of the pharmacy floor plan shows a square, denoting a balanced Brahminical site.[105] Additional recommendations advise a north-facing construction, erected upon sweet-smelling ground, white in colour and consecrated to *Brihaspati* (Jupiter) and *Brighuji* (Venus), the two principal planets associated with learned instructors, lawgivers and priestly advisors. Both planets offer these same credentials within the celestial court (see p.97, Planetary status and qualities of metals).

105 Square, sloping north, sweet fragrant soil and white = Brahmin. Short rectangle, sloping east, astringent/bitter soil and reddish in hue = Kshatriya. Long rectangle, sloping south, sour soil and yellowish in hue = Vaishya. Long narrow rectangle, sloping west, pungent soil and blackish in hue = Shudra.

In cases where most, or preferably all, of these requirements were met, the proposed site would be cleared and appropriate offerings made to the relevant deities. Ceremonies for the placation of deities included anointing the ground with jasmine flowers mixed with ghee, honey and jaggery. After consecration, construction could begin.

Naturally times have changed, and the availability of or access to land is now a big issue. As these recommendations show, the siting of a Rasashala was no small affair. From inception to completion, the whole project required immense planning and consideration. Of course, this is a major headache for anyone today who wishes to build such a facility in according to classically prescribed designs.

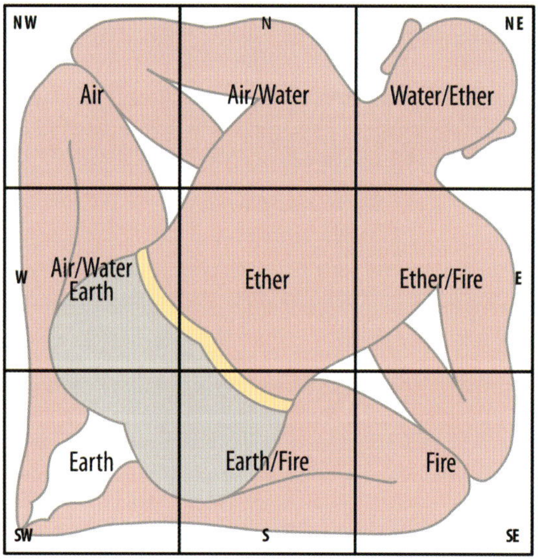

Left, Sthir Vāstu diagram showing cardinal points and corresponding elements. The centralised figure is Vāstu Purusha, lying with his head toward the north-east and facing downward; the demon is imprisoned on the earth's surface

4.2/ Sarpa (snakes)

Having met all of these criteria, the prospective plot for the pharmacy then needed to be inspected for omens of inauspiciousness, including foul smells, sourness of soil, fungal growths, excessive dead foliage, stagnant water, poisonous insects and, of course, the presence of snake lairs.

Note: The presence of a snake could also be as construed as a good omen,[106] however, to avoid their displeasure, it was considered good form to offer sweetened milk at the south-western boundary point and to mentally ask for their blessings. Forcible removal or the killing of a snake was considered inauspicious.

Resident serpents might also be further pacified by the use of Yantra technology. These were sacred diagrams, inscribed and hung upon the outer wall of the pharmacy. These protective Yantras could also be dyed or painted with *Hiṅgula* (powdered cinnabar). *Sarpa Bhaya Hara Asseeyaa Yantra* is one such example. This is a square, sixteen divisional, number-eighty-based Yantra (see illustration). By counting in any direction the number always totals eighty. Furthermore, eighty can be simplified to 8+0=8, eight being the number associated with longevity and death.

੩੨ 32	੩੯ 39	੨ 2	੭ 7
੬ 6	੩ 3	੩੬ 36	੩੫ 35
੩੮ 38	੩੩ 33	੮ 8	੧ 1
੪ 4	੫ 5	੩੪ 34	੩੭ 37

Left, Sarpa Bhaya Hara Asseeyaa Yantra (numbers in red represent the number values)

106 The presence of a snake or snakes can be auspicious as these creatures were seen as agents of the 'Nāga Folk'.

Essentials in the pharmacy

When it comes to pharmacy essentials, there are items and ingredients that should always be easily accessible, since they are crucial and may be required at any moment. The following lists identify and outline many of these essentials. Additionally, these tables offer viable alternatives, allowing for substitution if a particular item is unavailable or out of season. Moreover, they also assist in determining preferred choices. These have been highlighted by a (*) icon.

Pharmacy essentials	
Amlavarga (essential sour drugs)	Lemon/lime* (*Citrus limon*) and (*Citrus latifolia*) Amlavetasa (*Garcinia pedunculata*) Tamarind (*Tamarindus indica*) Pomegranate (*Punica granatum*) Malabar tamarind (*Garcinia cambogia*) Chickpea (*Cicer arietinum*) Jujube fruit (*Ziziphus jujube*) Karanda (*Carissa caranda*)
Amlapañchaka (5 sour drugs)	Tamarind* (*Tamarindus indica*) Pomegranate (*Punica granatum*) Jujube fruit (*Ziziphus jujube*) Kokum (*Garcinia indica*) Indian sorrel (*Hibiscus sabdariffa*)
Pañchamrittika (auspicuious clay mixture)	Powdered brick* Gaireeka (red ochre) Rock salt Plant ashes Termite cement
Pañchatikta (essential bitter drugs)	Neem* (*Azadirachta indica*) Guduchi (*Tinospora cordifolia*) Moringa (*Moringa oleifera*) Agastya (*Sesbania grandiflora*) Bitter gourd (*Momordica charantia*)

Madhuratraya (5 sweet drugs)	Honey* Jaggery Ghee Milk Cane sugar
Raktavarga (reddening drugs)	Saffron* (*Crocus sativus*) Red sandalwood (*Pterocarpus santalinus*) Cinnabar (HgS) Khadira (*Acacia catechu*) Lac (*Peepal shellac*) Manjistha (*Rubia cordifolia*) Arjuna (*Terminalia arjuna*)
Pancagavya (bovine drugs –cow/buffalo)	Milk* Ghee Yoghurt Urine Faeces
Ksharashtaka (astringent drugs)	Arka* (*Calotropis gigantea*) Apamarga (*Achyranthes aspera*) Sesame (*Sesamum indicum*) Tamarind (*Tamarindus indica*) Yava (*Hordeum vulgare*) Snuhi (*Euphorbia neriifolia*) Palasha (*Butea monosperma*)
Dugdhavarga (essential types of milk)	Cow* Buffalo Goat Ass Camel Elephant Human
Tailavarga (essential oils)	Sesame* (*Sesamum indicum*) Coconut (*Cocos nucifera*) Mustard seed (*Brassica juncea*) Castor seed (*Ricinus communis*) Bhallātaka (*Semecarpus anacardium*) Jayapāla (*Croton tiglium*)

Ksharas (latex drugs)	Arka* (*Calotropis gigantea*) Banyan tree (*Ficus benghalensis*) Snuhi (*Euphorbia neriifolia*) Karaveera (oleander, *Nerium indicum*)
Pancalavaṇa (5 essential salts)	Rock salt* Black salt (natural and artificial) Sea salt Sal ammoniac Romaka
Pancamitra (metal digestors)	Gunja* (*Abrus precatorius*) Jaggery Borax Honey Guggulu (*Commiphora mukul*)

4.3/ Modern interpretations

In modern cities like Colombo, space comes at a premium, and rent is high. As a result, most modern facilities don't have the luxury of accommodating these stringent Vāstu rules. That being said, I was able to visit a few pharmacies and take note of their position and construction during my time in Sri Lanka, and most of these pharmacies did try to incorporate some of these time-honoured designs to promote favourable energetics wherever possible while at the same time neutralising some potentially harmful features. This was primarily done through ritual, offerings, Yantra and prayer, for example in those cases where buildings were believed to be subject to malevolent or supernatural natural forces, you might see a small cloth bag of coloured fabric filled with pungent-smelling herbs to ward off any resident evils.

My nearest pharmacy was at the local Āyurvedic hospital, about a twenty-minute walk from my residence. This particular building was quite impressive, since it stood on one of those sites lucky enough to have had much thought and money spent on it. I was given a tour soon after arriving at the hospital and, even more fortuitously, was allowed to work in it later. The pharmacy itself was small, but as I was proudly informed by the hospital's administrator, it *"had been especially commissioned by the hospital and conformed to many, if not most of the accepted standards of traditional pharmacy construction."* Given that space was an issue on-site and that the newly commissioned building was an afterthought on the hospital's premises, they had done a great job of getting so much packed into a relatively enclosed space.

I got to appreciate how stress-free it was to have everything so close to hand when working there. It honestly made for smooth and speedy production. In the relatively short time I worked there, I surpassed in making remedies in

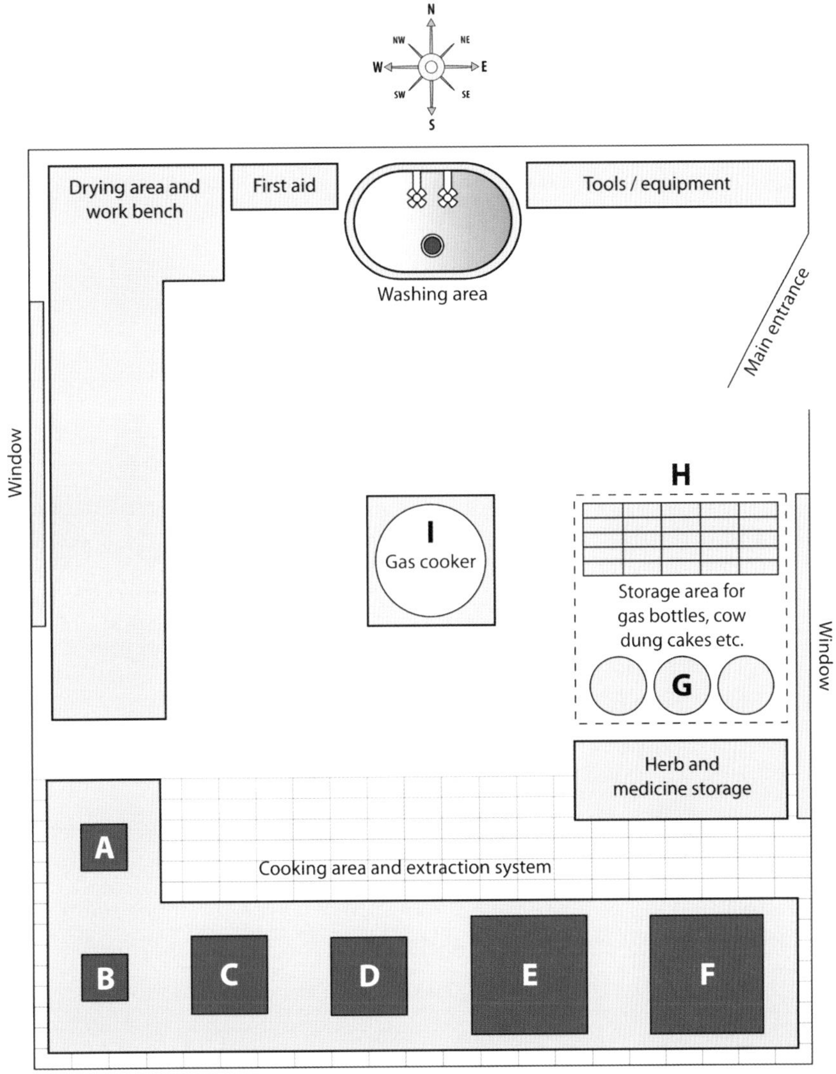

Left, floor-plan of Rasashala, Rajagirya Āyurvedic Hospital, Colombo

Key:
(A+B)= Laghu Puṭa;
(C+D)= Gaja Puṭa;
(E+F)= Maha Puṭa;
G= propane gas bottles;
(H)= dried cow dung fuel storage;
(I)= main gas burner

half the time. Not unexpectedly, the pharmacy also came with a resident cobra, although I only saw it once, vanishing under the building close to where they piled old wood and cloth sacks, in a south-west direction (of course). Luckily, that clinic wall was less accessible, however any foray in that direction suddenly made me extra mindful not to disturb dark corners.

As accurately as possible, I have reproduced the Rasashala at the Rajagiriya Āyurvedic Hospital Colombo. See the illustration on the previous page for more details.

4.4/ Celestial considerations

After selecting a pharmacy's siting, it is considered good practice to elect several suitably auspicious moments to lay one's foundation stones or officially open the building for use.

In Jyotish, electing a suitable time is called *Muhurta,* known as 'electional astrology' in the Western tradition of astrology. The astrologer is asked to select an appropriate horoscope within a given time frame that sees the nine planets occupying the most auspicious positions possible. That being said, determining a suitable window of opportunity is no easy task, especially when factoring in all the astrological dos and don'ts.

Using the rules of Muhurta, astrologers scan the months ahead and look for a number of auspicious times; these focus on moments when the Moon (especially important), the Sun and those planets having a direct bearing on the building's functionality are closely considered.

In the case of Rasashala it is Mars, Mercury and Jupiter that will be specially studied. Mars rules fire and heating, and Mercury and Jupiter rule intellect and adherence to good practices. In the case of our pharmacy consecration, attention would also be afforded to a favourable fourth astrological house, which details one's dwelling place or base. Planet Saturn holds sway over the land quality and the karma of that space in general. The tenth astrological house gives insight into the project's vocational strength and overall productivity and prosperity. Lastly, the sixth astrological house provides insight into the health-giving potential of the building and its overall medicinal potency – if you will.

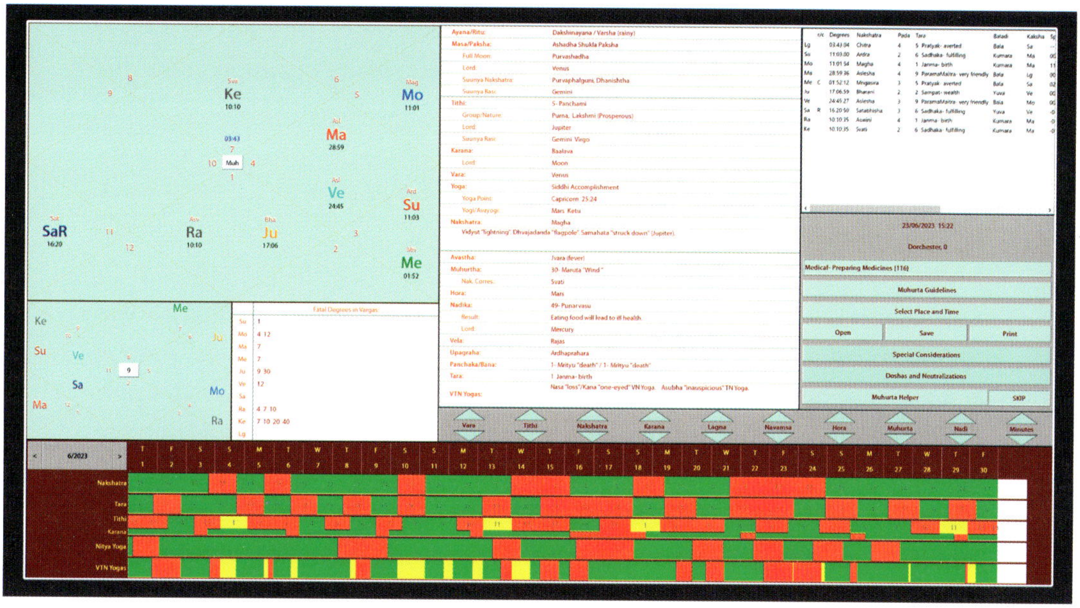

Above, modern Jyotish software allows easy calculation of Muhurta. What would take an accomplished astrologer to calculate in hours can be shown in the press of a button. Readers can see how many factors calculation can include, although, in practice, the actual selection of dates is generally far less sophisticated.

Additionally, an astrologer may also wish to consider the history of the proposed site, the horoscope of the person most directly responsible for its construction, and the nature of its intended use. Finally, and most importantly, the astrologer must work within the time constraints of that construction, taking into account the weather and seasonal variations; for instance, which seasonal variance is least likely to interfere with construction. Holidays and religious festivals and observances, of which there are many in Sri Lanka, also need to be taken into consideration. Then, armed with all this information, our astrologer must search for a suitable window in time to a point when all planets are positioned to give beneficial effects. With nine planets, twelve astrological signs, twelve zodiacal houses, holidays, seasons and festivals, there are many variables!

In India, astrologers commonly use Muhurta to elect all manner of auspicious events, including marriage, job applications, financial investments, home buying and family planning. The list is endless, and all can require astrological services. Electing a Mhurta when everything is perfect is impossible, and so, many times, it will come down to compromise and experienced judgement in the mind of the astrologer. Given a reasonable window of time, our astrologer can and will find something that lifts the energetic of the occasion and helps promote success.

4.5/ Final note

One final note on the science of Vāstu Shāstra[107] might be that it strongly favours the northerly direction for its cooling Himalayan breezes and rains. From a Vedic perspective, our planet's northern hemisphere was considered *Upari*, uppermost and auspicious. The northern hemisphere was also considered the abode of the Devas or gods. In contrast, the southern hemisphere or *Adhas* (lower realm), was thought to be the abode of hellish forces; hence the *Vāstu Purusha* figure (Sthir Vāstu diagram) faces downward toward that hemisphere.

These observations are not meant to denigrate those dwelling at southerly latitudes but to draw attention to the ancients' observation of the Earth and its energetics relative to its two poles and the qualities inherent to these upper and lower planes of existence. This observation extended to the edges of the cosmos, as a similar imaginary line, drawn across the heavens, demarcated *Deva-Bhāga* and *Asura-Bhāga*, the former representing a godly realm, the latter a worldly demonic realm. The ancients noticed that many major constellations and single prominent stars appeared to mirror one another within the complex star field that is the nighttime sky, and took this as confirmation of their parallel existence.

This reasoning also highlights the importance of one's geographical location relative to Earth's equator. When we compare ancient practices like Vāstu with the Chinese system of Feng Shui,[108] both variations of ancient geomancy paint a stark difference in opinion, particularly with the northern winds. For example, Feng Shui regards a northern wind as a harsh influence upon one's home and seeks to minimise exposure to it. Vāstu Shāstra, in contrast, encourages inviting a northerly breeze to enable a building to benefit from its cooling energetic.

107 Architectural system following design principles laid out in ancient texts based on integration with natural and celestial directional alignments.

108 The ordering of sacred space, from an ancient Chinese perspective.

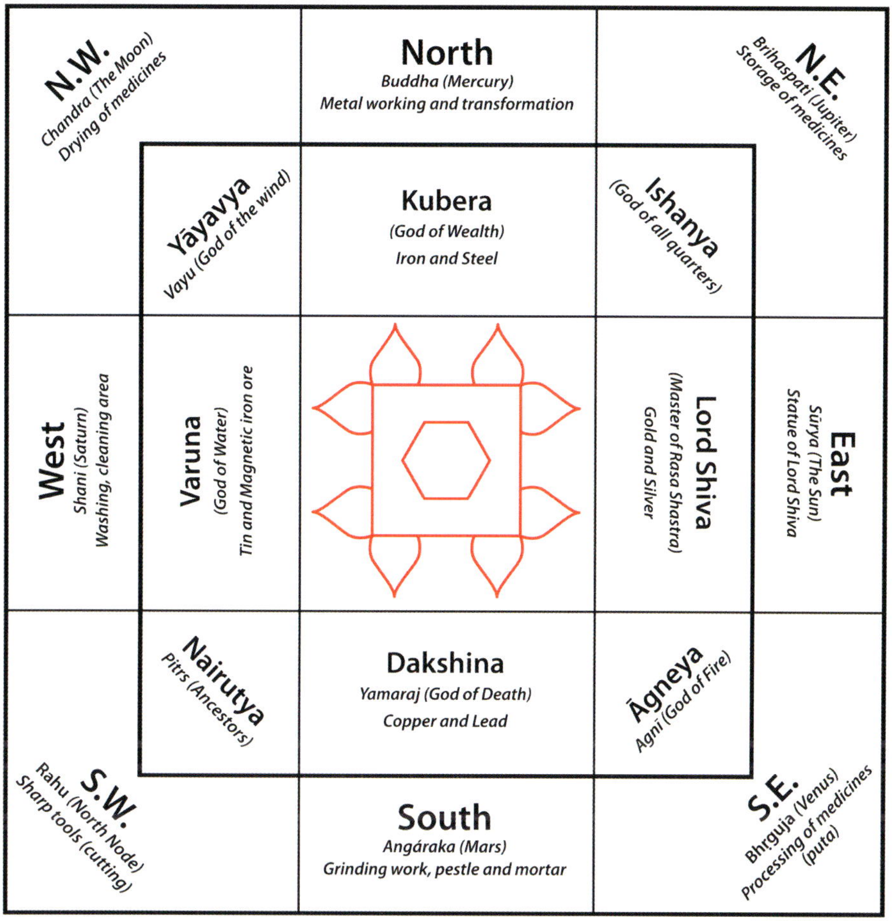

Left, eight cardinal points and respective planetary rulers. Taken from India by Al-Bīrūnī, cir.1030CE

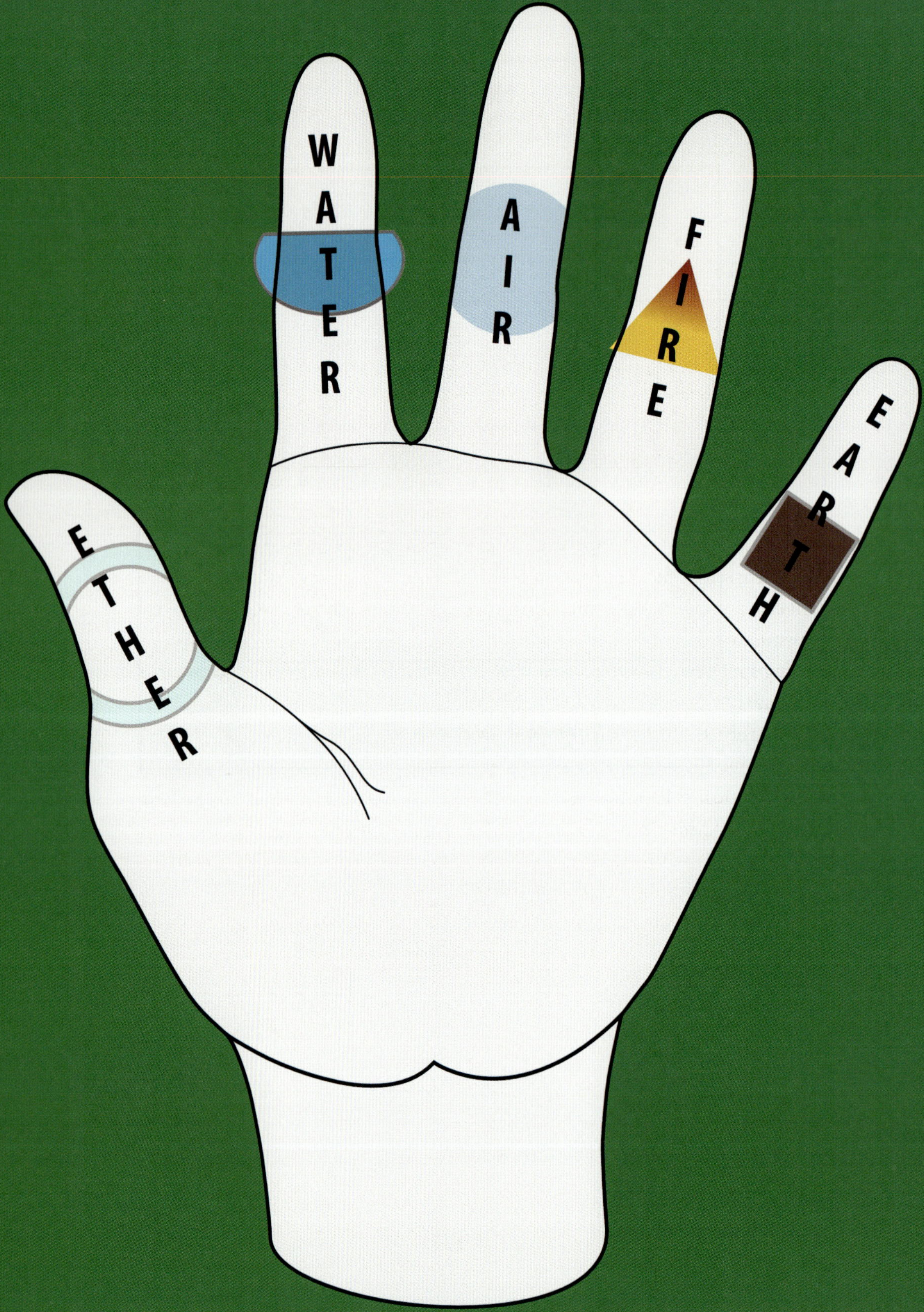
WATER
AIR
FIRE
EARTH
ETHER

Section 5

Selection of apprentices

5.1/ Alchemist and apprentice

> 'Metallurgy is a science which was taught by Lord Shiva himself. It is to be given to an earnest disciple by the preceptor in accordance with the usual procedure. In an auspicious moment, when the Moon and stars are strong, an earthen vessel full of water and containing gold, gems and fruits is to be placed in front of the Rasaliṅgam and covered with pieces of cloth.'
>
> *Rasa Jala Nidhi*

Training an alchemical apprentice was essential yet potentially hazardous, as imprudent selection would not only squander one's time but could be potentially dangerous to future patients. The choice was critical for ensuring the precise transmission of knowledge as well as the competency of practical endeavours. Making sure neither side invested unwisely, the ancient classical works provided a manual for the appropriate selection of a protégé.

To secure the transmission of knowledge and rituals across successive generations, the key to success is always a reliable selection process. While the theoretical side of alchemy could perhaps be transmitted to any willing student in a short time, mastering the practical side takes many years. With this kind of time-investment involved, it became imperative that would-be students pass a number of criteria. Rasa Shāstra texts recommend the following twelve traits for competency and reliability in the selection process.

Students to be:

1. Of good lineage and conduct;
2. Devoted to one's teacher;
3. Showing outstanding devotion to their studies;
4. Good-mannered and well spoken;
5. Alert and diligent, and free from laziness;

Far left, Pañcha Mahābhūta (five elements) as written upon the hand

6 Partake of wholesome food and drink;
7 Well conversant with the theories of Āyurveda (and Jyotish) where necessary;
8 A devotee to the gods and a believer in karma;
9 Not inclined to speak ill of others;
10 Honest;
11 Brave of heart;
12 Aspire to be an expert in their chosen profession.

The same text asks twelve questions of one's teacher/master to balance the selection process, as a commitment to one's master would engage the student for many years and so too would be a serious investment of their time. Essential qualities for one's tutor include the following.

One's master should be:

1 Of good lineage;
2 Well-skilled in the use of Āyurveda and the identification of disease;
3 A devotee of Lord Shiva and the Goddess Parvati;
4 Patient, thoughtful and compassionate;
5 Skilful in the medical-alchemical arts with diligence in experimentation;
6 Knowledgeable in the use of mantras;
7 Proficient in Vāstu (geomancy);
8 One who rises with the sun and retires at its setting;
9 Brave and honest at heart;
10 Partaking of wholesome food and drink;
11 Skilled in the use of Jyotish;
12 One who works with regularity and efficiency.

Lastly, there is some commentary on the process of selecting staff to work at one's pharmacy, albeit they would not be required to hold an exacting knowledge of alchemical processing and its medicinal applications.

Staff ultimately became resident and, in some ways, partly responsible for the end quality of the medicines produced; however, they were always to work under supervision. The quality of the staff was another integral part of the alchemical operation, and, just like an inappropriate student, poor selection of staff could seriously hamper production and efficiency and, most importantly, bring the pharmacy's reputation into disrepute. The choice of appropriate staff includes the following five guidelines.

Pharmacy staff should be:

1. Of an honest and truthful nature;
2. Aware of their duties;
3. Accustomed to cleanliness at all times;
4. Energetic, strong, brave and diligent;
5. Partaking of wholesome foods and daily regimes.

5.2/ Shat-Samudrik Shāstra

Bodily proportion, symmetry and physical strength greatly interested the ancients. One's size, weight and colouration were critical factors in the selection process. The body was carefully inspected for *Chinha* (symbols or auspicious signs). The most common place for these symbols to appear was the face, feet and hands.

Physical development through measurement, i.e. height, length, breadth and circumference, were all standardised using a measure known as an *aṅgulī*, roughly 2cm, or the breadth of an adult finger.

The core elements of this science are present in disciplines such as *Shat-Sāmudrika Shāstra*, a study that focuses on the six limbs or branches of the human body: head, torso, two arms and two legs. From the earliest age, careful observation and measurement were made of children's physical proportions and symmetry. Any impedance in growth rate or the appearance of strange bodily markings was treated with great seriousness.

Caraka Saṃhitā gives quite concise measurements for 'normal' foetal development, including the skeletal frame, joints and organs. Measurement also included relative distance and angles between the joints, appendages and organs. In addition, the examination of their parents was also used to assess a student's prospective temperament and health predispositions.

This same text goes on to several key tissues in the body and their inherited relationship to one's parents.

Father	Hair (on head and face), nails, fine body hair, teeth, bones, veins, ligaments, arteries, and semen (male)
Mother	Skin, blood, flesh, fat, heart, lungs, stomach, liver, small intestine, spleen, kidneys, umbilicus, bladder, colon, ovum (female), and rectum

5.3/ Hastā Rekha Shāstra

Nowhere were bodily signs taken more seriously than those written upon the palm, hands being so instrumental in our ultimate success or downfall.

Hastā Rekha (*Hastā* means hand and *Rekha*, the lines upon the hands), often proved a valuable tool in the selection process of apprentices. A close study of the hand and the lines upon it present the viewer with a living horoscope, written upon the skin and supported by flesh and bone. Palmistry requires no horoscopic calculation yet offers insight into the individual's true nature, as well as those planets having greater or lesser influence over the individual.

Planetary influences are read primarily through the four fingers and thumb and various fleshy projections upon the palm. These latter 'pads' are known as the planetary seats or *Parvata*. All are connected and fully interact with the Rekha. A study of the hand and palm can help determine the Āyurvedic constitution. Hand analysis, in many respects, can be quite an accurate assessment of your constitution, as the hand changes little throughout a lifetime.

In India, the practice of Hastā Rekha Shāstra and Jyotish are inseparable, and both are routinely studied to corroborate one another. In the West, these two disciplines went on to become more specialist in their own right, eventually diverging into Western astrology and Palmistry.

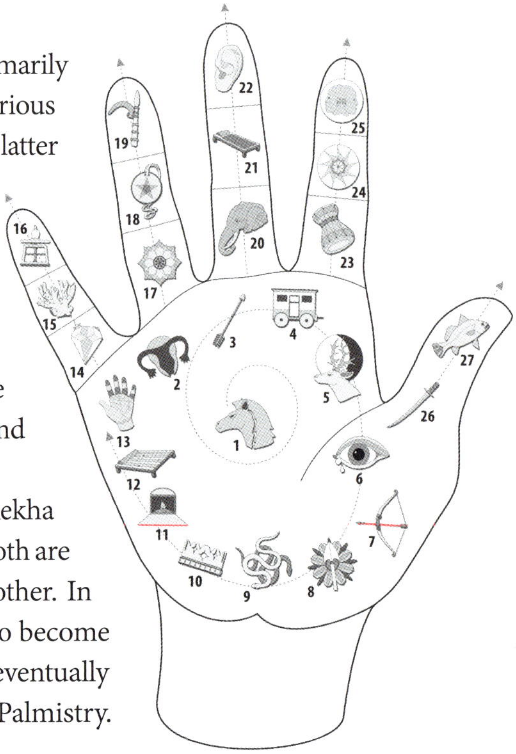

Far right, Nakshatra (lunar asterisms) as written on the hand

It was believed the hand was a microcosm of the universe, upon which the signatures of the planets were written. If one were so inclined to learn their language, the palms of a potential student would be a strong deciding factor in accepting them for tutelage.

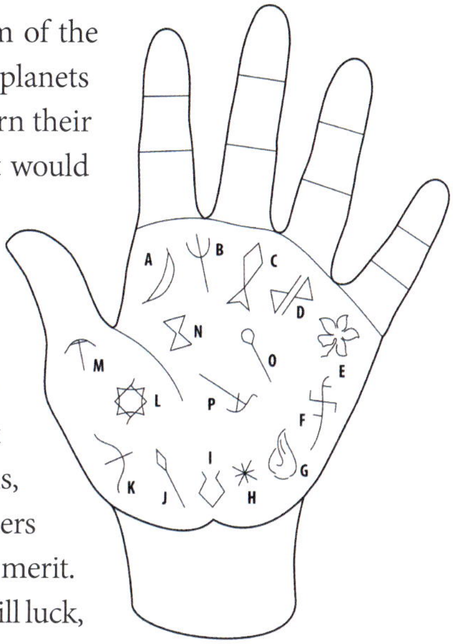

Left, auspicious symbols seen on the palm: A) crescent moon; B) trident; C) fish; D) flags; E) lotus; F) swastika; G) conch; H) six-pointed star; I) water pot; J) spear; K) bow and arrow; L) eight-pointed star; M) umbrella; N) two-headed drum; O) mace and P) plough

Symbols observed upon the palm

Chinha (symbols) upon the palm were particularly interesting and studied in great detail. Some were considered positive omens, others signs of impending good luck. Yet others were thought to indicate individuals of great merit. Conversely, some symbols were regarded as ill luck, detrimental to health and indicative of argumentative, morally deficient or troubled individuals.

Auspicious signs on the palm were said to include: *Padma* (lotus flower), *Matsya* (fish or fish's tail), *Shankha* (conch shell), *Trishula* (Shiva's trident), *Asta-chakra* (eight-sided disc), *Chandra* (the Moon) and *Nakshatra* (stars of less than six points). Inauspicious signs on the palm included: limbed squares (boxes with lines attached), *Sarpa* (snakes), *Shrinkhala* (chains), *Āvarita* (nets) and *Vi-shrinkhala* (broken lines). See the illustration above for some examples of Chinha.

Lines and planets upon the palm

The primary Rekha are: *Ayu Reka* or life line, *Matru Rekha* or head line, *Hṛdaya Rekha* (heart line) and *Karma Rekha* (fate line); see the illustration overpage. These four are laid down like foundation stones; collectively representing four significant life influences: longevity, health, intelligence and fate. While there are particular circumstances under which two lines can appear absent or merged, it is uncommon. For the most part, every palm will display four primary Rekha.

The next and most important assessment is the strength and prominence of a particular planet. The nine planets are each designated rulership of a specific finger or mound upon the palm. The individual strength and activity of each is assessed from the above fingers and mounds and their connections

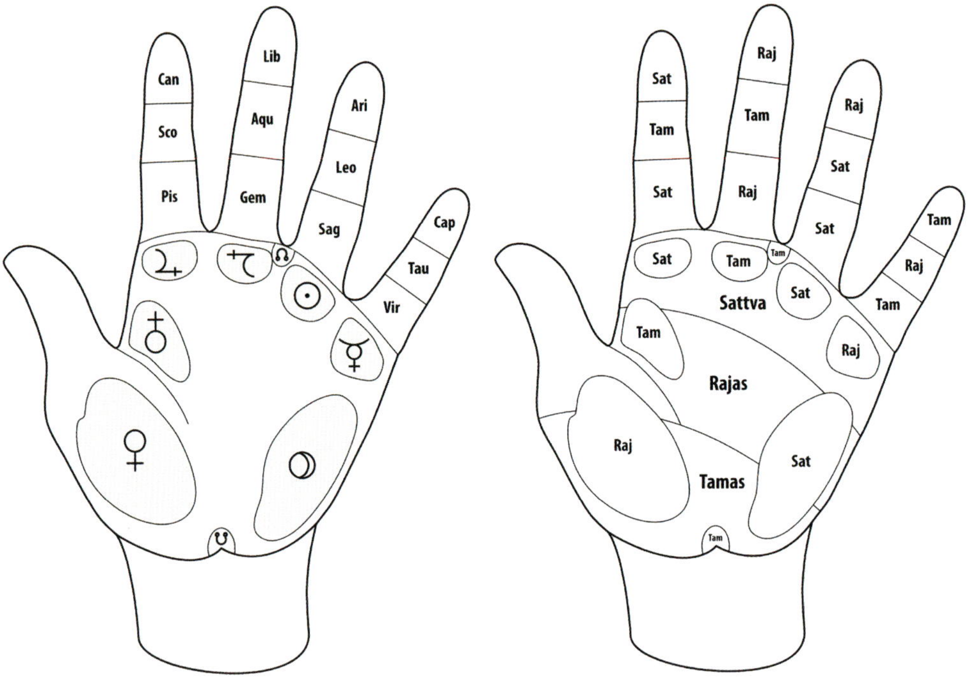

Right, planetary positions on the hand and the zodiacal signs on the fingers; far right, the same positions and their relationship to the three Maha Guna, i.e. Sattvic, Rajas and tamas

to the rest of the palm via the lines. Strong, clear and unbroken lines upon the palm were deemed auspicious; Rekha directly relating to a planet were seen as being of extra importance and elevated. The following is a short précis of the seven planets, two nodes, and the qualities they bestow.

The seven planets

Auspicious Rekha relating to **the Sun** makes one honoured, trustworthy and a master of men, leading with courage and perception. The native is brave and virtuous but also harsh and vindictive. He is prone to bilious (Pitta) diseases.

Far right, Sūrya: the Sun

Auspicious Rekha relating to **the Moon** makes one sensitive, caring and nurturing. The native is of a feminine disposition, youthful, fickle and prone to worry. Persons under the influence of the Moon suffer diseases of excess Kapha.

Left, Chandra: the Moon

Auspicious Rekha relating to **Mars** indicate those of a charitable disposition, masters of fire, forge and metalworking. Mars is quarrelsome and cruel but is a master of technical works and engines of war. The native, under the influence of Mars, suffers diseases of excess Pitta.

Left, Kuja: Mars

Auspicious Rekha relating to **Mercury** indicate those of an enquiring, questioning mind. Mercury is quick-witted, humorous and enjoys practical jokes; it readily assumes the attributes of stronger planets through mimicry and mockery. Mercury is also the planet most associated with the healing arts. The native under the influence of Mercury may suffer diseases of any Dosha.

Left, Budha: Mercury

Right, Guru: Jupiter

Auspicious Rekha relating to **Jupiter** makes one learned, wise and authoritative. The native attains wealth, status and health and is honoured by kings. Persons under the influence of Jupiter suffer diseases of excess Kapha.

Right, Shukra: Venus

Auspicious Rekha relating to **Venus** marks one with congeniality, diplomacy and great beauty. The native enjoys much luxury, pleasant company and fine clothes. The native is refined and learned in many types of scripture but uses knowledge for personal gain. Those under the influence of Venus suffer watery (Kapha) types of diseases.

Right, Shani: Saturn

Auspicious Rekha relating to **Saturn** marks one with patience, steadfastness and longevity. The native gains many fixed assets (such as land) through hard work and diligence. Saturn is melancholic in nature, rigid in stature and withdraws from worldly matters. Saturn frequently delays but never denies. The native under the influence of Saturn suffers airy Vāta types of diseases.

Rāhu and Ketu

Generally, Rāhu and Ketu tend to act as malefic forces; however, they do occasionally produce auspicious effects, particularly when connected to favourable planets via Rekha. Rāhu is positioned on the palm between the base of the ring and the middle fingers. The Sun rules the ring finger and Saturn the middle finger. Both the Sun and Saturn have deep connections to Rāhu. Firstly, Saturn and Rāhu share many qualities, in some ways reinforcing one another in their effects. Secondly, Rāhu (and Ketu) represent the eclipse points, both having the power to obscure sunlight. Ketu, who acts similarly to Mars, is positioned between the Moon and Venus mounds at the palm's base. Between their two localities, the nodes metaphorically cleave the palm into two halves; this 180° axis is astronomically mirrored, as the two eclipse points are always directly opposite one another.

Rekha relating to Rāhu mark one with foresightedness, imagination and innovation. This planet is typically associated with occult knowledge; it is worldly and highly perceptive. Rāhu is famed for its unpredictable, sometimes violent energy. It powerfully amplifies the effects of any planet it is closely connected to. The native under the influence of Rāhu typically suffers from airy Vāta types of disease.

Rekha relating to Ketu mark one with sharpness and a Martian temperament. Ketu can be impulsive and reactive yet otherworldy and strongly intuitive. He is the planet said to bestow longevity, enlightenment and Moksha (liberation from suffering). A native under the influence of Ketu similarly suffers airy Vāta types of diseases.

Auspicious signs for an acolyte

Auspicious signs are seen in the area ruled by the planet Mars (between the thumb and forefinger). Called the Line of Enemies or *Maṅgal Rekha*, it can denote one skilled in the art of metalworking. The mound of Mars should be inspected for prominence, but if it is heavily featured, i.e. reddish, scarred or angry looking, it provokes the negative Martian attributes like arguments, irritation and short-temperedness.

The Venus Belt or *Shukra Mekhala* (sometimes referred to as the Venus Girdle) is ostensibly another favourable sign on the palm, making one respectful, knowledgeable and convivial, however, should this belt descend to touch the *Hṛdaya Rekha* (heart line), the individual becomes too preoccupied with material gain, luxury and the opposite sex.

Budh Rekha (the Mercury line) is considered a beneficial sign as it stimulates one's interest in the healing arts and, most importantly, connects one to Pārada (mercury) and, by association, the alchemical arts. However, if overly strong, this line enhances and promotes mercantile qualities in a student, perhaps making them too business-like in their activities.

Confused or broken lines fragment one's energy; a weak fate line might make one work-shy or feel somewhat lost in life; Karma and Saturn are strongly connected to the fate line.

Planetary lines entangled with Rāhu or Ketu were generally seen as portents of bad luck, addictions or those seeking power; however, if the nodes are held in check and balanced by stronger planetary forces, they might prove helpful in the taming of poisons, working with and purifying toxic minerals or plants. After all, Rāhu and Ketu were once united as a single body and partook of the nectar of immortality; see Sagar Manthan and the Churning of the Milky Ocean legend.

Other lines of interest

Other lines of interest are those descending from or encircling the fingers; these include the Sun line or *Keerthi Rekha*, the Jupiter line or *Dīksha Rekha*, the Jupiter Crescent or *Guru Chandrika* and the fortune line or *Punya Rekha*. Any of these, as mentioned earlier, was seen as a highly positive sign. Keerthi Rekha denoted individuals of outstanding performance; trustworthy, honourable and able to command. Dīksha Rekha indicated a student well versed in the sciences, natural law and religious observances. Guru Chandrika emphasised one wise and thoughtful, devoted to the gods, while Punya Rekha bore the fruits of previous good karmas and past life merits.

Lastly, the twelve finger joints on the four fingers relate to the twelve astrological signs of the zodiac, so proportions, malformations or prominent signs and Rekha were seen as reflexive representations of the twelve zodiacal signs.

Note: Just as the planets remain in motion and constantly change their positions, so too can Rekha appear to morph, collide and vanish from view. The hand is a ledger, which updates as new karma is created and discharged throughout life. The course of the four primary lines is less subject to change, but even they may be modified over time, however just like the course correction of a great river is measured in centuries as it wanders through the landscape, the primary Rekha moves little during the course of a single lifetime.

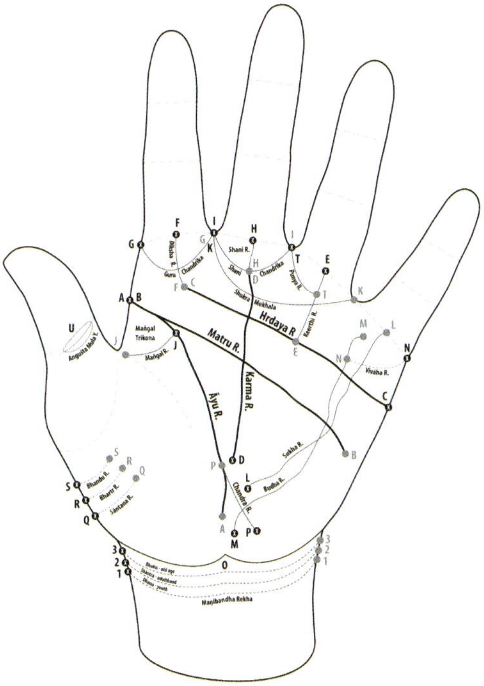

Left, primary lines:
A) Life line/Āyu Rekha;
B) Head line/Matru Rekha;
C) Heart line/Hṛdaya Rekha;
D) Fate line/Karma Rekha

Secondary Lines:
E) Sun line/ Keerthi Rekha;
F) Jupiter line/ Dīksha Rekha;
G) Jupiter crescent/ Guru Chandrika;
H) Saturn line/ Shani Rekha;
I) Saturn crescent/ Shani Chandrika;
J) Line of enemies/ Maṅgal Rekha;
K) Venus belt/ Shukra Mekhala;
L) Contentment line/ Sukha Rekha;
M) Mercury line/ Budh Rekha;
N) Union line/ Vivāha Rekha;
O) Bracelets/ Maṇibandha Rekha;
P) Moon line/ Chandra Rekha;
Q) Progeny line/ Sāntana Rekha;
R) Siblings line/ Bhartr Rekha;
S) Kinship line/ Bhanḍu Rekha;
T) Fortune line/ Punya Rekha;
U) Thumb Barley/ Anguśta Mula Yava

Note: the black dots with an 'x', mark the start of Rekha and the grey dots, their end point.

For more information about the practice of Indian Palmistry, see my earlier work: *Vedic Palmistry ~ Hastā Rekha Shāstra.*

5.4/ Shodhana (purification)

Shodhana is the name used in Rasa Shāstra to describe the process of purification whilst reducing gross elements to subtle ones through trituration (a grinding process) and, during the same process, removing toxic content. Shodhana employs various plant juices, herbal decoctions and salts, along with lots of intensive grinding and sieving. These actions help reduce particle size and provide mediums for the aforementioned grinding to be more effective.

Unlike modern purification, which systematically removes elements to obtain purity, Shodhana ironically adds a variety of substances to materials. The effect of this approach is to loosen and coerce toxic elements to conjoin with the medium (usually liquid); this can then be removed during the washing and evaporation process.

Loss of material is inevitable during Shodhana – the labour-intensive process of washing, grinding, drying and heating slowly, and imperceptibly, erodes volume. Even though purification requires adding other ingredients, the overall trend is toward *lightening* and *lessening*. Depending upon the material in question, it can be expected that at least one-third to half the total volume will vanish during most purification processes.

Shodhana is a cyclical process, slowly moving the initial material toward a highly refined state. If followed correctly, the result should be a very fine powder, easily assimilated by the human body without fear of damaging or deranging the Dhātus.

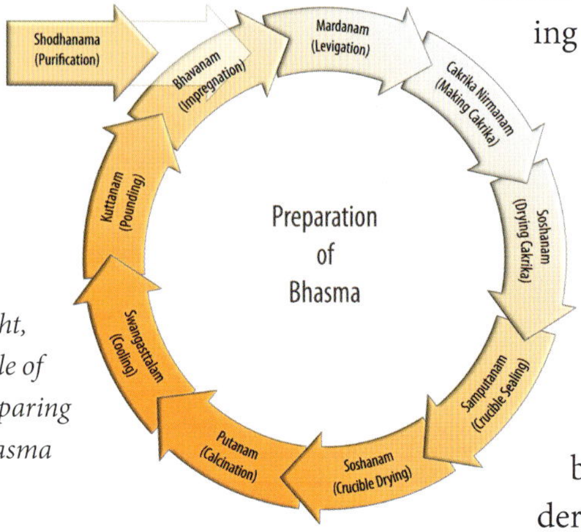

Right, cycle of preparing Bhasma

The following outlines the ten-step process of Bhasma preparation:

1 **Impregnation** After powdering the base material, a liquid or gel medium is stirred well into the powder to make a coarse paste.

2 **Levigation** Both materials are ground for one hour or until the liquid content has begun to evaporate. During this process, the material should become a fine paste. Note: different types of mortar will be appropriate due to the

different consistencies of ground material. See Section 6.3/ Khalwa Yantra (pestle and mortar) for more information.

3 **Making cakrika** *Cakrika* means a disc or disc-shaped. Here the drying paste material is formed, by hand, into small flat pills. No specific size is recommended, and examples vary widely; however, 2–3cm is quite common.

4 **Drying cakrika** Once formed, cakrika are allowed to dry in the open air under sunlight.

5 **Crucible sealing** The dried cakrika are placed inside a flat bottomed crucible, called a *Sharaava*, a term that also refers to cooking plates. Effectively this type of crucible is usually two deep earthenware plates bound together using long strips of cotton cloth smeared with clay.[109]

6 **Drying crucible** Once sealed, the crucibles are dried before adding another layer of protective cloth and clay. The process of wrapping the crucible is repeated four to seven times, depending on the quality of the cloth used and the thickness of the clay. Once sealed, contents are effectively airtight, reducing oxidisation when heated to temperatures over 800°c.

7 **Calcination** Sealed crucibles are placed into various sized puṭa (earthen pits) and filled with dried cow dung (above and below the crucible). Once ignited, the dung burns for about three to four hours.

8 **Cooling** The puṭa are allowed to fully cool for twenty-four hours before attempting to retrieve the crucibles. For more information, see Section 6.7.

9 **Pounding** Upon retrieval, the crucibles are carefully opened, and their contents are recovered. The cakrika are collected from within and ground in a mortar.

10 **Impregnation** The cyclical processing begins anew, with the powdered cakrika again mixed with liquid and stirred well to make a new paste. This paste is then formed into new cakrika, and the processing continues. Depending upon the material being prepared, this cycle can be repeated 3–4 times or 30–100 times.

---

109 Powdered termite cement is favoured in Sri Lanka. There are a number of different sealing materials mentioned in Rasa Shāstra texts; these are various clays, chalk, rock salt, iron oxides and various animal dung, mixed with milk.

This cyclical process of reduction and removal acts to cleanse the material of impurities and potentise the remaining material. This magnification of potency partly explains why the dosage of Rasa medicine tends to be minuscule, and why its effects are felt almost immediately. Conversely, improper processing during manufacture tends to magnify the toxicity. In Rasa Shāstra, there are no shortcuts – reward only comes from adhering to the prescribed processes. Quality is quite literally determined by the time invested, and cutting corners will not only reduce efficacy but invite the risk of creating a compound that could be potentially lethal for the end user. With these constraints in mind, it is understandable why some Āyurvedic manufacturers shy away from Rasa medicines. All alchemical medicines mean an investment in specialised equipment, labour-intensive effort and, finally, testing and analysis, all of which add considerable cost to the final product.

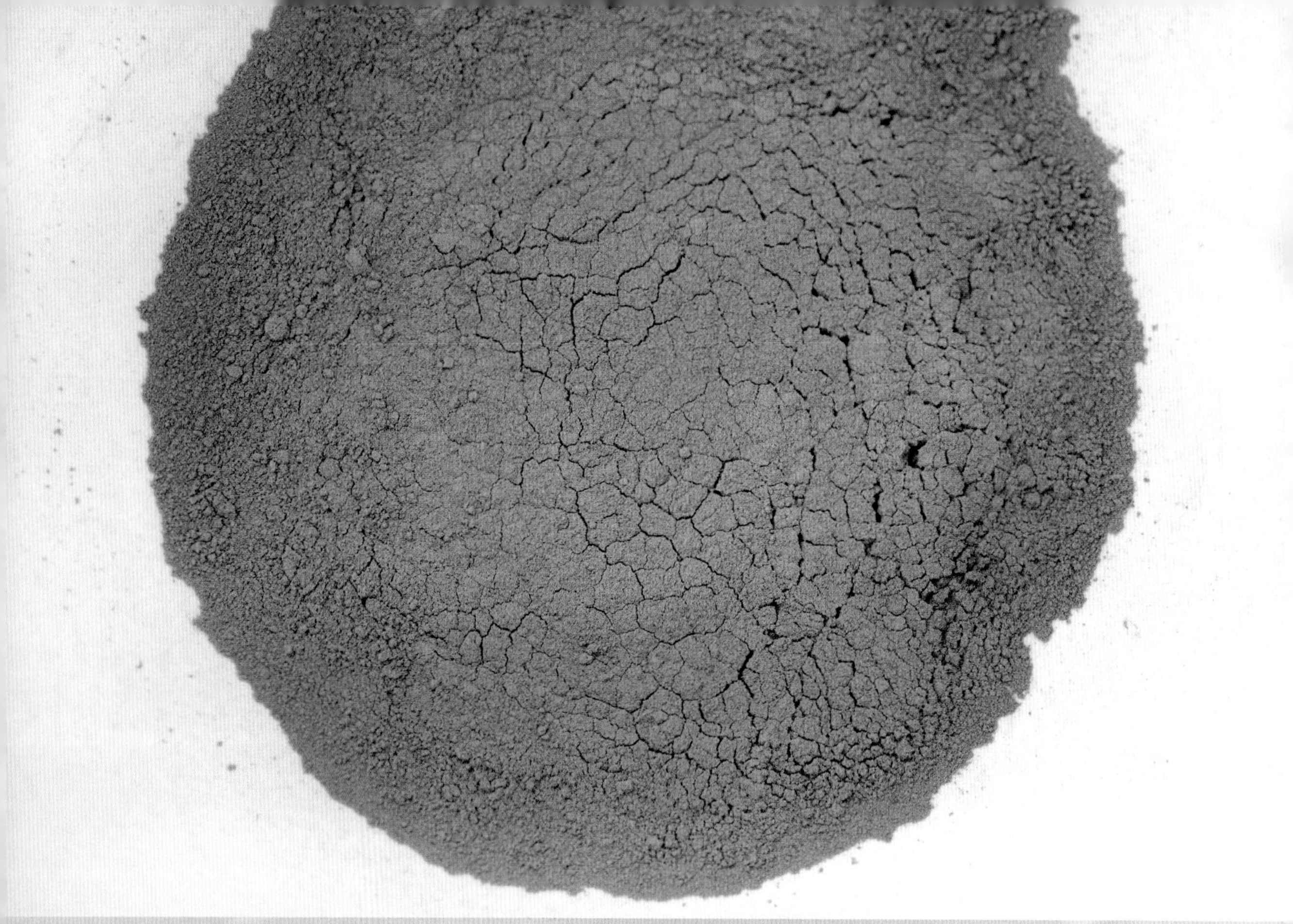
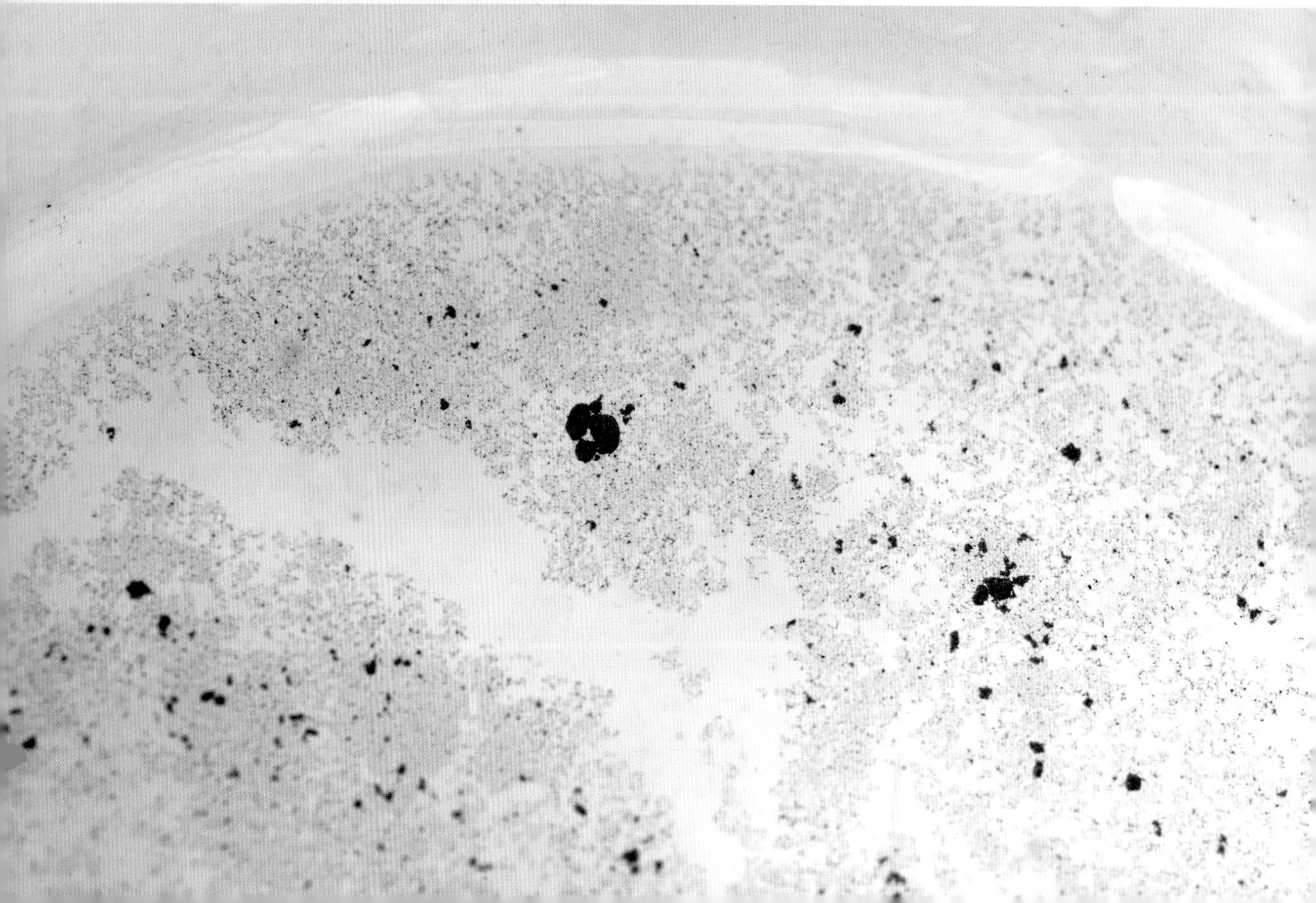

Section 6

Preparation of medicines

6.1/ Testing of alchemical preparations

The ancients went to extraordinary lengths to fully prepare these medicines, developing a set of elaborate protocols that involved the slow reduction of materials by exposure to high temperatures and systematic grinding. This processing is described as 'killing' the base material, reflecting its inability to be resurrected to its former state, unlike the way metals are able to reconstitute to their previously metallic form.

To this end, Rasa Shāstra provides seven qualitative tests for finished medicines.[110] Although a far cry from modern analysis, these tests remain a valuable indicator of consistency and purity. They establish a baseline, quickly allowing you to see if something has gone wrong during the manufacturing process.

The table on the following page outlines seven methods used to determine the purity of Rasa medicines.

Once passing these tests, Bhasma could be stored almost indefinitely. This is one reason why these remedies proved popular, particularly with travelling physicians who could not access herbal medicines all year round. In addition, the longevity or shelf-life of these medicines was just one feature that made Rasa Shāstra preparations more advantageous than their herbal counterparts.

Five additional reasons to include Rasa medicine in your portable medicine kit might well be:

1 Only small doses are required, and they act quickly;
2 Administering Rasa medicines is very easy, and they have no unpleasant taste;
3 Due to their micro-particle size, Rasa medicines are absorbed quickly and penetrate the deepest and densest tissues, even to a cellular level;

Far left, testing Bhasma for its lightness by floating it on water

110 Bhasma = alchemical ash.

4 Rasa medicines act as a metabolic catalyst and so have a more holistic effect, which is why they can appear to carry on working, even when no longer being administered;

5 The efficacy of Rasa preparations increases over time; they have no expiry date.

Bhasma testing	
Subtlety	A small amount of freshly prepared Bhasma should be placed between the index finger and thumb and rolled about in a circular motion until the ash becomes engrained into the fingerprints. This should be done until the excess material naturally falls away. The remaining material will stay embedded in the fingerprints.
Purity	A small amount of Bhasma, cast into a naked flame should remain inert. Improper processing and/or purification is indictated by smoking powder, pungent smells or some level of reconstitution.
Taste	Bhasma, placed on the tongue, is sampled. Correctly prepared, it should be free from any taste.
Colour	While there are permissible colour variations in Bhasma, most batches will attain a similar hue. Unexpected colouration is a sign of improper processing and/or purification, for example if the expected result is grey Bhasma, but the end product looks reddish.
Lustre	Correctly prepared, Bhasma should have an even matt finish, without any kind of shine or lustre. The presence of minute shiny particles is a sign of improper processing and/or purification.
Texture and lightness	Bhasma should have a smooth quality; it should feel velvety to the touch. It should also be light, without the feeling of scratchiness. When placed in water, it should be supported by water tension, allowing it to float.
Stability	Once appropriately stored, Bhasma should remain unaffected by the passage of time. Moulds, crystallisation or compaction are signs of improper processing and/or purification. Correctly stored, the potency of Bhasma will only increase with age

6.2/ Mercury (a special case)

One of the most controversial materials to find its way onto the shelves of the Rasa pharmacy is the liquid metal mercury. At the mention of this, jaws tend to drop, and eyebrows rise. I want to demystify its use in this book, bringing the reader closer to understanding its time-honoured use and necessity. The fact is that this material forms the mainstay of many Rasa medicines, hence the prefix *Rasa* itself being just one of the many names for this liquid metal.

Left, liquid mercury

Can mercury be made safe for human consumption? Does mercury relinquish some or all of its toxicity following purification? These questions are not easy to answer. What can be said is that mercury is seldom consumed without first combining it with the element sulphur. These two, together, form one of the most stable chemical bonds in nature, and the resulting black sulphide form, known as Kajjali, is non-water soluble. It is this stable HgS form that goes on to become the base of many Rasa medicines.

Left, extraction of mercury from cinnabar using Damaru Yantra

As expected, this particular element requires considerable levels of purification[111] before use. It is considered

111 From a modern scientific standpoint mercury is considered highly toxic, known to cause irreparable damage to the body. Mercury toxicity can accrue through direct inhalation or via absorption as $HgCl_2$ (mercury chloride), or CH3Hg (methylmercury).

the ultimate panacea, but mercury does not willingly give up its medicinal benefit without a fight. Purification of the liquid metal can be both involved and complex. There are also differing opinions on how to best harness its medicinal qualities, with many texts expressing the dangers of incorrect or partial purification. Many of these same commentaries speak about contamination, adulteration and inferior grades of mercury sold by merchants. There are also lengthy discussions on elimination techniques should these impurity states manifest after internal consumption.

Note: Much of mercury's historical fame is heavily intertwined with gold-making and longevity elixirs, each requiring extended and complex purification practices to awaken its transformative powers. In this context, it was the iconic mercurial ore cinnabar that became associated with the transmutation of base metals into gold and the rejuvenation of the tissues through the use of immortality drugs. For more information see Section 7.11/ Hiṅgula (cinnabar).

6.3/ Khalwa Yantra (pestle and mortar)

Alchemically speaking, this is perhaps the most important piece of equipment in Rasa Shāstra and it comes in a myriad of forms. During the purification process, nearly every base material necessitates the use of Khalwa Yantra. In this section, we take a look at the common types of yantra and discuss their individual merits and downfalls.

Right, cast iron mortar

Cast iron

If we were to categorize all mortars into three fundamental classes, namely stone, metal and ceramic, then iron would stand out as the ideal mid-range mortar. It is hard enough to break stones without damaging its integrity, yet smooth enough to grind a reasonable grade powder. Cast iron Khalwa are therefore the most useful and versatile type, well suited to the initial reduction of hard and brittle materials like gemstones and some minerals. This is especially true when they have worn interiors, their iron walls smoothed. They can also help pound malleable metals such as tin, lead, and zinc. Iron mortar helps break up the more fibrous parts of plants, especially bark and seeds. They also find favour in mercurial works due to irons' inability to form an amalgam with mercury.

The downside to iron is oxidisation – they get rusty. Prolonged surface corrosion eventually leads to pitting, contamination and inefficiency. Iron mortars must be kept scrupulously clean and regularly oiled; linseed is especially good. If left to the mercy of nature, iron Khalwa quickly rusts and contaminates and colourises its contents. During the processing of plant materials, it is not uncommon for the juice to develop a reddish hue on a mild interaction with rust. From a therapeutic point of view, this interaction might be negligible, perhaps even beneficial under some circumstances; however, it is still bad practice.

Overall, iron khalwa is highly durable, it can see many years of active-duty work, withstanding high temperatures, pounding and other abuses. My own iron mortars, the rounded variety, frequently double as quenching receptacles for molten copper and are none the worse for wear.

Brass and bronze

Brass and bronze are hard non-ferrous alloys used to manufacture some types of Khalwa Yantra. Although durable, they are susceptible to tarnishing and surface corrosion over extended periods of time. Both alloys were considered to have curative properties in their own right; often associated with relieving diseases relating to the blood, liver, spleen and skin.

Due to the malleability and availability of these metals, they can be quite commonplace in Indian pharmacies. Brass and bronze were thought to be auspicious due to their practical application as plates and vessels from which to eat and drink

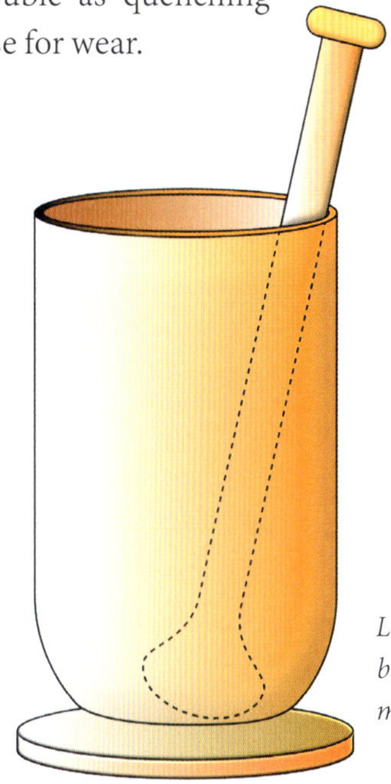

Left, typical brass pestle and mortar design

(when allowed to stand in a bronze jug, water was thought to be beneficial for health). The two alloys are often used to cast ceremonial items such as statues and bells; the ringing of these still being used to ward off malevolent spirits.

Like iron mortars, brass and bronze mortars are highly functional, easy to clean and durable. Unfortunately, the deep-throated design is about the only version of Khalwa readily available. Having used this style to prepare materials I have to say they are best for pulping raw plant material. They do an adequate job for these kind of softer minerals, but the pestle quickly becomes damaged when trying to grind harder materials. Just like iron, brass and bronze mortars benefit from regular cleaning. In Sri Lanka, they were polished with gypsum, or limestone powder with lemon juice; this mild abrasive and acidic type of polish really does make this metal gleam.

Granite

Perhaps the oldest and most iconic material in any alchemical pharmacy, granite mortars are useful beyond words. They quickly devour dried plant materials and shatter coarse compact minerals. They allow gums and resins (such as Guggulu) to be pounded easily with minimum effort, using the gravity and weight of the pestle to do all the hard work. High grades of granite are composed mainly of silicates, such as feldspar, and some black varieties have a Mohs hardness of around 6.5–7. All of these qualities make for a pretty impressive material, resistant to reasonably high temperatures, corrosive acids and alkalising powders and liquids.

Right, granite pestle and mortar (circular variety)

Indian granite comes in several hues, from red to green, white to bluish-grey, and the highly coveted rich black variety. Chemically, all granites are a mineral assemblage of elements such as iron, mica, obsidian and quartz, their final mixture determining colour and hardness. Some highly sought-after varieties are thought to contain small flecks of copper, producing a specular sheen in cut and polished stonework (see note below). This effect is especially eye-catching in black granite. This last category of granite is particularly favoured for Pārada (mercurial) works, especially when cut into the iconic mortar design known as a *Naukakruti Kharal*.[112] Its boat-shaped profile allows for long back-and-forth strokes, essential for mixing the liquid metal into various herbo-mineral-metal ingredients.

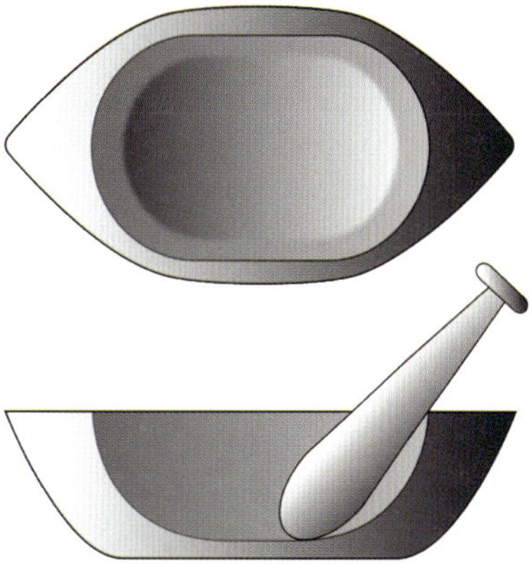

Left, Naukakruti Kharal mortar; the traditional apparatus is preferred for the purification of mercury. It is believed that black granite imparts special properties to the processing

Left, Naukakruti Kharal mortar, Purbeck granite

This type, or rather shape of mortar, is prevalent in India; most pharmacies have them in varying sizes. Obtaining these outside India can be troublesome. Due to their weight and size, shipping can be cost-prohibitive. Not so long ago, I gained access to a Stonemasons workshop and decided to have a go at preparing a naukakruti kharal. It was not the easiest of tasks, but, with help and encouragement, I crafted a half-decent rendition. To make life a bit easier, I opted for a softer material someplace between a regular marble and granite, in this case, Purbeck granite, also known as Purbeck marble. The end result was quite useful but unfortunately a bit too soft for day-to-day work.

112 Known also as Tamari Pathar Yantra.

Note: Due to the formation processes of granite, the presence of copper is unlikely due to the tendency of this metal to bond with more compatible elements like sulphur. The coloured flecks observed in the granite are more likely to occur due to manganese or magnesium oxidation. The different grades (and colours) may impress some subtle quality upon the materials being processed, but as these constituents vary in their percentages, it would be difficult to standardise or quantify.

Ceramic/porcelain

Ceramic is a highly robust material that is very useful for fine grinding materials; its ultra-smooth surface is well suited to producing fine powder. The tough and durable surface is achieved using kaolinite or china clay, a finely levitated mineral, rich in aluminium silicates. When heated to approximately 1300°c, it produces a hard, white and slightly translucent surface. Other materials associated with the assemblage of kaolinite include feldspar, iron oxide and gypsum. Due to this composition, its vitrified surface is impervious to liquid and does not require glazing. Repeated grinding in ceramic mortars will eventually abrade the surface, forming a key upon which contaminates can adhere. Regular light abrading with an ultra-fine grade of wet and dry paper will keep its surface clean. Although hard, this material is highly brittle and should only be used during the final processing of Bhasma.

Above, ceramic pestle and mortar

One additional note on ceramic mortars is their inherent ability to gauge material purity. This measurement falls into a 'qualitative' category of analysis, assessing the ability to clean the mortar after use. If its surface is easily cleaned, i.e. it leaves no residue or stain upon the ceramic surface, then it can be assumed that processing has reached a refined state. If the reverse is true, i.e. the mortar's surface has become heavily stained and now requires significant cleaning to restore the ceramic interior, it can be assumed that the material will require further processing.

In this respect, you might think of a ceramic mortar as a kind of one-stop lab-testing kit whose surface perhaps reflexes the body's digestive tract and how each material is likely to interact with the mucus membranes. During the digestion process, is the material easily assimilated and distributed, or will it quite literally stick to the bodily tissues?

Glassware

Although easily available, glassware is not favoured over more traditional materials. While it does have advantages, such as transparency and hardness (glass is a little harder than marble), its surface will become scratched with prolonged use. Much like some semi-transparent rock crystal mortars, the ability to see sediment is advantageous; for instance, during the processing of Rasaka (calamine), the powdered raw material is ground with cow's urine and allowed to settle over several hours. Once settled, the unwanted liquid content is carefully poured off and replaced with fresh urine. Seeing sediment below the liquid makes it very easy to pour off the excess.

Other stone mortars

Sometimes less affordable and, in some cases, impractical, these mortars are manufactured from jade, quartz (rock crystal), onyx, agate, and marble. Most have a similar Mohs hardness of around 6–7.0. Both black and white varieties of Marble have a Mohs hardness of about 3–5. Some of these alternative mortar materials are deemed auspicious in preparing certain materials, i.e., rock crystal mortars are highly useful for potentising materials such as pearl and coral Pisti under moonlight.

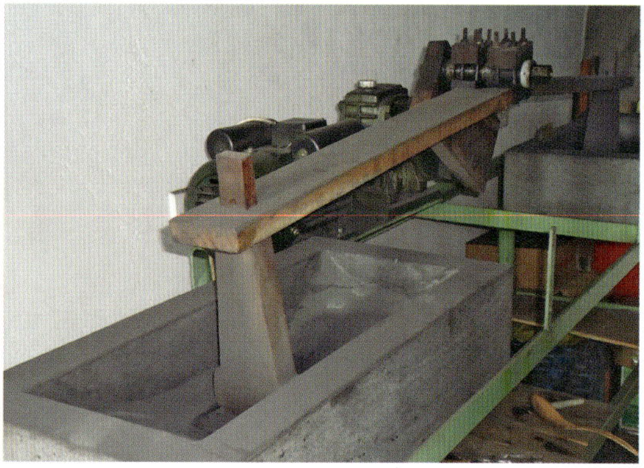

Above, and above right, types of improvised stirring machines: above, stirring mercury in its liquid state; above right, mixing mercury as a powder

6.4/ Man vs. machine – automation in Rasa Shāstra

Much of the traditional equipment used in Rasa Shāstra is still hugely relevant today, although certain techniques have evolved. The advent of mechanisation has permeated modern pharmacies, streamlining once arduous tasks into efficient and straightforward processes. As pharmacies increasingly adopt mechanisation, the ingenuity of their mixer/grinder/furnace systems evolves to cater to specific tasks and available materials. Refer to the images above for notable examples of innovation in the Rasa pharmacy.

The impact of these machines on the quality of medicine remains uncertain. However, Āyurveda maintains that intent is paramount, suggesting that introducing profit-driven machinery as a replacement for human involvement contradicts its fundamental principles. Conversely, proponents argue that these same machines free up time and enable individuals to engage in more productive endeavours. Hence, is this development truly unfavourable?

6.4/ Man vs. machine – automation in Rasa Shāstra

Among the alchemists in Sri Lanka and India, there is a prevailing consensus that something is lost when the human touch is removed and replaced by a mechanised arm.

During one of my weekly excursions to the city, where I stocked up on equipment and materials, I found myself at a chemical supplies shop in a bustling complex. It was teeming with vendors, some confined to shop spaces while others spilled out onto the walkways. This particular dealer specialised in supplying chemicals to schools. The proprietor had become accustomed to my visits and even engaged in light conversation as I sifted through the assortment of unmarked boxes and wax paper packets. On this occasion, I was in search of an additional pestle and mortar, as well as sulphur and copper salts. The proprietor was engrossed in conversation with another customer when I entered, but he seemed content to let me explore without assistance.

After finding the items I needed, I approached the makeshift counter to settle up. As the money and goods exchanged hands, the gentleman I had initially assumed to be a customer inquired if I was working in Sri Lanka, speculating that I might be a teacher. I said that I was a student studying at a small Āyurvedic hospital, where I was learning the art of preparing traditional remedies.

Upon hearing about my studies in Āyurveda, he displayed mild interest and revealed that he had also dabbled in the subject. He mentioned having written a paper on an experiment he and his fellow students conducted during their study days. Although he seemed to be a teacher now, he had once contemplated a career in Āyurvedic medicine. Intrigued, I inquired about the focus of his paper, and he explained that they had sought to determine whether hand-prepared remedies were more effective and safer than their commercially produced counterparts. Naturally, I was eager to hear his conclusions. He disclosed that their experiment involved administering larger doses of *Rasa Sindoora* (a preparation of red mercuric oxide that I will delve into in more detail later) to mice.

They procured several handmade batches from a local physician renowned for his artisanal medicines, together with several of the same remedy purchased from a large pharmacy and prepared by machine. The mice were then given the doses and closely monitored. Daily urine samples were analysed, and post-mortem studies were conducted to assess any internal damage, such as to the kidneys and brain.

By the end of the study, they concluded that the handmade medicines exhibited lower detectable levels of toxicity in both urine samples and post-mortem examinations, causing minimal harm to the test subjects. On the other hand, the mice treated with the commercially prepared medicine showed higher levels of mercury in their urine and tissue damage upon post-mortem analysis. Some level of toxicity was expected in both groups due to the very high doses administered.

As a student of Āyurveda, he found these results gratifying, since they aligned with Āyurvedic theory. However, years later, now teaching modern biochemistry, he found it difficult to explain why the two approaches yielded such disparate outcomes. He admitted that the study was not conclusive and that, by modern testing standards, their results would be considered inconclusive or inviable. Ultimately, he concluded that the differences observed could be attributed to poor manufacturing practices, contamination, variations in test parameters, and so on. He mentioned that without replicating the experiment with different samples of handmade and commercially prepared medicine, it would be impossible to draw any definitive conclusion. He added that all the medicines had undergone chemical testing in a laboratory, primarily revealing HgS (mercury sulphide) and a few trace elements – so nothing unexpected.

Unfortunately, I was unable to pursue further discussions or access the written paper. Nevertheless, the teacher's comments remained etched in my memory, and I share them here, although they are purely anecdotal.

6.5/ Mediums used for Bhavana

> 'Pārada (purified mercury), ground with the root of white Eranda (*Ricinus communis*) and Kanta Pashana (magnetic iron ore), is to be heated along with silver, in a closed crucible.'
>
> *Rasanavakalpa*

Before calcination, the process of purification and assimilation begins with *Bhavana* – adding a liquid to aid in the grinding process. Popular mediums were not randomly chosen but carefully selected because of their history of medicinal use. Common Bhavana liquids include lemon and Ambuldodam (sour orange), apple juice, aloe gel, raw milk, and castor oil. Decoctions of Triphala, Bhringaraj (false daisy) and lotus seed are also popular and find their way into many alchemical recipes. Rosewater is slightly unusual in that it is the de-facto Bhavana used in the conversion of gemstones to prepare Pisti – see See Section 6:8/ Pisti (Agnitapta Bhasma), a highly prized medicine in both Rasa Shāstra and Unani Tibb.

Above, Manah śhilā (realgar), Bhavana with ginger juice

Over centuries of practical experimentation, all have found a unique application in the manufacturing of Rasa medicine, most notably during this initial pre-digestion/assimilation stage of preparation, and all of the above have a specific energetic that they bring to this union.

In this short section on Bhavana, we take a closer look at a handful of the more popular Bhavana and their individual energetics.

Lemon (*Citrus limonum*), Nimbu

Lemon, though initially acidic, has a long-term alkalising nature. This makes it a wonderful and abundant medium to begin the process of purification and assimilation. According to *Shad-Rasa* (six tastes), lemon juice is sour and astringent, light, dry and penetrating; it also evaporates quickly, making it ideal for the preparation of cakrika. Lemon stimulates and excites Agni (digestive capability) and helps reduces phlegm, having excellent decongestant properties. It also has an overall anti-inflammatory action on the bodily tissues. Lemon and lime help reduce all three Dosha; however, in greater quantity, lemon is more likely to aggravate Pitta Dosha. Either lemon or lime juice can be substituted for most Bhavana purposes; its universal action and abundance make it the go-to material in the absence of other specified Bhavana. If lemon or lime is absent, the juice of Ambuldodam (sour orange, *Citrus aurantium Linn*) can also be substituted. Due to the latter's abundance in Sri Lanka, I used it far more than any other juice. You were never further than twenty paces from an orange tree in the pharmacy where I worked.

Triphala decoction

The volume of literature on Triphala is enormous; you could literally write an entire book on the merits of this time-honoured Āyurvedic mixture. I will keep this brief as I have gone into some detail on its use in other medicines. Triphala, or three fruits, is a mixture of amalaki (*Phyllanthus emblica*), Bibhītaki (*Terminalia bellirica*), and Harītakī (*Terminalia chebula*). Each is a powerful medicine in its own right; in combination, they produce a synergistic effect that enhances all three. Triphala has excellent Rasāyana properties and supports the immune system while helping to detoxify and support liver functionality. It has good anti-inflammatory and antioxidant properties, is a digestive tonic, and relieves constipation while promoting the absorption of nutrients. When used as a decoction, the three fruits (dried and of equal quantity) are boiled in water, eight parts to one part fruit, then simmered down to roughly one quarter of the original volume, cooled and filtered.

Aloe gel (*Aloe barbadensis*), Kumārī

The *Aloe vera* plant has been used medicinally for thousands of years. *Kumārī* means 'young maiden' or 'virgin', so it was traditionally seen as a tonic that imparted youthful vigour. The gel, freshly extracted from its saturated leaves, can easily be applied externally to sores, burns and wounds. Aloe balances all three Doshas and services all seven Dhātus. According to Āyurveda, aloe gel is cooling, sweet, slimy and astringent. Kumārī gel is a popular Bhavana for many Rasa preparations; its ability to promote digestion, eliminate and destroy toxins, and rejuvenate the blood and skin make it a much sought-after material. Due to its astringent property, it is fast-drying, which also aids in the production of cakrika. The gel is readily pasted and combines with all manner of minerals and metals. Aloe gel is particularly popular in the manufacture of zinc and serpentine Bhasmas.

Milk / Godugdha

Āyurveda considers milk a necessity of life; it is sattvic, tonifying and nourishes all seven tissues. Using this vital liquid in the Bhavana process increases the rejuvenating property, as milk not only enhances strength, it also feeds Ojas. The properties of milk are sweet and cooling; they pacify Vāta and Pitta while increasing Kapha. Milk is an aphrodisiac, a laxative, and strongly counteracts the effects of all poisons. It is used to purify metals, minerals, and most animal products. Milk is particularly beneficial in the purification of toxic plant material.

Castor oil (*Ricinus communis*), Eranda

Considered the king of oils, the use of castor oil is centuries old, and its healing properties are legendary. The best medicinal grades were those with a reddish or yellow hue; however, other variations of colour were acceptable although they were considered less effective. The oil was said to be best when extracted on Sundays, under the performance of rituals that honour the Sun.

Left, lotus seed decoction and rosewater are both highly useful in the preparation of gemstone Bhasma

Eranda was particularly useful in reducing those materials considered resistant to conversion and that very often required some level of extended purification. These materials included mercury, copper, iron pyrite and copper sulphate. Castor oil is sweet, pungent and heating in action; while it reduces Vāta, it increases Pitta and Kapha. It also has analgesic as well as nervine qualities.

Lotus seed (*Nelumbo nucifera*), Kamala

Lotus seed decoction is a potent medium to begin the transformation of *Ratna* (precious gems) and *Uparatna* (semi-precious gems). The qualities of lotus seeds are light, sticky, oily, cooling, sweet and astringent. It reduces Pitta and Vāta but increases Kapha. Lotus seeds have an affinity with the heart, urinary system and large intestine (inducing peristaltic motion). They also promote the complexion while aiding in the preservation of the skin, reducing wrinkles and maintaining youthful looks. Typically, rosewater is recommended for the preparation of Pisti (see Section 6:8: Pisti or Agnitapta Bhasma); however, lotus seed decoctions can be used instead. Like horse gram decoction, lotus seeds similarly have the power to break stones, helping to reduce the hardest of minerals. This is another reason why this plant is recommended to help break up kidney stones and urinary calculi.

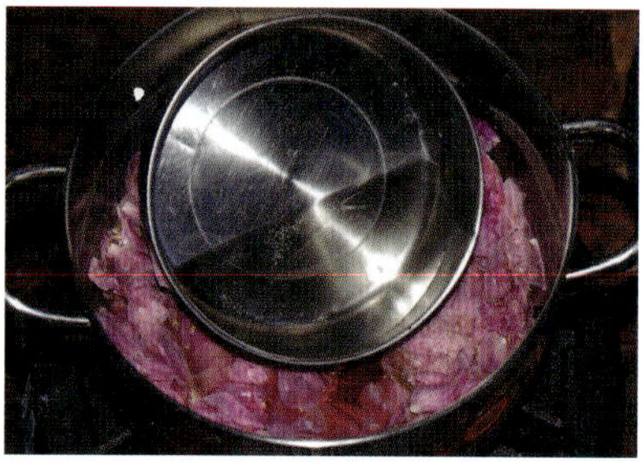

Above, preparation of rosewater, using simple distillation equipment.

Rosewater (*Rosa damascene*), Arq gulab

Rosewater has a strongly cooling energetic, making it perfectly suited for ailments characterised by excess heat, like high fever and inflammation. It also proves beneficial in conditions related to blood imbalances, such as those tied to the female reproductive system. Rose is light, bitter and astringent in taste; it reduces Pitta and Kapha while slightly increasing Vāta. It can contract tissues, thereby reducing bleeding; its post-digestive sweetness helps tonify and strengthen those same tissues. Additionally, Arq gulab supports plasma, blood and the heart; its cooling and sweet nature nourishes both the circulatory and nervous systems. It also has a sattvic quality that calms the mind and senses. Rosewater is particularly important when preparing Pisti and is used almost exclusively to prepare all gemstones (except diamond).

Apple juice (*Malus sylvestris domestica*), Rubb-e-seb

Apple juice is a highly favoured ingredient in a number of Unani formulations, used as both Bhavana and Anupāna. It is considered a hepatic-cardio-tonic as well as having excellent antioxidant properties. It is cooling, astringent and slightly sour, and is known to aid in healing mucous membranes, helping to bind stools while reducing intestinal bleeding.

6.6/ Mārana (calcination)

This section deals with *puṭa*, a word meaning 'to cook'. This is usually done by digging earthen pits fill with dried cow dung cakes. The high temperature produced is used to calcinate ingredients sealed within crucibles.

Calcination is an essential and interesting part of alchemical manufacturing. The term *Mārana* means to kill or extinguish life. It essentially signifies the irreversible transformation of a material – be it a metal, mineral, gemstone or otherwise – to a state where reconstitution to its original form is not possible. Until this is achieved, the materials will remain unprocessed and unsuitable to use.

Saṁpuṭa means to envelop or contain. This important piece of technology enables the alchemist to protect the alchemical works whilst they are undergoing the intense process of heating. In essence, Saṁpuṭa is the enveloping, cooking process, during which the material being processed is heated yet kept separate from the fire. The most commonly used items for this purpose are Sharaava (cooking plates and pots). Two of these Sharaava are secured about their rims with cloth and clay, protecting and accompanying the purified materials as they go into the heart of the fire. These crucibles are therefore known as *Sharaava Saṁpuṭa* (enveloping containers).

PUTA (BURNING PITS)

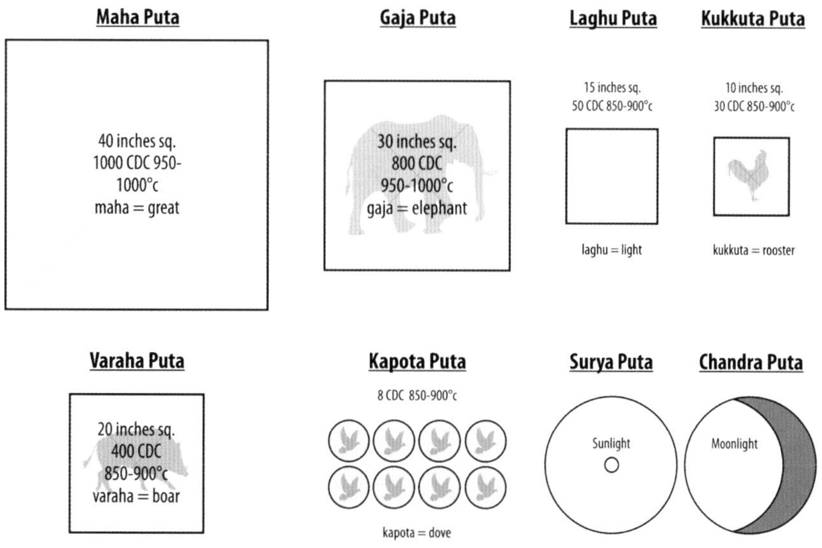

Left, different sizes of puṭa and their uses

Above, Maha Puṭa; above right, Varaha Puṭa

Different sizes of *puṭa* (earthen pits) were then specified as units of heat, as a way of approximating the temperature required for certain types of Rasa materials. To this end, different pit sizes were dug, based on dimensions many could easily visualise without a tape measure. So, for example, *Gaja Puṭa (gaja* means 'elephant') is loosely based upon the depression in the ground left by an elephant's belly. *Varaha Puṭa* (*varaha* means 'wild boar') would be the depression left by a boar. The dimensions of *Kukkuta Puṭa* (*kukkuta* means 'rooster') would equate to the hop of a rooster, and so on. See the diagram on the previous page, showing eight different types of puṭa and their symbology.

Sūrya and *Chandra Puṭa* harness the respective forces of the two great luminaries they are named after. They are deployed during the Bhavana process, wherein the natural drying action of the Sun – in the case of Sūrya Puṭa – is used to evaporate the liquid between triturations. On the other hand, Chandra Puṭa is reserved for the preparation of specific Pisti (see Section 6:8/ Pisti (Agnitapta Bhasma)), when the energetic of coolness is advised. Typically, Chandra Puṭa is reserved for materials such as pearl, mother-of-pearl and red coral, materials selected for their inherent 'coolness', since the use of Chandra Puṭa magnifes these properties.

In classical texts, most materials requiring puṭa will advise the size required. The largest puṭa recommended was *Maha Puṭa* (*maha* means 'great' or 'grand'), and was reserved for those more heat-resistant materials such as mica and iron. In contrast, smaller puṭa, such as Kukkuta and Kapota, were usually preferred for certain mercurial or Amṛītakarana[113] processes.

113 *Amṛītakarana* = to instill the quality of Amṛīta (divine nectar) into a substance.

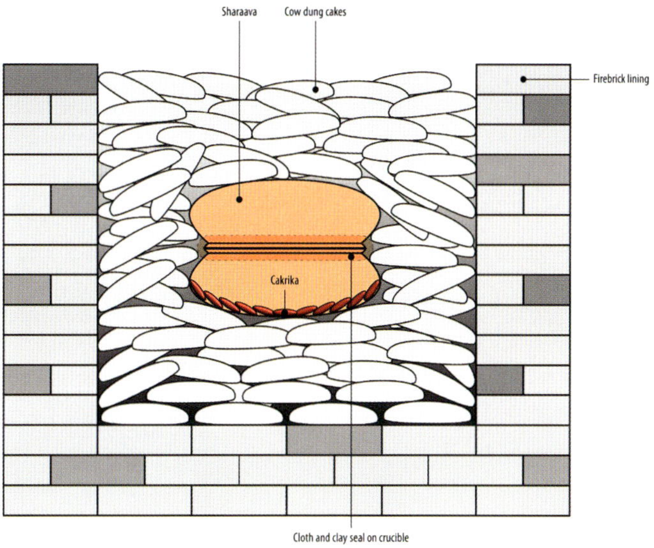

Left, graphic showing what happens inside a puṭa.

As the classical dimensions of puṭa were somewhat open to interpretation, modern interpretation of their actual size is also a bit in question. To make things more complicated, some texts can differ in their heating recommendations, so all of these factors should be kept in mind.

Returning to Rasashala and the idealised layout of the alchemist's pharmacy, puṭas were excavated to the required size in the southeast corner of the premises. These earthen pits could be and were frequently used for multiple firings. However, due to the inevitable impact of natural erosion and rain, these pits would eventually need to filled and re-sited. Modern facilities do not have the luxury of re-siting due to space limitations and so opt for permanent concrete structures; these are either shuttered concrete pits or earthen pits lined with firebrick. Wherever possible, they will be constructed in a south-easterly direction, but many now use ventilation hoods to channel toxic smoke up and away from factory workers.

Cow or buffalo dung?

Traditionally, both cow and buffalo dung, called *Vanopala*,[114] was recommended as the preferred fuel source for puṭa. Presumably, this fuel was abundant and gathered at the forest edge or on grasslands, naturally dried by the heat of the

114 Also called Gomaya

day and stored in sheltered areas. While some research has been undertaken to ascertain if there is any advantage to using cow dung over buffalo dung, or vice-versa, the results do not seem to show any significant differences.

Additionally, when considering the large number of cakes advised in some sizes of puṭa (Maha Puṭa requires 1500 cakes), it makes it difficult to imagine these quantities were being rigorously adhered to. From my own observations, puṭa are filled to one-third, at which point the Sharaava are positioned close to the centre before being filled to capacity.

Once ignited, using ghee-saturated cloth, the puṭa slowly burns downward, its mass diminishing over the next three to four hours until only ash remains. The next day, the puṭa will be cleared, and the Sharaava will be recovered and opened.

6.7/ Puṭa paka (temperature)

The temperature and duration of heating are essential components in the alchemical process. The term most commonly used to describe this process is *Mārana*, a word meaning 'to kill' or, in this case, to be calcined.

In practice, internal temperatures of most puṭas are close to 1000°c. These high temperatures begin the pre-digestion process of metals, minerals, gemstones and animal products. The heat renders their structures more brittle, further facilitating their reduction into powder. Metals are the most heat resistant, so they have their own preferred practices before the Mārana process; see Section 2.3/ Reduction and conversion of metal. Their eventual conversion to Bhasma is assured; however, repeated exposure to high temperatures can be a longer process.

Once converted, all materials lose weight but energetically retain residual heat from the puṭa processing. This *Agni*, or inner fire, ultimately allows for better assimilation and penetration into the subtle tissues of the body.

Puṭa temperature analysis: right, Varaha and far right, Kukkuta. Image on the right (Varaha) shows a pyrometer being used to monitor temperature

Most real alchemy takes place at the heart of a puṭa, where materials subjected to intense heat are energetically altered, allowing new properties to be impressed upon the heated mass while discarding less desirable ones. This fascinating area of research has, until fairly recently, received little attention in the available literature of Rasa Shāstra. Apart from a few research institutes studying the effects of puṭa, questions regarding peak and optimal temperatures seem to have been largely skirted around.

Above, and above left, standardisation of cow dung cake size using a circular template

I found this study intriguing and decided to reproduce a number of puṭa, namely Gaja, Varaha, Laghu, Kukkuta and Kapota, measuring each with a combination of pyrometric cones and pyrometers to better understand this critical heating process. After spending a warm summer preparing thousands of cow dung cakes (an unenviable task, I can assure you) I was able to satisfactorily reproduce all five puṭa and document their individual temperature ranges.

Note: The number of individual cow dung cakes given for each size of puṭa is based on a standard 9cm (diameter) cake. See the image above for the standardisation of cake size.

On the following pages I have reproduced the graphs of their heat signatures. These hopefully demonstrate not only the intensity of heat generated but also how the temperature difference is not that significant for all of their variations across the different sizes of puṭa.

Right, Gaja Puṭa: approximate dimensions thirty squared inches; content approx. 800 c.d.c.

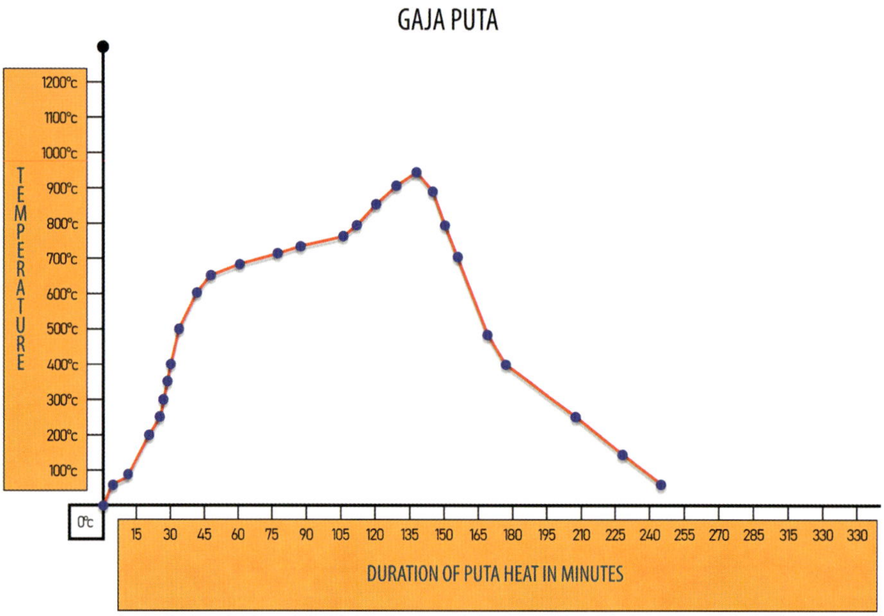

Right, Varaha Puṭa: approximate dimensions twenty squared inches; content approx. 400 c.d.c.

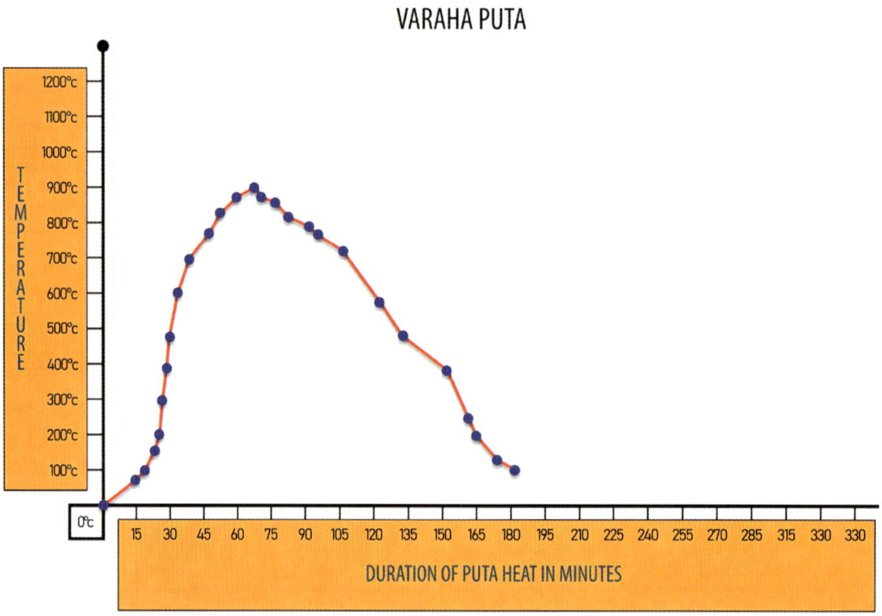

LAGHU PUTA

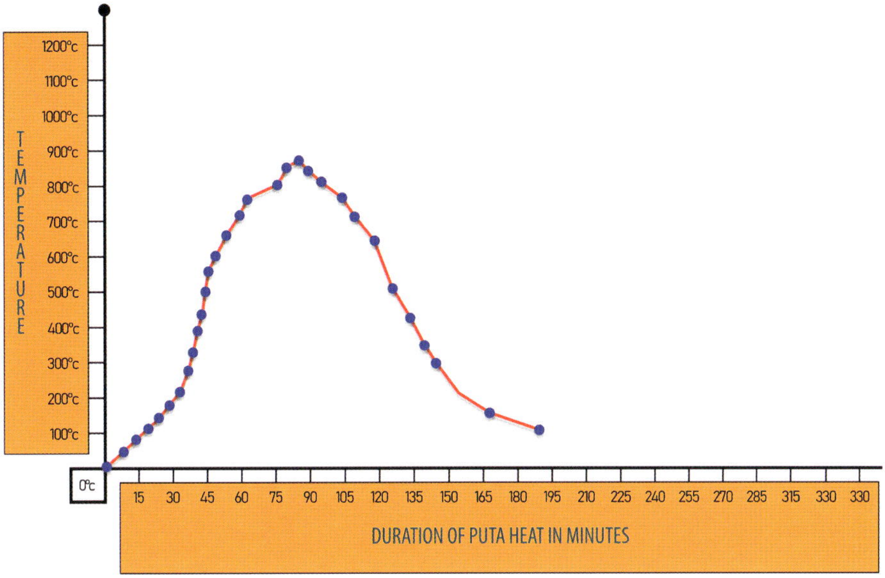

Left, Laghu Puṭa: approximate dimensions fourteen squared inches; content approx. 200 c.d.c.

KAPOTA PUTA

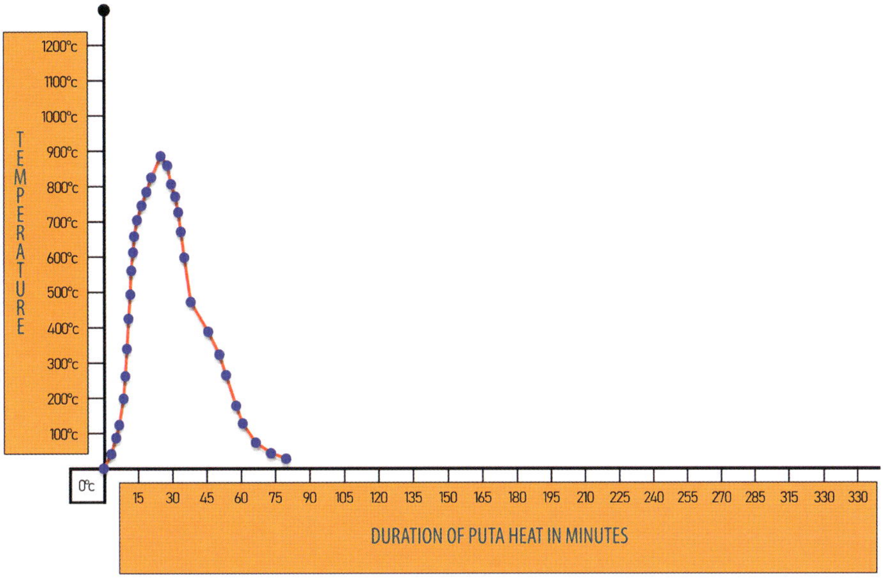

Left, Kukkuta Puṭa: content approx. 30 c.d.c.

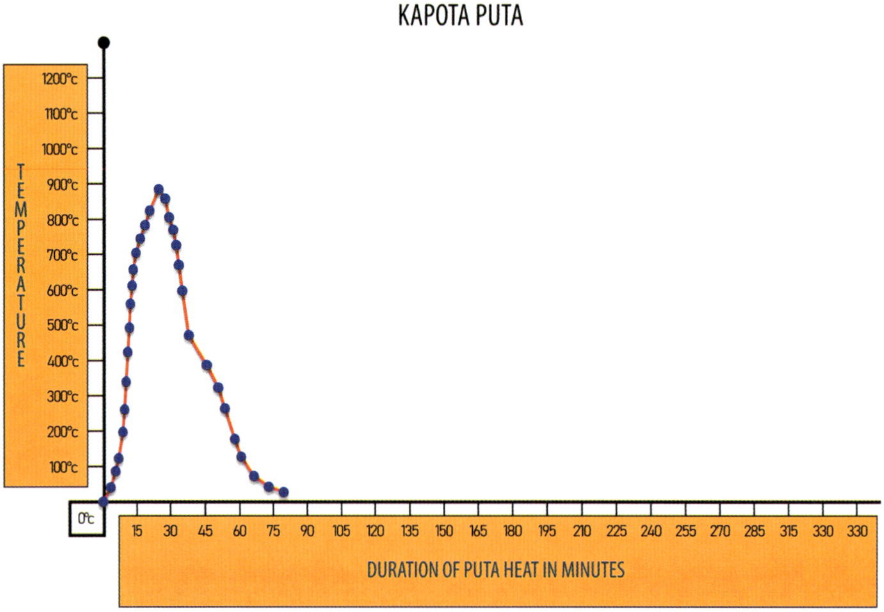

Right, Kapota Puṭa: content approx. 8 c.d.c.

Puṭa temperature table		
Puṭa type	**Max. temp**	**Approximate duration**
Gaja Puṭa	960°c	255 min
Varaha Puṭa	910°c	219 min
Laghu Puṭa	910°c	210 min
Kukkuta Puṭa	500°c	255 min
Kapota Puṭa	896°c	90 min

As seen in the results, the peak temperature of all puṭa was confined within a 60°c range, and were achieved within a comparable time frame – about fifty minutes, with kapota as the sole exception. These results would indicate little difference in the actual overall temperature, with a slight variation in duration. The main difference is when maximum temperature is achieved and maintained, this being a little longer in the larger puṭas.

As I was able to reproduce the same results more than once, I am confident that these graphs are reasonably accurate. On one occasion, whilst preparing Kukkuta Puṭa, I recorded a surprisingly low-temperature reading of 500°c for a 260 minute duration of heat. This result was obtained only once, so I can only think the cakes were not sufficiently dry. The UK (it would seem) is not the best environment in which to manufacture cow dung cakes. The Indian climate is far more agreeable for the manufacture and use of this fuel – one of the reasons why it is available on most street corners (see image below).

Left, cow dung cakes drying in the streets of Faridabad, India

6.8/ Pisti (Agnitapta Bhasma)

Pisti, pronounced *pishtee*, is yet another method of preparation that does not use heat. Materials suitable for Pisti are mostly gemstones; these can be *Ratna* (precious) or *Uparatna* (semi-precious) categories. Pisti is also recommended for many animal products like pearl, mother-of-pearl and red coral. The preparation of Pisti is much simpler, and its potency is generally considered less than that of the Bhasmas.

Right, red coral (Pravala) being prepared in a Rasa pharmacy in Faridabad, India

Pisti is usually advised for those conditions where excessive heat is a problem, for example fever and inflammation. Pisti is relatively simple to prepare in comparison to Bhasma, so it is much more popular with manufacturers.

Preparation method

1. Materials are first washed in salted water.
2. They are then dried and powdered using an iron mortar. The contents are then repeatedly sieved through finer meshes to remove large pieces.
3. The resultant powder is triturated with freshly prepared rosewater. The first trituration happens in granite mortars, then the process moves to ceramic mortars, as the powder becomes very fine.
4. As this mixture dries out, it is exposed to sunlight (Sūrya Puṭa), fresh rosewater is added, and trituration is continued. The mixture will be exposed to Moonlight (Chandra Puṭa) during the preparation period to increase its cooling potential. Finished Pisti will have a residual rosewater scent and not feel gritty in the mouth. Usually, preparation requires around fifteen to twenty hours of grinding.

6.9/ Anupāna (vehicle of delivery)

Anupāna is a term used a lot in Āyurveda.[115] It describes a unique system that helps ensure a medicine arrives at or is at least is pointed toward a specific location in the body. There is no equivalent for this in allopathic medicine, where it is believed all ingested medicines follow the same digestive pathway. However, as previously explored in Sapta Dhātu, Āyurveda perceives the body as a series of interconnecting systems and pathways, enabling the directed transportation of specific medicines to the areas needing them the most. The understanding of this 'personalised delivery system' of Anupāna sets Āyurveda apart.

Many Anupāna are used alongside Rasa medicines; some are quite common, like milk and honey, others less so. The following selection gives a brief overview of some popular Anupāna and the practical and energetic reasoning behind their use.

Honey

Above, honey being combined with herbal powders to make herbal honey pills

Generally, some preference is given to the use of honey *(Madhu)* as an Anupāna. This mainly stems from honey's pre-digested state and slightly heating nature. Honey has a cleaning, scraping quality, along with some astringency, and is characterized by its subtle yet penetrating nature. It is rapidly absorbed by all Dhātu. Āyurveda generally advises against heating honey, considering it a pre-digested food and medicine in its own right. It was believed that the application of heat disrupted its inherent heating and astringent properties, impairing its digestibility, and, therefore, could cause an imbalance in the Dosha.

Many ancient societies revered the honey bee and the medicinal power of its nectar. While the honey of feral bees was considered somewhat superior in medicinal value, domesticated varieties were more than acceptable.

Whether using honey as an Anupāna or just including it in formulas, it is always

115 The principle of Anupāna is not exclusive to Āyurveda; Unani and Classical Chinese medicine also make use of a similar principle, described in their own terminology.

worth sourcing locally-produced honey. Most neighbourhoods will have a few beekeepers willing to sell excess stock. As a rule of thumb, it is always worth avoiding mass-produced brands, as these are nearly always blended, making it impossible to know their original sources. Due to the global bee crisis (CCD)[116] sourcing quality honey is becoming ever more problematic, so when in doubt, buy from a local beekeeper. Additionally, many types of commercially produced honey are heat-treated, making blending easier and improving their visual appearance.

Due to the natural crystallisation of honey's sugars, granulation will occur at different times following its extraction from the comb.

When clear honey is required (usually for preparing honey pills) you may apply very low heat to granulated honey. The heating should be less than 35°c. Jars can be placed in a cardboard box with a 40W light bulb to clear crystallised honey and left for about a week. As the box warms up, the honey will slowly de-crystalise.

Dairy

Āyurveda recognises dairy products as having Rasāyana properties and being tonifying, nutritive, and sattvic, thereby promoting physical strength. Moreover, dairy products are acknowledged for their Vājīkarana potential, often associated with aphrodisiac effects.

Milk is highly regarded in Āyurveda for its cooling, sweet, and emollient properties. It can nourish Ojas and, similar to honey, it quickly permeates all the Dhātus (tissues) in the body. It exhibits a special affinity towards *Shukra Dhātu* (reproductive tissues) effectively nourishing at a rapid pace. Milk as an Anupāna is particularly favoured with various Rasa medicines, such as pearl, mother-of-pearl, deer horn, Shilajit, Bhallātaka, serpentine, aconite and Guggulu.

Ghee holds significant importance as an Anupāna in Āyurveda, and although it is commonly used in everyday cooking, it is considered more of a medicinal substance than a mere food. The healing properties of ghee encompass a range of benefits, from strengthening the heart and soothing duodenal ulcers to reducing fevers, repairing nervous tissue, improving difficult menstruation, and alleviating burning sensations. Ghee possesses a sweet and cooling taste and,

116 Colony Collapse Disorder

when consumed moderately, it reduces Vāta while balancing Pitta and Kapha Doshas. It provides nourishment to all tissues, acts as a Rasāyana, and helps maintain a healthy Agni (digestive fire). Both the internal consumption and external application of ghee are effective. Medicated ghee, formulated with specific herbs, is particularly potent and available in various formulations, such as Bṛāhmi Gṛtha (see medicated ghees).

Butter While less commonly utilised as an Anupāna, unsalted butter remains an effective carrier for various substances in Āyurveda. It serves as a vehicle for materials such as deer horn, zinc, eggshell, copper sulphate, topaz, garnet, and chrysoberyl. Butter possesses a sweet and cooling taste, mildly decreasing Vāta and Pitta while increasing Kapha. It provides nourishment to all tissues and exhibits remarkable aphrodisiac qualities. Butter is favoured as an Anupāna because it nourishes tissues, particularly when building Shukra Dhātu (reproductive tissue) and Ojas.

Triphala decoction Triphala is undeniably one of the most well-known remedies in Āyurveda, renowned for its diverse medicinal properties. It serves as a gastrointestinal tract cleanser, an antioxidant, a blood cleanser, a laxative, and a rich source of bioavailable vitamin C. Triphala is often favoured as an Anupāna due to its potent anti-Visha (toxin) and antioxidant effects. In addition to tonifying the tissues, it is commonly administered as a decoction. Typically, Triphala is combined with Loha (iron) preparations like Kasisa and Loha Bhasma, to aid their absorption. When used as an Anupāna, Triphala serves two primary functions: facilitating the absorption of nutrients in the blood and acting as their carrier to the head, brain, and eyes.

Jaggery Kithul jaggery holds a prominent place as an Anupāna in Sri Lanka, where it is commonly consumed either grated or in small pieces, rapidly dissolving in the mouth. Unlike other varieties of whole cane sugars, kithul jaggery is derived from the sap of the date palm, which is believed to possess superior therapeutic qualities. Ideally, kithul jaggery should be well-formed, dry, and brittle, with a dark colouration indicating higher quality. While most jaggery is heat-inducing, date palm jaggery has natural Pitta-reducing properties, making it less likely to aggravate this Dosha.

Kithul jaggery is enriched with essential minerals such as iron, phosphorus, potassium, calcium, and magnesium. It is favoured as an Anupāna because it is nutritive, unctuous, and diuretic. It is particularly effective with Makara Dwaja, Swetha and Sheetal parpati, as well as Bhallātaka nut, in fact the combination of Bhallātaka nut and kithul jaggery is especially significant, as pounding the raw nut with date palm jaggery neutralises its Pitta-aggravating properties. For further information on the preparation and uses of plant material, See section 12.4/ Bhallātaka (bhilawan Nut).

Right, kithul jaggery

Warm water Water holds significant importance in Rasa Shāstra, where its properties are extensively explored. Various water sources, including mountain streams, lakes, rivers, rainwater, dew, and dew collected during a full moon, are considered viable. Once collected, water stored in copper or bell metal vessels is believed to possess purifying and potentising qualities.

Water obtained from mountain streams under moonlight is particularly esteemed for its highly cooling and rejuvenating attributes. Warm water infused with a small amount of lime juice is considered an excellent Anupāna (vehicle) for administering substances such as red coral, cowrie shells, chalk, salt petre, soapstone, Jews' stone, Guggulu, and milk hedge.

Different types of water play a vital role in Rasa Shāstra and are recognised for their ability to enhance the efficacy and therapeutic properties of various substances, particularly when used as an Anupāna.

Part III

Materials, Formula and Processing

Rasa materials

Quick reference guide

The following tables present a comprehensive list of ninety-two materials for creating diverse alchemical preparations. While it is important to note that this compilation does not encompass all available materials, it serves as a handy reference guide for pharmacists. For more in-depth information regarding each material, please refer to the Appendix: Catalogue of materials and their use.

The table's categorisation predominantly follows a traditional arrangement, with a few unconventional categories here and there. The reasons behind this arrangement remain uncertain, but availability or therapeutic potency may have strongly influenced its development. For example, Abhraka (mica), abundantly mined in north-east India, appears in the Maha Rasa category. Conversely, rare substances like cinnabar, not commercially mined in India, feature in the Sadharana Rasa[117] category. Interestingly, Gandhaka (sulphur), an essential element for processing and stabilising mercury, appears as an Uparasa.[118] This might appear peculiar given its significance and intimate bond with mercury. Scholars have proposed a potential explanation for this arrangement, suggesting that it highlights mercury's affinity for each material. However, this theory has its issues since the combination of mercury and sulphur is known to be one of the most chemically stable in nature, which should logically place sulphur in the Maha Rasa category.

Identity issues

The identity of certain Rasa Shāstra materials has been lost to us, and in these cases, I have highlighted the material with an (*). Chapala and Kankusta are presently associated with bismuthinite (Bismuth Sulphide) or *Garcinia cambogia* (Malabar Tamarind), respectively. These substances are included in the table for reference, however, it is important to note that their associations are still subject to investigation and verification.

Far left, ingredients to prepare a silver-like alloy.

117 *Sadharana Rasa* refers to common materials that fall between Uparasa and Dhātu.

118 *Uparasa* are secondary (mineral) drugs that are used in connection with Pārada (mercury).

Rasa materials – quick reference guide * No consensus on identification				
No.	**English**	**Sanskrit**	**Category**	**Indications/effects**
1	Mercury	Pārada Hg	Mercury (metal)	Yoga vāhin, Rasāyana, tonic, extends life, heals wounds, and cures all disease
2	Mercuric sulphide	Kajjali HgS		
Maha Rasa				
3	Mica	Abhraka $K(MgFe)_3(AlSi_3)$ $O10(OHF)_2$ (biotite) $KAl_2\ (AlSi_3O_{10})\ (F,OH)_2$ (muscovite)	Maha Rasa (mineral)	Tuberculosis, asthma, persistent cough, heart disease, diabetes, and anaemia
4	Fluorite	Vaikrānta* CaF_2	Maha Rasa (mineral)	Poisoning, anaemia, high fever, skin diseases, asthma, and urinary infections
5	Copper pyrite	Swarna Maksika $CuFeS_2$	Maha Rasa (mineral)	Eye diseases, tuberculosis, anaemia, skin diseases, abdominal diseases, an immune stimulant
6	Iron pyrite	Vimala FeS_2	Maha Rasa (mineral)	Eye diseases, tuberculosis, anaemia, skin diseases, abdominal diseases, an immune stimulant
7	Bitumen	Shilajit $C_{14}H_{10}N_4O_5$ (+)	Maha Rasa (mineral/plant)	Kidney stones, urinary calculi, irregular blood sugar, and high Vāta conditions
8	Copper sulphate	Sasyaka Cu_5FeS_4	Maha Rasa (mineral)	Poisoning, diseases of the eye, heart disease, diabetes, and consumption
9	Bismuth sulphide	Chapala* Bi_2S_3	Maha Rasa (mineral)	Tuberculosis, gonorrhoea, menorrhagia, high fever, excess mucus, obesity
10	Calamine	Rasaka $(Fe,Mn)ZnCO_3$	Maha Rasa (mineral)	Tuberculosis, diabetes, trachoma, eye disease, anaemia, consumption, leprosy
Uparasa				
11	Sulphur	Gandhaka S	Uparasa (mineral)	Leprosy/skin diseases, high fevers, mercury poisoning, and the removal of āma (toxins)

12	Red iron oxide	Gaireeka Fe_2O_3	Uparasa (mineral)	Heart disease, anaemia, poisoning, snakebite, vomiting, impaired vision, abdominal bloating
13	Ferrous sulphate	Kasisa $FeSO_4·7H_2O$	Uparasa (mineral)	Tuberculosis, poisoning, skin diseases, kidney stones, painful urination, greying or falling hair, and loose teeth
14	Alum	Kanksi $KAl(SO_4)_2·12H_2O$	Uparasa (salt)	Leucoderma, bleeding gums, malaria, and stomatitis (inflammation of the mouth and gums)
15	Arsenic trisulphide	Haritāla As_2S_3	Uparasa (mineral)	Cancers, leprosy, poisoning, high fevers, Pitta type skin diseases, gout
16	Arsenic disulphide	Manah śhilā As_2S_2	Uparasa (mineral)	Tuberculosis, skin diseases, high fever, leprosy, intestinal parasites, gout, chronic bronchitis
17	Antimony sulphide	Anjana Sb_2S_3	Uparasa (mineral)	Improves vision, reduces blood toxicity, reduces obesity, and is a general Rasāyana
18	Galena	Nilanjana PbS	Uparasa (mineral)	Eye diseases, ulcers, haemorrhaging, and bone fracture
19	Malabar tamarind	Kankusta*	Uparasa (plant)	Skin diseases, abdominal bloating, enlargement of the spleen, excess mucus
Sadharana Rasa				
20	Monkey face fruit	Kampilla	Sadharana Rasa (plant)	Skin diseases, boils, parasites (it has powerful anti-fungal properties), and reduction of excess phlegm
21	Arsenic trioxide	Gauri pashana As_2O_3	Sadharana Rasa (mineral)	Cancer, heart disease, consumption, leprosy, scorpion bites, filarial, bronchitis, asthma, high fever
22	Ammonium chloride	Narasara NH_4Cl	Sadharana Rasa (salt)	Mouth ulcers and bleeding gums, heart disease, eye diseases, poisoning, headaches, and acid reflux
23	Cowrie shells	Kapardika $CaCO_3$	Sadharana Rasa (animal)	Diminished eyesight, cataracts, hearing loss, high fever, heart disease, urinary stones, tuberculosis, hyper-acidity
24	Ambergris	Agnijara	Sadharana Rasa (animal)	Tetanus, reduction of Vāta, irregular digestion, loss of appetite, colitis, convulsions, paralysis. It is also an aphrodisiac

25	Mercuric oxide	Giri Sindoora HgO	Sadharana Rasa (mineral)	External wounds, skin infections, burns, poisoning, leprosy, eye diseases, digestive dysfunction, ulcers, and herpes
26	Cinnabar	Hiṅgula HgS	Sadharana Rasa (mineral)	Leprosy, old age and debility, hepatitis and splenomegaly, rheumatism, chronic fevers, diabetes, impotency
27	Lead monoxide	Mrddara Śrnga PbO	Sadharana Rasa (mineral)	Bone fractures, greying or falling hair, eczema, Improves fertility
Dhātu				
28	Gold	Swarna Au	Dhātu (metal)	Heart disease, lowered immunity, tuberculosis, diminished eyesight, skin diseases, and diabetes
29	Silver	Rajata Ag	Dhātu (metal)	Heart disease, Rasāyana, stomach disorders, diabetes, vertigo, anaemia, and senility
30	Copper	Tamra Cu	Dhātu (metal)	Consumption, eye diseases, liver disorders, anaemia, abdominal diseases (gastritis and colic pain), chronic fever, leprosy, asthma
31	Iron	Loha Fe	Dhātu (metal)	Blood disorders, anaemia, leprosy, eye diseases, abdominal disease, enlarged spleen or liver, dysentery, and colic pain
32	Rust of iron	Mandura Fe_2O_3	Dhātu (metal)	Childhood anaemia, intestinal parasites, jaundice, abdominal disease, enlarged liver and spleen, dysentery, colic pain, and diabetes
33	Tin	Vanga Sn	Dhātu (metal)	Diabetes, urinary disorders, premature ejaculation, infertility, anorexia or emaciation, and brain dysfunction
34	Lead	Nāga Pb	Dhātu (metal)	Diabetes, rheumatism, digestive disorders, piles, poisoning, skin diseases, abdominal bloating
35	Zinc	Yasada Zn	Dhātu (metal)	Diabetes, eye diseases, skin diseases, reproductive and urinary disorders
36	Brass	Pittala Cu+Zn	Dhātu (alloy)	Hepatitis, anaemia, intestinal parasites, blood disorders, skin diseases, enlargement of the liver and spleen
37	Bronze	Kansya Cu+Sn	Dhātu (alloy)	Skin and eye diseases, blood disorders, obesity, poor digestion, and intestinal parasites

38	Three Metals	Tri-Loha Au+Ag+Cu	Dhātu (alloy)	Heart disease, low immunity, senility, and fertility issues. Enhances cognitive function
39	Five Metals	Varta Loha Cu+Sn+Zn+Fe+Pb	Dhātu (alloy)	Skin diseases, reduced vision, insufficient digestive power, intestinal parasites, constipation, and urinary disorders
		Ratna		
40	Ruby	Maanikya Al_2O_3	Ratna (precious gemstone)	Heart disease, brain disorders, eye diseases, tuberculosis, reproductive issues, digestive weakness
41	Pearl	Mukta $CaCO_3$	Ratna (precious gemstone)	Heart disease, poisoning, stomach disorders, Asthi (bone) disorders, impaired digestion, inflammation, and diminished eyesight
42	Coral	Pravala $CaCO_3$	Ratna (precious gemstone)	Bleeding/blood disorders, tuberculosis, atrophy of Māmśa (muscle) and sinew, diminished eyesight, cough, poisoning
43	Emerald	Tarksya Al_2O_3	Ratna (precious gemstone)	Lung diseases, asthma, bronchitis, immune dysfunction, CNS diseases, senility, poisoning
44	Topaz	Pushparaga $Al_2SiO_4(F,OH)_2$	Ratna (precious gemstone)	Poisoning, inflammation, skin diseases, vomiting, digestive disorders
45	Diamond	Hiraka C	Ratna (precious gemstone)	Cancer, tumours, immunodeficiency diseases (HIV), consumption, diabetes, anaemia, impotency, diminished eyesight, and urinary tract infections
46	Sapphire	Nilama Al_2O_3	Ratna (precious gemstone)	Tuberculosis, arthritis, poisoning, brain dysfunction, diseases of the CNS, infertility
47	Garnet	Gomeda $Ca_3Al_2(SiO_4)_3$	Ratna (precious gemstone)	Consumption, skin diseases, anaemia, digestive disorders, high fever, learning difficulties
48	Chrysoberyl	Vaiduryam $BeAl_2O_4$	Ratna (precious gemstone)	Blood diseases, heart disease, senility, loss of brain functionality, general debility, constipation

Uparatna				
49	Agate	Akika SiO_2	Uparatna (semi-precious gemstone)	Heart disease, internal bleeding, infertility, insanity, menorrhea, general debility, urinary calculi
50	Sunstone	Sūryakanta $KAlSi_3O_8$	Uparatna (semi-precious gemstone)	Heart disease, blood disorders, brain dysfunction, poor digestion, senility, and V-K imbalances
51	Moonstone	Chandrakanta $KAlSi_3O_8$	Uparatna (semi-precious gemstone)	Stomach disorders, blood diseases, heart disease, digestive disorders, high fever, and Pitta imbalances
52	Lapis lazuli	Rajavarta $3(NaAlSiO_4).Na_2$	Uparatna (semi-precious gemstone)	Diabetes, consumption, urinary disorders, blood disorders, excess Pitta, tuberculosis, anaemia
53	Jade	Sangeyasab $NaAl(Si_2O_6)$	Uparatna (semi-precious gemstone)	Heart disease, blood diseases, infertility, stomach disorders, gastric pain, dysentery, urinary calculi, internal bleeding
54	Turquoise	Pirojaka $Al_6(PO_4)_4(OH)_8.4(H_2O)$	Uparatna (semi-precious gemstone)	Heart diseases, palpitations, eye diseases, poisoning, renal calculi, duodenal ulcer, infertility, constipation, learning difficulties
55	Tourmaline	Vaikrānta (Na,Ca) $(Mg,Fe_2+,Al,Li)_3B_3$ $(Al,Fe_{3+})_6O_{27}(OH,F)_4$	Uparatna (semi-precious gemstone)	Skin diseases, tuberculosis and weakness of the lungs, asthma, bronchitis, anaemia, high fever, diabetes and urinary disorders
56	Amber	Kaharuba $C_{10}H_{16}O$	Uparatna (semi-precious gemstone)	Heart disease, blood circulation, internal bleeding, bleeding piles, dysentery, menorrhea, and gastro-intestinal weakness
57	Quartz	Sphatika SiO_2	Uparatna (semi-precious gemstone)	Low immunity, blood disorders, high fever, burning sensations, internal bleeding, infertility, and memory loss
Miscellaneous (mineral)				
58	Lead tetroxide	Nāga Sindoora Pb_3O_4	Miscellaneous (mineral)	Eczema, bone fractures, ulcers, herpes, scabies, leprosy, swellings
59	Serpentine	Nāgapashana $Mg_3Si_2O_5(OH)_4$	Miscellaneous (mineral)	Poisoning, excessive vomiting, snake bite, heart disease, infertility, learning difficulties

60	Sodium borate	Tankana $Na_2B_4O_7{\cdot}10H_2O$	Miscellaneous (salt)	Bleeding disorders, severe itching, ulcers, insect/animal bites, CNS disorders, high fevers
61	Jew's stone	Badarasma $CaSiO_3$	Miscellaneous (mineral)	Kidney stones, polycystic kidney disease, renal calculi, poisoning, snake bites, skin diseases
62	Lodestone	Kanta Pashana Fe_3O_4	Miscellaneous (mineral)	Heart disease, consumption, diabetes, anaemia, obesity, parasitic infestation, nephritis, oedema, and leucorrhoea
63	Asbestos	Kauseyasma $Mg_3Si_2O_5(OH)_4$	Miscellaneous (mineral)	Diabetes, epilepsy, anaemia, urinary disorders, burning sensations, blood disorders, pyorrhoea, and gum disease
Miscellaneous (animal)				
64	Peacock feather	Mayūr piccha $CaCO_3$	Miscellaneous (animal)	Chronic bronchitis, chronic hiccups, asthma, breathing difficulties, poisoning, and general weakness of the lungs
65	Laccifer lacca	Laksha	Miscellaneous (animal)	Sciatica, osteoporosis, jaundice, asthma, epilepsy, and heart palpitations
66	Bezoar stone	Gorochana	Miscellaneous (animal)	Heart palpitations, poisoning, pain, and stress
67	Deer horn	Mrga Śrnga $Ca_{10}(PO_4)6(OH)_2$ with $(CaCO_3)$	Miscellaneous (animal)	Heart disease, pleurisy, pain in the sides of the chest, eye diseases, sinus and migraine conditions
68	Deer musk	Kastūrī	Miscellaneous (animal)	Inflammation, nerve debility (paralysis), reduced vision, Rakta Pitta (bleeding), skin diseases, excess vomiting, poisoning
69	Conch shell	Shankha $CaCO_3$	Miscellaneous (animal)	Gastritis, indigestion, peptic ulcer, irritable bowel syndrome (IBS), diarrhoea, duodenal ulcer, and eye diseases
70	Eggshell	Kukkutanda $CaCO_3$	Miscellaneous (animal)	Osteoporosis, sciatica, asthma, leucorrhoea, bronchitis, diarrhoea, and rickets
71	Cuttlefish bone	Samudra Phena $CaCO_3$	Miscellaneous (animal)	Osteoporosis, mineral deficiency (silica), low phosphoric acid. Heals bone fractures
72	Mother-of-pearl	Sukti $CaCO_3$	Miscellaneous (animal)	Heart disease, poisoning, splenic disorders, colic, urinary stones, asthma

73	Goat bone	Ajasthi $CaCO_3$	Miscellaneous (animal)	Osteoporosis, rickets, bone fractures, calcium deficiency during pregnancy and in early childhood
Sudhā Varga				
74	Gypsum	Godanthi Haritāla $CaSO4{\cdot}2H_2O$	Sudhā Varga (mineral)	Tuberculosis, asthma, anaemia, rickets, osteoporosis, high fever
75	Calcite	Surama sapheda $CaCO_3$	Sudhā Varga (mineral)	High fever, anaemia, lung diseases, diarrhoea, and excessive thirst
76	Chalk	Khatika $CaCO_3$	Sudhā Varga (mineral)	Excessive bleeding (external and internal), wounds, eye diseases
77	Limestone	Sehunda $CaCO_3$	Sudhā Varga (mineral)	Hyperacidity, reduction of high Pitta, duodenal ulcers, intestinal parasites, painful joints
78	Soapstone	Dugdha Pashana	Sudhā Varga (mineral)	Skin diseases, high fever, leucorrhoea, gonorrhoea, chest pains, and bleeding disorders
Visha				
79	Cobra venom	Sarpa-Visha	Visha (animal)	High fever, high blood pressure, senility, and poor digestion. Enhances virility
80	Aconite	Vatsanābha	Visha (plant)	Heart disease, consumption, rheumatism, skin diseases (such as leprosy), mercury poisoning, high fever, inflammation
Upavisha				
81	Crown flower	Arka	Upavisha (plant)	Growths (such as tumours), syphilis, whooping cough, asthma, oedema, enlarged spleen or liver, rheumatism, poisoning, and stiff limbs
82	Milk hedge	Snuhi	Upavisha (plant)	Rheumatic disorders, gout, enlargement of the liver, leprosy, poor digestion, abdominal disorders
83	Dhattura	Dhattura	Upavisha (plant)	High fever, leprosy and other skin disorders, rheumatic pain, neuralgia, sciatica, and intestinal parasites
84	Flame lily	Langali	Upavisha (plant)	Blood impurity, skin diseases, ulcers and boils, inflammation, haemorrhoids, colic, and abdominal parasites

85	Indian oleander	Karaveera	Upavisha (plant)	Heart disease, toxic blood, high fever, skin diseases, ulcers, and boils
86	Indian liquorice	Gunja	Upavisha (plant)	Leprosy, boils, itching skin, infertility, alopecia (hair loss), vertigo, asthma, excessive thirst, and intestinal parasites
87	Opium poppy	Khasabeeja	Upavisha (plant)	Pain, nerve spasms/convulsions, inflammation, restlessness, asthma, and arthritis
88	Cannabis indica	Vijaya	Upavisha (plant)	Loss of appetite, poor digestion, poor blood circulation, intestinal spasm, and heart disease
89	Croton	Jayapāla	Upavisha (plant)	Constipation, jaundice, skin diseases, abdominal pains, intestinal parasites, haemorrhoids, and water retention
90	Marking nut	Bhallātaka	Upavisha (plant)	Haemorrhoids, splenic disorders, leprosy, skin diseases, abdominal bloating, obesity, and intestinal parasites
91	Poison nut	Kuchala	Upavisha (plant)	Poor digestion, infertility, low urine flow, constipation, poor menstruation, hyperacidity, asthma, and paralysis
Miscellaneous (plant)				
92	Indian bdellium	Guggulu	Miscellaneous (plant)	Heart conditions, blood toxicity, liver diseases, diabetes, obesity, thyroid conditions, and joint problems

Dhātu – subcategories of iron

Iron (Loha) can be categorised into various types based on different criteria. Here is a detailed categorisation of the various types of iron used in Rasa Shāstra.

Subcategories of Dhātu (iron)	
Munda loha (pig iron)	Mrdu: malleable; it has a shiny surface and melts quickly
	Kuntha: malleable, but requires much hammering to form it
	Kadara: brittle, un-malleable and breaks easily, revealing a blackened surface
Tikshna Loha (wrought iron)	Khara: inflexible with a lined surface; coarse in texture
	Sara: similar to steel; shows fine lines/grain. Very hard
	Hrnnala: yellow/black colouring; a distinctive beak pattern on cut surfaces
	Tarapatta: glazed smooth blackish finish; fine curved lines upon its surface. Highly durable and rust resistant (like stainless)
	Vajra: glazed, smooth blackened finish with deep fine lines; bright flashy surface like a diamond
	Kala: blue-black colouring and heavy; brittle if struck by other higher grades of iron
Kanta loha (magnetic iron ore)	Bhramaka: magnetic repelling (strong lodestone)
	Cumbaka: magnetic attracting (strong lodestone)
	Karsaka: magnetic attracting (weak lodestone)
	Dravaka: used for alloying purposes
	Roma: used for piercing the skin; encourages hair growth
Kanta loha (colour grades)	Yellow variety: used in the art of gold-making
	Red variety: used for the fixation of mercury
	Black variety: used for medicines

Section 7

Mercury

Disclaimer: It should be noted that the following section and subsequent sections detailing the practice of mercury purification were performed and documented prior to 2016. Since then (UK) law no longer permits individuals to retain/store liquid mercury without a license; see gov.uk: 'Licensing for home users of poisons and explosive precursors 2022.' Readers are advised to refer to any law or regulation on the handling/storage of mercury pertinent to their own country.

7.1/ Use of mercury-based medicines

> 'Rasa (mercury) is superior to all medicinal substances due to its effectiveness. Even in minute dosage, it produces no unwanted taste upon consumption. It can be used in all difficult conditions, returning health in no time at all. Scholarly physicians prescribe medicines only for curable diseases, but Pārada can be administered even in cases of incurable disease. Hence Rasa (and its preparations) are superior among all medicines.'
>
> *Rasendra Sāra Saṅgraha*

The prefix *Rasa* is a synonym for mercury, so it is perhaps unsurprising that mercury has been widely used as a fundamental component in Rasa Shāstra formulations. This element was recognised by ancient scholars for its potent medicinal properties while also acknowledging its significant toxicity. It is often remarked that mercury possesses remarkable healing potential, which, unfortunately, can be matched by its ability to harm the body if improperly prepared.

Far left, preparation of Pārada/ mercury (Hg), mixed with Vaikrānta/ Tourmaline

Mercury, the eightieth element on the periodic table, possesses a captivating allure. Its chemical symbol, Hg, (derived from *Hydrargyrum* (Latinised Greek), 'meaning watery silver') perfectly captures its essence. With a boiling point of

356.73°c and a freezing point of -38.8°c, mercury exhibits unique properties. In its natural state, it readily bonds with sulphur, forming various inorganic, non water-soluble compounds. Combined with carbon, it forms a fascinating organometallic compound known as methylmercury.[119]

At room temperature, mercury displays a striking appearance, gleaming with a bright silver-white metallic sheen. When exposed to heat, however, it rapidly transforms, vaporising into an odourless, colourless gas. The sublimation process of mercury in its sulphide form, such as cinnabar, demonstrates its ability to separate from its companion element, sulphur, at moderate temperatures and completely evaporate at higher temperatures. Upon cooling, it condenses back into a liquid state. This process, employed by ancient alchemists, served as a relatively crude yet effective method for liberating sulphur's hold and removing metallic impurities like tin and lead, as well as metalloids such as arsenic and stibnite.

One of the remarkable properties of mercury is its affinity for combining with various metals, forming amalgams. Interestingly, when in the presence of iron, mercury does not readily combine, which has led to the widespread use of iron flasks as reliable containers for this metal.

Mercury in modern gold production

The acquisition of gold and silver has historically generated significant interest in the mining industry, particularly in extracting mercury from various ores. Mercury's unique ability to bind with these precious metals made it an invaluable tool in the reclamation process. However, the mercury/gold amalgamation process, although outdated, has gradually declined over time and is now primarily employed in small-scale mining operations. It is crucial to note that this process is crude, inefficient, and highly toxic, posing a significant risk to workers in the immediate vicinity.

In modern times, efforts have been made to refine this practice, incorporating water baths and retorts (see illustration opposite) to enhance the

119 CH_3Hg (mono-methylmercury) readily combines with chlorine, nitrates and sulphur, making it bio-accumulative, bonding with amino acids and proteins in living matter. Methylmercury can be also produced by anaerobic aquatic organisms, breaking down inorganic mercury pollutants.

process while mitigating health hazards. Nonetheless, poor mining practices in developing countries remain a concern, as the allure of high gold and silver prices tempts individuals seeking short-term profits.

To address this issue, local governments in affected areas have taken steps to regulate these mining activities. They provide essential equipment such as retorts and water baths to mining crews while also promoting the traditional practice of gold panning. Although panning for gold is a time-consuming process, when combined with finely ground gold slurry and heated with sodium borate, it can yield substantial amounts of the precious metal.

After separating the fine gold particles from the slurry, they are mixed in equal quantities with borax and subjected to high temperatures. The inclusion of borax serves two purposes: it lowers the melting point of gold due to its natural fluxing capacity, and offers a safer alternative to the conventional gold-mercury amalgamation process. This method ensures that the obtained gold metal is free from the harmful effects of mercury. For further purification, the gold can undergo cupellation, a process that reduces it to its purest form and enables accurate assaying.

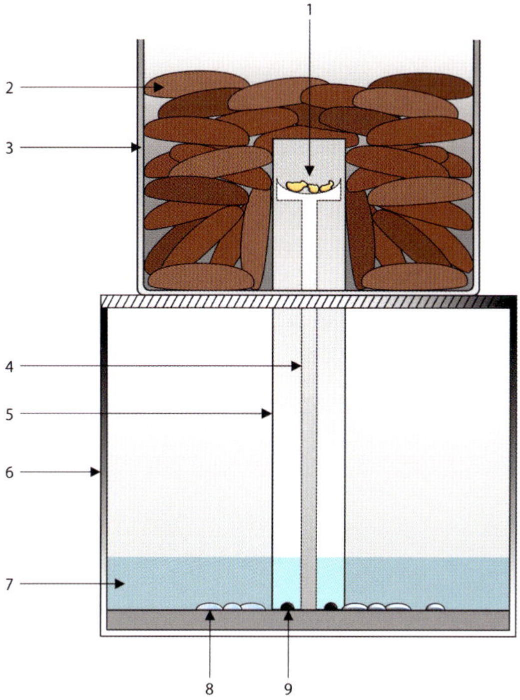

Above, retort and water bath for gold extraction:

1) mercury/gold amalgam held in an iron cup;

2) cow dung cakes;

3) top container to hold burning cow dung;

4) central iron support and cup;

5) iron sleeve with welded top and open bottom;

6) lower container to hold water bath;

7) water;

8) reclaimed mercury droplets;

9) vents at the base of the iron sleeve that allow condensed mercury to escape.

7.2/ Origins of mercury and its impurities

'Deep in the Himalayas, Lord Shiva and Parvati engaged in an intense union, causing tremendous commotion and disturbance across the three worlds. The gods, driven by the desire to have a son from this union, who could ultimately vanquish the demon Tārakāsura, decided to intervene. They dispatched Agni, disguised as a bird, to disrupt the couple. Upon seeing Agni, Lord Shiva grew furious and abruptly ceased his actions. He collected his ejaculated semen in his hand and hurled it at Agni. Overwhelmed by its intense heat and power, Agni cast it into the River Gaṅgā in an attempt to cool it. However, even the mighty river couldn't withstand its force, and it expelled the substance onto the shoreline, eventually cooling and transforming into Pārada (mercury). Other metals gathered around it, later manifesting as imperfections within Pārada.'

Rasendra Sāra Saṅgraha

Rasa Shāstra explains that mercury's toxicity is attributed to its inherent impurities, specifically poison (Visha) and wastes (mala). These impurities generate a burning quality within the metal, manifesting in the body as heat, leading to inflammation and fever. In contemporary analysis, mercury is frequently found to contain contaminants like tin, bismuth, lead, zinc, and arsenic.

Right, Pārada being purified in Triphala decoction

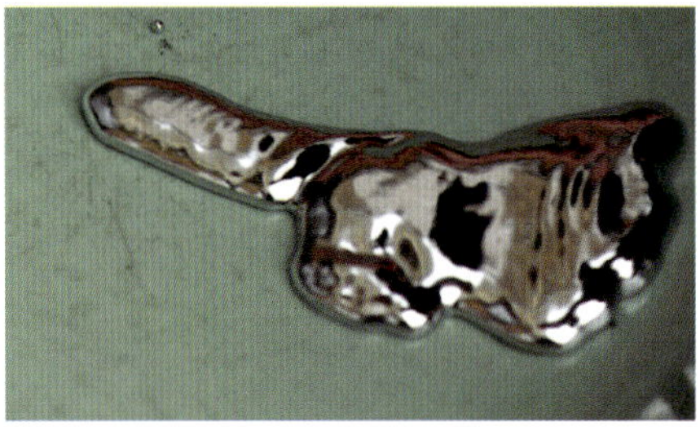

Right, liquid mercury after extraction from cinnabar

The ancients recognised five[120] different types of mercury and developed specific qualitative tests to help ascertain its purity. Experts in this science could discern the presence of Dosha by directly observing the colour and blemishes upon the sheath (skin) of pooled mercury.

120 Pārada was also said to be of four colours: white (a curer of disease), red (a remover of senility), yellow (facilitates the transmutation of metal) and black (allows one to attain flight, to move through the air).

The following tables outline these properties and their subsequent effects if contaminated.[121]

Types of Pārada	
Rasa	Blood-red and free from blemish. Highly rejuvenating (Rasāyana)
Rasendra	Blackish in colour, enhances the body and keeps one free from senility. Its nature is quick and drying
Sūta	Yellowish in colour, contaminated, yet useful for either Deha or Loha siddha. It is particularly useful for making gold
Pārada	Bluish-white in colour, contaminated, yet useful for Loha siddha. It is particularly useful for making silver. It is capable of curing all disease after suitable purification
Miśraka	Displaying colours similar to a peacock's feather. It is capable of curing all disease after suitable purification

Natural impurities of Pārada	
Visha	There is an inherent poison contained within unprocessed mercury which can cause death
Vahni	Due to its firey and heating quality, Pārada causes high fever and a burning effect when ingested
Mala	Natural wastes (mala) and residual metals, intermixed with mercury, promote exhaustion and lethary. The presence of metals such as tin, lead and bismuth produced skin diseases and impotence
Giri	Mineral contents such as clays, sand, and silicon give rise to fainting and coma
Chapala	*Chapala* means 'always in motion'. This unsteadiness gives rise to sterility

121 It was noted that intake of an impure mercury produced the following conditions: abdominal pain, vomiting, burning sensations in the body, ringing in the ears, impaired vision, excess salivation, headaches, drowsiness and excess sleeping, convulsions, depression, high temperatures, aching limbs, teeth falling out, receding gum lines, bleeding gums and swelling of lymph nodes.

Typical metal impurities in Pārada	
Vanga	Tin (Sn)
Nāga	Lead (Pb)
Yasada	Zinc (Zn)
Chapala	Bismuth (Bi)

Kanchuka Dosha (sheaths) surrounding Pārada	
Parpati	Roughened outer layer, looking like coarse skin. Parpati dries bodily tissues, and obstructs the elimination of urine and stool
Patini	Broken surface resembling ruptures or dimples. Patini causes dryness and cracking of the skin
Bhedi	Imperfections and indentations resembling small holes. Bhedi causes strong purgation such as diarrhoea
Draarvi	Areas of inconsistency that resemble degeneration or decomposition. Draarvi promotes water retention and obstructs urine
Malkari	A disturbance of Vāta–Pitta–Kapha. Malkari causes tri-Dosha disturbances in body function
Andhkari	Dullness, a loss of reflection, appearing blind. Andhkari causes blindness
Dhwankshi	Darkness or blackness upon the surface. Dhwankshi causes discolouration of the skin and dryness of the mucous membranes

7.3/ Transmutation of base metals

> 'If Rasa (mercury) is processed by adding sulphur or Swarna Makshika (copper pyrite), its Bhasma/alchemical ash so prepared would be a remedy par excellence for curing all ailments. The patient suffering from kustha (obstinate skin diseases including leprosy) should take this recipe . . . [Similarly, mercury] processed with diamond and Shilajit, or Yogarāja,[122] cures all ailments. The patient suffering from kustha should take this recipe every day.'
>
> *Caraka Saṃhitā, Cikitsāsathānam*

Is it possible to transform lower metals into those of great value, such as gold and silver?

Gold-making recipes are certainly an important aspect of this ancient science. I am frequently asked, "Is it possible to make gold?" To partially answer this, it's important to note that it was believed to be possible, which attracted much interest. For the ancients, the concept of metal transmutation was valid. The *Pañcha Mahābhūtas*, or the five great elements, were understood to combine and fashion themselves into all the various forms we see in the physical world. Therefore, it didn't require a great leap of faith to imagine that metals such as lead, tin, and copper were perhaps only a step away from becoming gold.

The transmutation of metal in Rasa Shāstra was seen as one step along the pathway to immortality and obtaining enlightenment. To the alchemist, the ability to manifest gold from lower metals was nothing more than a sign that things were going well; it was not a means to an end. In addition, there seems to be a tradition that those seeking the formula for gold-making embark on a similar quest to those seeking Somā or Shangri-La. It appears to reveal itself only to those not actively looking for it.

There is a long-standing tradition of aurification (preparing imitation gold), and various classical texts provide guidance on the process. One example is the *Rasaprakāśasudhākara*, in which the author advises alchemists on how to create gold and silver using mixtures of copper, brass, bronze, zinc, tin, arsenic, and

122 Yogarāja: *Yoga*=combination, *Raja*=royal, therefore 'royal combination' means it combines readily with all medicines to improve potency and effectiveness.

small amounts of gold and silver. The recipes within this text can range from relatively simple to highly complex. Still, they yield substances that somewhat resemble the appearance of these precious metals.

If you're interested in exploring these methods further, I recommend visiting the following link:

http://www.ayuryog.org/blog/aurifiction-imitating-gold

The Ayuryog Project extensively studied the histories of Yoga, Āyurveda, and Rasa Shāstra (Indian alchemy and iatrochemistry) from the tenth century onwards. I had the opportunity to contribute to some of the videos on their channel, making it a valuable resource for those interested in aurification and argentifaction. For more detailed information, please refer to the Ayuryog Project (Entangled Histories of Yoga, Āyurveda, and Alchemy in South Asia) section.

Historical accounts of gold making and the preparation of mercury

Throughout history, there have been intriguing reports of accomplished alchemists showcasing their skills by seemingly creating gold in front of astonished onlookers. One such account involves visits to a temple in Delhi in 1942, where two renowned Rasa Vaidya Sāhtrīs[123] performed the act of gold-making. They claimed to produce pure gold, from mercury combined with a certain amount of previously prepared mercury. While these narratives are captivating, they lack conclusive evidence and may appear too incredible to sceptical individuals.

As a wise friend once pointed out to me, if someone possessed the ability to create gold, it is unlikely that they would willingly disclose their secret to their neighbours or broadcast it to the world. The secretive nature of alchemy suggests that those who possessed such knowledge would keep it closely guarded and reveal it only to a select few.

The advanced practices within alchemy, like the ones mentioned, are typically classified into two main categories: the refinement of metals through

123 A master of the Rasa science. For more information see Dash, V.B. (1996). *Alchemy and Metallic Medicines in Āyurveda*. Delhi: Concept Publishing.

transmutation and the refinement of the human body to counteract decay and ageing. The process of metal transmutation is referred to as *Lohasiddhi*, while the pursuit of immortality and longevity is known as *Dehasiddhi*.

In the context of alchemy, the goal is to achieve a state where the body, much like gold, remains untarnished, unaffected by the passage of time, and effectively attains immortality.

Just as gold does not tarnish, pale, or fade, an alchemical body is believed to possess qualities of endurance, perpetuity, and immutability, standing the test of time for eternity.

Mercury was a crucial component in both branches of alchemy, and was also categorised into two distinct types: *Samanya*, which was its common use, and *Rasāyana*, which was associated with rejuvenation, longevity, and the eradication of diseases. To enhance mercury's potential for the latter purpose (Rasāyana), eight additional practices or *Saṁskāras* were recommended to awaken its powers for rejuvenating the body.

Known as *Ashtasaṁskāra*, these eight practices constituted a complex process, often resulting in a significant loss of material during processing. To compensate for this loss, most texts advised preparing a larger quantity of the liquid metal, anticipating a reduced yield. These losses were primarily attributed to the unpredictable movements, referred to as *Gati*, intrinsic to the liquid metal. This inherent volatility made the process challenging and prone to inefficiencies.

Above, from top down, aurifiction: imitating gold

The table below outlines these losses in more detail.

Loss of Pārada	
Jalgati	The dissolution of mercury into liquid mediums
Hansagati	A loss of mercury during trituration
Malagati	The loss of mercury's natural impurities during processing
Dhoomagati	The evaporation of mercury during the heating procedures
Jeevgati	The loss of mercury due to its inherent nature (unaccountable losses)

The following descriptions of Ashtasaṁskāra are based upon the methods given in the alchemical classic *Rasahṛdayatantra*, cir.10th century CE.

Descriptions of Ashtasaṁskāra based upon the methods given in *Rasahṛdayatantra*	
Svedana (steaming)	In the initial stage, mercury is subjected to Svedana (steaming). It is combined with specific proportions of mustard greens, rock salt, Trikatu, Chitraka, fresh ginger, and radish, along with sour gruel. This mixture is then heated over a gentle flame for three consecutive days using a Dolā-yantra (a specialized steaming apparatus). This primary Saṁskāra effectively eliminates any impurities or waste materials from the mercury.
Mardana (trituration)	The trituration of mercury involves it being mixed with various ingredients such as molasses, burnt wool, rock salt, soot, brick dust, and mustard greens, in proportions of a sixteenth part each to the mercury as a whole. This mixture is then combined with sour gruel and undergoes the trituration process for three days. Sometimes, this process takes place in a moderately heated mortar using a specialised apparatus known as *Taptakalva Yantra*.
Mūrchā (thickening)	The three inherent faults of mercury are called dirt, fire and poison. Through dirt it produces fainting, through fire, burning sensations, through its poison – death. Aloe gel removes dirt, Triphala (the three myrobalans) removes fire, and leadwort (Chitraka), removes poison. Therefore, mercury should be thickened seven times using a mixture of these. Here, mercury is ground with aloe gel, Triphala and Chitraka until becoming fixed.

Utthāpana (raising)	Through the process of trituration, the mercury is now washed in sour gruel to relieve it of any lingering waste materials. This will include the dirt, fire and poison collected during the last process. The mercury is now brought to a rise from the sour gruel decoction using a condensation apparatus. Having performed Utthāpana, it (mercury) becomes cleansed from any contamination of the metals Nāga (lead) and Vanga (tin).
Pātana (causing to fall)	Once a coppery paste (Cu/Hg amalgam) is prepared, a condensation process is performed to eliminate any remaining traces of lead or tin that may still be present. Through this procedure, the purified mercury condenses. To further ensure the contaminating tin, lead, and copper are removed, a sublimation technique is employed. This process may require multiple iterations using devices known as *Adhahpatana* (upward) and *Urdhvapatana* (downward), each facilitating the upward and downward movement of mercurial vapour. These repeated sublimations contribute to the purification of the mercury by separating it from any remaining impurities.
Nirodha (reviving)	Following the sublimation process, mercury undergoes a state of impotence and requires revitalization to regain its original potency. This revitalization is achieved through a process of revivification, wherein watery emissions such as cow's urine, saline water, or Kanji (a fermented rice drink) are used. These substances are employed to rekindle the lost potency of the mercury, ensuring its efficacy is restored.
Niyama (fixing)	The mercury is now enlivened but also highly unstable, requiring the application of Svedana (sweating technique) to bring it under control. To achieve this, a bolus consisting of macerated herbs such as betel, garlic, salt, false daisy, spiny gourd, and tamarind is utilised as a restraining agent for the mercury as it is steamed in a Dolā-yantra (steaming apparatus). The combination of these herbs effectively ensures that the mercury remains stabilised and contained throughout the steaming process.
Dīpana (stimulating)	Svedana is also employed to activate and stimulate the mercury. During this process, which lasts for three days, the mercury is combined with various substances, including alum, ferrous sulphate, borax, black pepper, salt, mustard greens, Moringa, and sour gruel. This Svedana induces a sense of hunger in the mercury so it transforms into a state where it becomes 'one desiring a morsel.' Once the mercury has been prepared in this manner, it can be ingested to assimilate both minerals and metals while maintaining a constant weight. Following this step, the mercury is washed in lemon or lime juice, which ensures its cleanliness and purity. It is then rinsed with water and left to dry under sunlight, allowing any residual moisture to evaporate.

Left to right, from top:
1) Svedana stage of Pārada Purification;
2) recovery of Pārada after Svedana;
3) Mardana processing;
4) Mūrchā processing;
5) Utthāpana processing;
6) Nirodha processing;
7) Pātana processing;
8) Niyama processing;
9) Dīpana processing (start);
10) Dīpana processing (end)

Remaining 10 Saṁskāra

After completing the Ashtasaṁskāra, those who desired to activate mercury fully were advised to undergo ten additional Saṁskāra. However, accomplishing these supplementary steps would be no easy task.

The remaining ten Saṁskāra are:

1	Abhrakagrāsamāṇa	Measuring a morsel of mica
2	Cāraṇa	Feeding mica to mercury
3	Garbhadruti	Enclosed liquefaction
4	Bāhyadruti	External liquefaction
5	Jāraṇa	Amalgamation/digestion
6	Raṃjana	Dyeing/colouring
7	Sāraṇa	To set in motion, or potentiation
8	Krāmaṇa	To advance its power of penetration
9	Vedha	Transmutation
10	Sevana	Final ingestion

Regime whilst taking mercurial drugs

There are specific lifestyle and dietary guidelines, known as *Pathyas* (indications) and *Apathyas* (contraindications), to be followed when taking mercurial drugs. These guidelines aim to support the body during the treatment.

They are:

Pathyas

1 Preparation of the body through Pañcakarma therapies (refer to Part I).

2 Consumption of milk, ghee, Sāli rice, ginger, patola/pointed gourd (*Trichosanthes dioica*), Kakamachi herb/black nightshade (*Solanum nigrum*), Katuja herb/bitter oleander (*Holarrhena antidysenterica*), bringjal/eggplant (*Solanum melongena*), and the flower of the Kadali/banana tree (*Musa paradisiaca*).

Apathyas

1. Avoiding daytime sleeping.
2. Cold water bathing should be avoided.
3. Walking in cold winds or cold weather should be avoided.
4. Avoiding consumption of pungent or fatty foods, alcohol, and excessive indulgence in sex.
5. Engaging in conversations with women or in direct confrontations with others should be avoided.
6. Managing anger and avoiding rage.
7. Excessive physical activity or exhaustion should be avoided.
8. Intense fasting should be avoided.
9. Avoiding sour, salty, and pungent tastes in the diet.

These guidelines help create a favourable environment for the effectiveness of mercurial drugs while minimising potential adverse effects. It is essential to adhere to these recommendations for optimal outcomes during the treatment.

7.4/ Mercury – planet and metal

From a celestial standpoint, a strained relationship exists between gold, symbolised by the Sun, and mercury, represented by the planet Mercury. Mercury's proximity to the Sun exposes its surface to scorching temperatures of up to 427°c,[124] strikingly close to the boiling point of the liquid metal.

In Jyotish (Vedic astrology), Budha[125] (the planet Mercury) embodies the prince of the planetary court, with the Sun, his eternal guardian, never more than one astrological sign away. Due to his remarkable swiftness through the celestial realm, Mercury earned the titles of celestial messenger and planet of communication. Although Mercury's sidereal year spans a little over 87 Earth days, his journey around the zodiac, as perceived from Earth, closely mirrors that of the Sun.

From our vantage point on Earth, the Sun traverses its annual orbit in the

124 The boiling point of mercury is 357°c.

125 *Budha* means intelligence and discretion, coming from 'Buddhi' or enlightenment.

heavens, making contact with each planet during the course of its solar year. Mercury, being close to the Sun, appears to dart back and forth between the Sun and the approaching and receding planets. It frequently races ahead then appears to chase back after planets that have already passed. This rapid movement was also observed in Pārada (mercury) as it entered the body. It seemed to have the ability to permeate and connect with all seven bodily tissues, which had been associated with the qualities of the seven planets. In this context, mercury, both as a planet and a metal, is portrayed as fast-moving and all-pervading.

Astrologically, the planet Mercury is considered the planet of healing, emitting a subtle green ray that permeates all bodily tissues, diligently seeking out ailments in the deepest recesses of the body. He is also associated with Rasa Dhātu, the pervasive medium that nourishes and sustains the other six essential tissues, functioning as a fluid that communicates and regulates bodily metabolism.

Above, Budha: Planet Mercury

The Ayuryog project

(Entangled histories of Yoga, Āyurveda and alchemy in South Asia)
In 2019, I was approached by Professor Dagmar Wujastyk, the principal investigator of the area of the Ayuryog project that focused on alchemical research. She requested my assistance in carrying out a series of alchemical reconstructions based on various alchemical texts, starting with Rasahṛdayatantra. The objective was to gain practical insights into the procedures employed by ancient practitioners of Rasa Shāstra. It was a tremendously enjoyable experience for both Dagmar and me, and we learned a great deal. In fact, this initial project led to the undertaking of two additional smaller projects. All this work can be viewed at:

http://ayuryog.org/content/alchemy-reconstruction

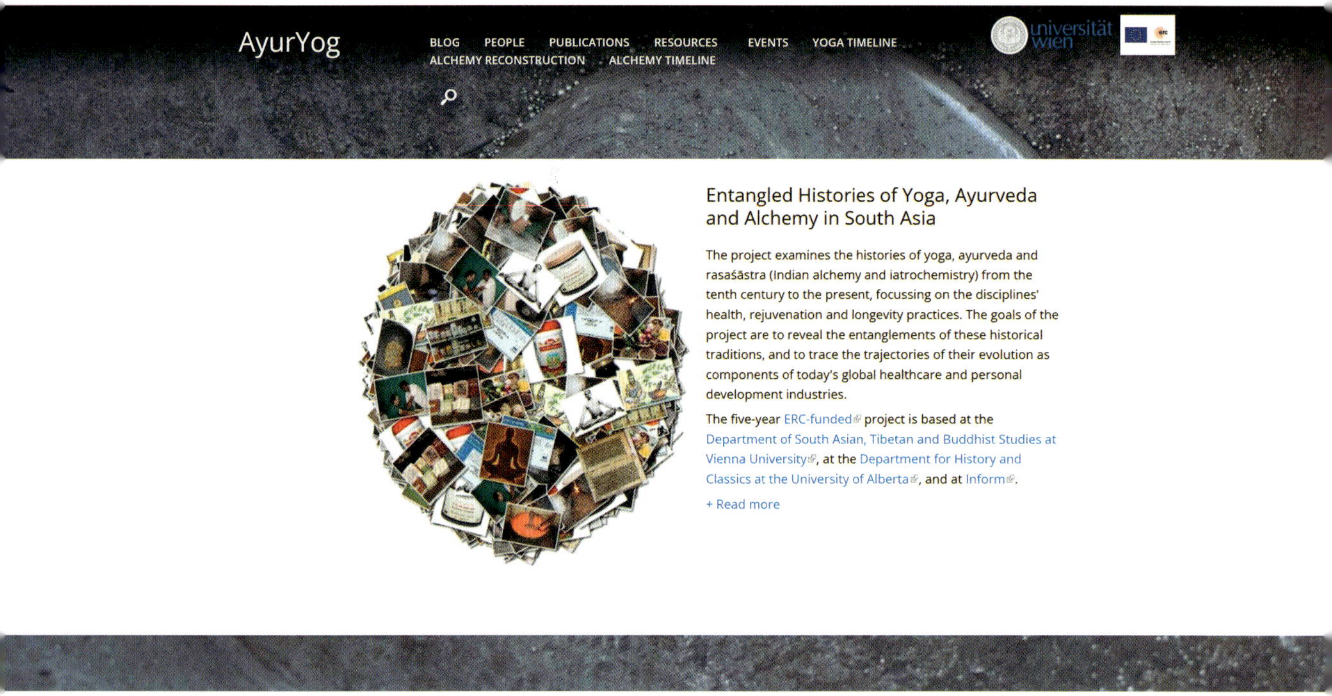

Above, banner taken from Ayuryog.org

I highly recommend watching it as the videos provide a visual guide that complements my explanations, and observing an actual process can enhance understanding. Additionally, I encourage you to read Professor Wujastyk's blog posts, as they contain valuable additional details. In addition, there are also interactive alchemy and Yoga timeline pages, which are excellent resources for those interested in the historical evolution of alchemy in India.

To provide a brief summary of the Ayuryog project, here are Professor Wujastyk's own words, taken from Ayuryog.org:

> 'The Ayuryog Project examined the histories of Yoga, Āyurveda and Rasaśāstra (Indian alchemy and iatrochemistry) from the tenth century to the present, focussing on the disciplines' health, rejuvenation and longevity practices. The goals of the project were to reveal the entanglements of these historical traditions and to trace the trajectories of their evolution as components of today's global healthcare and personal development industries.
>
> Practices aimed at achieving longevity, physical rejuvenation and transmutation constitute a key area of exchange between the three disciplines, preparing the grounds for a series of important pharmaceutical and technological innovations

and also profoundly influencing the discourses of today's medicalised forms of globalised Yoga as well as of contemporary institutionalised forms of Āyurveda and Rasa Shāstra.

Drawing upon the primary historical sources of each respective tradition as well as fieldwork data, the research team explored the shared terminology, praxis and theory of these three disciplines. We examined why, when and how health, rejuvenation and longevity practices were employed; how each discipline's discourse and practical applications relates to those of the others; and how past encounters and cross-fertilizations impact contemporary health-related practices in Yogic, Āyurvedic and alchemists' milieus.

The five-year ERC-funded project was based at the Department of South Asian, Tibetan and Buddhist Studies at Vienna University, the University of Alberta, and Inform, King's College London.

Regarding our alchemical experiments, our aim was to reconstruct the information given in Sanskrit alchemical works. To gain a deeper understanding of the various alchemical treatises' presentation of procedures, A. Mason and I have been working on reconstructing some of the alchemical operations described in the texts. These procedures, called Saṁskāra, are supposed to ready mercury and also other materials for their use in mercurial elixirs. The first eight of the Saṁskāra (of a total of eighteen) are aimed at ridding mercury of unwanted contaminants and flaws.'

Professor Dagmar Wujastyk

7.5/ Ancient sources of mercury

'Gold is found in Tibet in very large quantities, and often uncommonly pure. In the form of gold dust it is procured in the beds of rivers, attached to small pieces of stone, and sometimes it is found in large masses, lumps, and irregular veins. Cinnabar, containing a large portion of quicksilver, is a production of Tibet, and might be advantageously extracted by distillation, if fuel were more plentiful.'

Walter Hamilton, The East India Gazetteer

Historical studies into the origins of Rasa Shāstra have raised intriguing questions about the sourcing of mercury during that era. They suggest that materials may have been imported from neighbouring regions that were all connected through the Silk Road, including Tibet and China to the north or Afghanistan to the northwest. The imports might have contained contaminants like tin, lead, and arsenic, which, whether mixed in intentionally or unintentionally, would have resulted in the need for purification.

Although India is not known for substantial mercurial reserves, its northern mountainous terrain has a notable history of volcanic activity and geothermal phenomena. This suggests the potential formation of stable ores containing pockets of free mercury, which might have been mined over time. However, a significant portion of India's mercury used for alchemical purposes was probably imported from regions beyond its current borders.

Tibet has always been renowned for its abundant mineral deposits, earning the title of the world's treasure trove. Among its vast resources are copper, iron, mica, potash, rock salt, borax, azurite, graphite, lapis lazuli, lead, zinc, cinnabar, and, notably, gold. Many of these deposits are known to be associated with mercury minerals. Tibet's neighbor Kyrgyzstan, which is located to the northwest, boasts the world's second-largest reserves of mercury, surpassed only by the Guizhou Mines in southeast China.

Earth's natural alchemy

From the images opposite, it can be seen that magmatic processes involve the mobilisation of magma from the earth's mantle and its expulsion at the oceanic ridge (1), creating a nascent oceanic crust. In the specific scenario under consideration, two diverging oceanic plates are being propelled towards a landmass. The denser oceanic crust (SiMa) is compelled to subduct beneath the archipelago due to its higher density relative to the continental crust. The intense pressure and friction resulting from the dynamic interaction between the continental and subducting oceanic crusts generate substantial heat (2) and seismic activity. At regular intervals, these plates experience 'jamming' events, causing the accumulation of pressure. When the jam is eventually released, it triggers an earthquake.

The frictional forces acting on the subducting oceanic plate, as it sinks below the continental plate, instigate the melting of the oceanic plate, giving rise to magma formation (3). This molten material ascends through the continental crust and manifests as volcanic activity on the Earth's surface. The volcanic

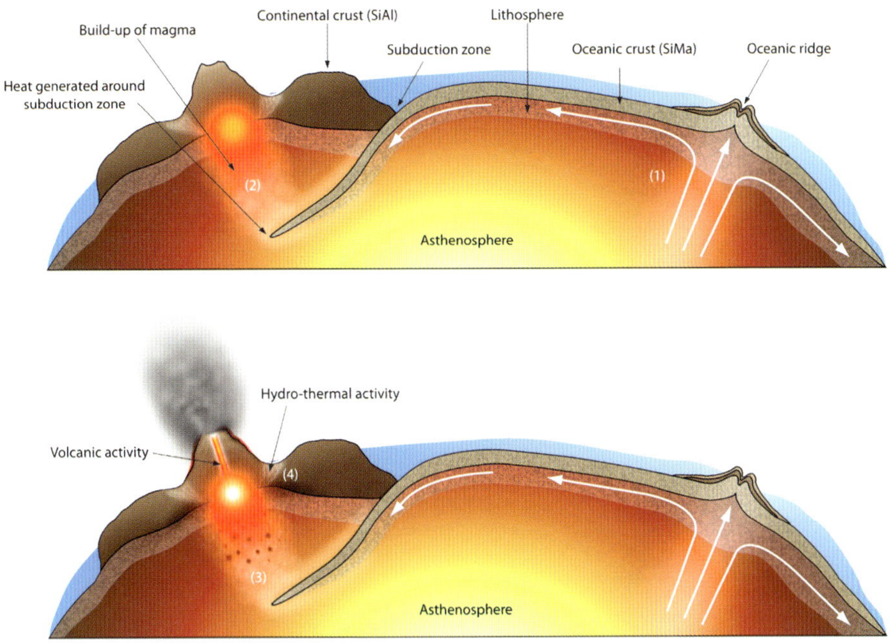

Left, Earth's natural alchemy

processes associated with this phenomenon also give rise to hydrothermal and hot spring activity (4), which are responsible for diverse and intriguing mineral assemblages, such as cinnabar, marcasite, and stibnite.

7.6/ Common purification methods for mercury

> 'Mercury is to be mixed with Trikaṭu[126] and boiled in Dolā-yantra with the juice of kārpāsa leaf[127] for one week. Thereafter it should be washed in exceedingly soured kāñjī and duly heated.'
>
> *Rasārṇava*

Prior to any application, mercury has to be purified, and several techniques are used in this regard. The following outlines two common and condensed methods of purification, both of which I was able to employ on a number of

126 Pippali (long pepper), Maricha (black pepper) and Sunthi (dried ginger)

127 Indian cotton plant (*Gossypium herbaceum*)

occasions. In practice, there are a number of variations on these methods, where different plants and minerals are sometimes substituted, but overall, they are quick, clean and safe to use in place of more complex procedures such as those outlined earlier in this book.

Far right, Kashmiri single-clove garlic

Mercury purification – method 1

Ingredients

500g Mercury
250g Kashmiri single clove garlic
500g Betel leaves (*Piper betel*)
Q.S. Triphala Kwatha[128]

1 The garlic is washed and peeled.
2 The garlic is then mortared into a pulp and squeezed through a cotton cloth to extract the juice.
3 The garlic juice is added to 500g of unpurified mercury and triturated for nine hours. The texts advise the contents to be mixed in one direction for the whole process.[129] During the mixing process, it may be necessary to add more garlic juice.
4 After nine hours, the mercury is 'washed' with a solution of vinegar to help break and disperse any impurities and the garlic residue. The liquid metal is repeatedly washed until only 'brightened' mercury remains.
5 The final removal of impurities is achieved by filtration through a fine cotton cloth. The mercury will pass through the cloth with a little squeezing, and the cloth will absorb any remaining debris. Any excess vinegar is allowed to evaporate before continuing.
6 500g of betel leaves (*Piper betel*) are selected and juiced. This is best

128 Made from equal quantities of Āmalaki = *Phyllanthus emblica*, Bibhītaki = *Terminalia bellirica* and Harītakī = *Terminalia chebula*.

129 Before undertaking the purification of Pārada, one decides upon the direction of mortar action (clockwise or anti-clockwise). This is then maintained throughout the processing.

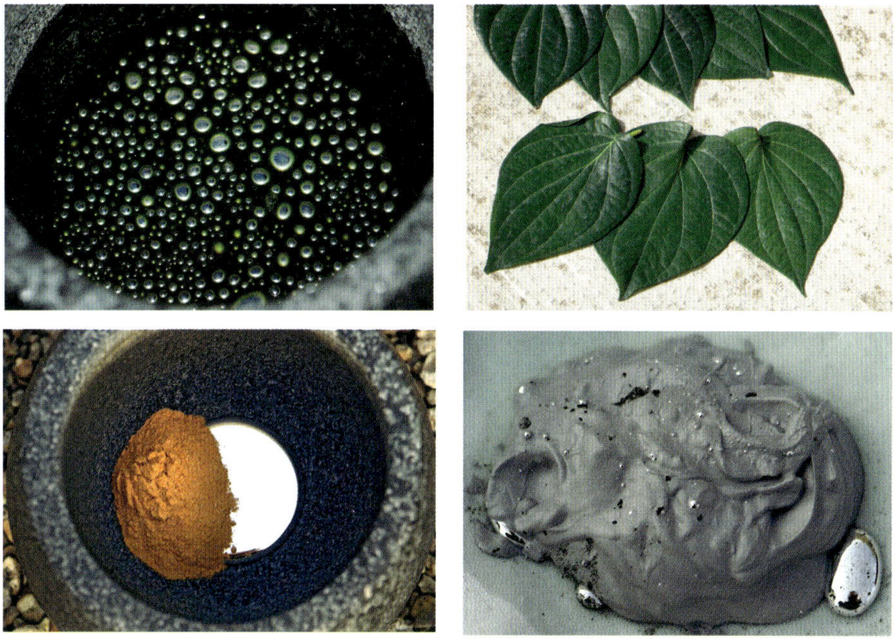

Top left, Pārada ground in betel leaf juice

Top right, fresh betel leaf

Bottom left, Pārada ground with turmeric and brick dust

Bottom right, Pārada loses all brightness and becomes paste-like

achieved by pounding fresh leaves in an iron mortar and using a cotton cloth to squeeze the juice from the macerated pulp.

7 The betel leaf juice is added to the filtered mercury and triturated for nine hours. During this period, it is usually necessary to top up the juice levels.

8 After completing the trituration, the mercury is again washed to remove any residue and debris.

9 As before, the remaining purified mercury is passed through a cotton cloth to clean, while any excess vinegar is allowed to evaporate before continuing.

10 The next stage of processing requires Triphala Kwatha. Kwatha is prepared by adding 60g of powdered Āmalakī, Harītakī and Bibhītaki in equal quantities to about eight cups of water. This is reduced over a mild heat, evaporating three quarters of the contents. When complete, the Kwatha is added to mercury and again triturated for nine hours.

11 Following trituration, the mercury is again washed and filtered through a cotton cloth. The excess vinegar is allowed to evaporate before storing the purified mercury in a glass bottle. Purification is now complete.

Mercury purification – method 2

Ingredients

500g Mercury

Q.S. Turmeric (powdered)[130]

Q.S. Brick dust[131]

1 Equal quantities of turmeric powder and brick dust are added to 500g of mercury and stirred for at least twenty-four hours.
2 Upon completion, the mercury is again washed in vinegar and filtered through a cotton cloth. The excess vinegar is allowed to evaporate before storing the purified mercury in a glass bottle. Purification is now complete.

Right, Pārada ground with turmeric and brick dust; far right, Pārada after washing

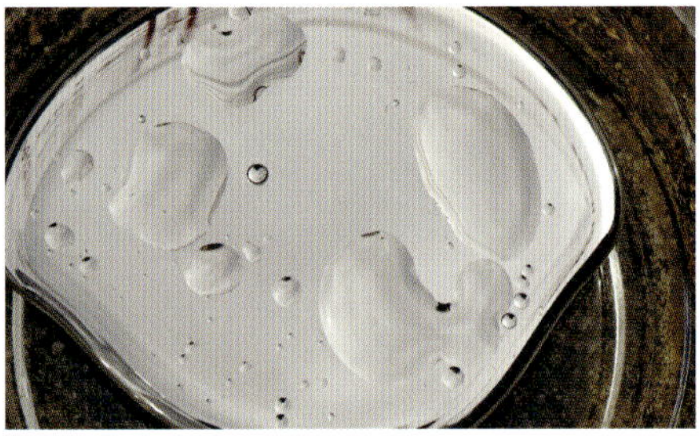

7.7/ Kajjali (HgS) black sulphide of mercury

Above, Kajjali is prepared from Purified Pārada and Gandhaka

After the purification of mercury (as outlined in Section 8.2), it is combined with purified sulphur to create a black sulphide compound. This process is referred to as *Kajjali Bandha*. Kajjali, thus formed, serves as a fundamental component in numerous Rasa medicines.

The preparation of Kajjali itself is a simple procedure. Equal quantities of mercury and

130 In some cases, fresh turmeric is recommended.

131 Brick dust is usually some sort of fire brick, or at least bricks that have been subjected to repeated heating. When the bricks of a puṭa enclosure are eventually replaced they can be kept for this process.

sulphur are meticulously ground together for approximately eight to nine hours, or until the two substances blend harmoniously, resulting in a smooth, lustrous black powder. It is important to note that the final powder should exhibit a matte finish and lack any lustre. If any shine remains, it indicates the presence of free mercury within the powder.

Far left, Purified Pārada and Gandhaka; left, Kajjali Pārada and Gandhaka, after five hours of mixing by hand

The final Kajjali should adhere to the following descriptions:

Kajjali – Acceptable signs	
Slakshnatvam	Smooth and non-irritating to mucous membranes
Kajjalabhas	Black, looking like soot
Rekhapurnata	Filling the minute spaces on the fingertips
Nischandratvam	Loss of mercurial lustre
Anjana Sannibha	Minute particle size, suitable for collyrium or Anjana
Loha Pariksha	Mixed with lemon juice and applied to gold sheet, mercury does not form an amalgam
Agni Pariksha	Fumes upon heating, leaving no ashes
Jala Pariksha	Floats upon water

During the grinding process, this material becomes very powdery, often rising and distributing outwards. When mixing Kajjali, it is advised to wear a mask to avoid inhaling dust. Some texts recommend lightly sprinkling water over Kajjali to dampen down dust particles.

Ratio of mercury to sulphur

Various alternative mixtures are recommended regarding the mercury/ sulphur ratios in Kajjali. While the most commonly used ratio is 1:1, other ratios, such as 2:1 and 10:1, are employed in more intricate and diverse manufacturing processes. Additionally, when Kajjali is subjected to heating in Kūpīpākwa,[132] the ratio may be adjusted to 1:3 based on the desired recipe and effect. Generally, higher sulphur content in Kajjali corresponds to higher dosages and an extended duration of consumption without concerns about side effects. This consideration becomes particularly relevant when determining the appropriate dosage and treatment duration.

Benefits of Kajjali

Kajjali serves as a fundamental component in numerous Rasa medicines and is generally not taken alone. However, there are exceptional circumstances where this may be recommended. Typically, Kajjali is combined with other herbal ingredients and subjected to additional Bhavana (processing), or it is prepared in various forms such as Parpati or medicines made using the Kūpīpākwa technique.

Right, purified Pārada and Gandhaka are triturated for approximately nine hours or until the black sulphide of the mercury is without lustre

The benefits of Kajjali are manifold. It enhances the potency of other drugs and possesses robust anti-ageing properties. Furthermore, it acts as an excellent Yoga vāhin[133] or catalyst and a potent Rasāyana substance.

Suitable Anupānas for administering Kajjali include honey, ghee, butter, and egg white. These Anupānas not only facilitate the delivery of Kajjali but also contribute to its effectiveness and may help minimise potential side effects.

---

132 A glass bottle specially adapted for the heating process. See Section 7.9/ Makara Dwaja

133 *Yoga vāhin* = a vehicle for other medicines

7.8/ Rasa Parpati (HgS)

One of the widely known applications of Kajjali involves its use in preparing Rasa Parpati. *Parpati* refers to a delicate wafer or flake-like form that is achieved by rapidly cooling and solidifying a liquefied medium, typically by compressing it between Eranda or banana leaves. Kajjali, commonly known as HgS, is the preferred substance for creating Parpati.

Parpati is of four varieties:

1. Kajjali only (Pārada and Gandhaka);
2. combined with previously prepared Bhasma;[134]
3. a prepared Parpati is ground with various herbal ingredients;
4. Without the presence of Pārada and Gandhaka, as in Swetha Parpati.

In comparative studies investigating the effects of Kajjali and Rasa Parpati, both of which are composed primarily of HgS, notable distinctions have been observed, particularly regarding the impact on stool consistency in patients undergoing treatments. Kajjali induces softer or looser stools, whereas Parpati tends to have a stool-hardening effect. It is worth mentioning that Rasa Shāstra advocates the application of heat to enhance the potency of mercurial drugs, as exemplified by Makara Dwaja and similar formulations. However, it is important to note that this heating process is not always necessary, as most popular Rasa preparations rely solely on Kajjali.

Above, several different types of Rasa Parpati

134 Herbs and Bhasma may be added to Kajjali; for example, copper Bhasma or iron Bhasma to produce Tamra Parpati or Loha Parpati.

Rasa Parpati preparation method

Ingredients

Kajjali
Ghee (clarified butter)
Banana or Eranda leaves
Gorvara (cow dung)

1 An iron Kadai is heated, and a little ghee is melted.
2 A small amount of Kajjali (usually one tablespoon per heating) is placed into the Kadai.
3 Kajjali will quickly melt but produces toxic smoke. (Note: Heating this material is extremely hazardous, exposing one to mercuric/sulphur vapour). The softened mass is swiftly removed and dropped onto a waiting banana cushioned with cow dung. Upon contact with the leaf, a secondary tamping bundle (made of cow dung wrapped in banana leaf) is firmly depressed on the molten Kajjali, pressing it into a thin wafer.
4 The resultant disc is recovered from the leaf, ground into a powder and stored.

There are three *pakas* (levels of heating) in Parpati preparation:

1 Mrudupaka: low heat, Parpati remains soft and black after heating;
2 Madhyapaka: medium heat, Parpati becomes brittle and reveals a blackish interior with slightly peacock colouration when broken. This grade is the most common;
3 Karapaka: high heat, Parpati looks reddish and dry with a loss of lustre.

Benefits of Rasa Parpati

Rasa Parpati offers a comprehensive array of benefits, effectively addressing a wide range of ailments, including skin disorders, anaemia, fever, IBS/malabsorption, piles, diarrhoea, dyspepsia, ascites (excessive peritoneal fluid), hepatitis, splenomegaly (enlarged spleen), asthma, persistent cough, ulcerative colitis, intestinal parasites, and hyperacidity.

To optimise its efficacy and absorption, it is recommended that you consume Rasa Parpati alongside an appropriate Anupāna. Among the various available choices, honey, milk, and ghee are considered the most suitable options. These Anupānas act as carriers for the medicine, facilitating its delivery and potentially mitigating any side effects.

7.9/ Makara Dwaja (HgS) red sulphide of mercury

> 'Pārada and Gandhaka are turned into nectar and poison according to the purpose of their use. When they are taken to rule they act as nectar, but when they are used without observing any rule they act like poison.'
>
> *Rasanavakalpa*

A commonly used method to prepare mercuric sulphide involves heating it gradually for several hours within a specially adapted glass bottle called a *Kūpīpākwa*. This bottle is buried in a Vālukā Yantra with its neck protruding, and it is fortified by multiple layers of cloth and clay to shield the delicate glass from extreme external temperatures. This traditional technique produces many Rasa medicines, particularly those based on mercury compounds, including Makara Dwaja.

This specific Rasa preparation holds a deep-rooted legacy in Āyurveda. It is highly esteemed for its immune-system-boosting and Rasāyana qualities. Once Kajjali, the black mercury sulphide, is prepared, it transforms into its red sulphide form. The resulting red crystals are carefully collected from the neck of the bottle and finely ground into a red powder. It is believed that the efficacy of this and similar remedies remain intact for an extensive duration, with some proponents even suggesting that they improve further with time.

Above, several different types of mercurial drugs, including Makara Dwaja, being prepared in glass bottles

Makara Dwaja offers a means of infusing the energetic properties of gold into a remedy without the necessity of reducing it to a Bhasma (ash). This is accomplished by blending a mixture of mercury/gold amalgam with sulphur, then subjecting the powdered substance to a temperature of approximately 500°c for twelve to fourteen hours. All three elements become intricately bonded through this alchemical process, forming a dark reddish crystal. The resulting crystal can be finely powdered and consumed directly or further formulated into other medicinal preparations.[135]

135 One such remedy being *Siddha Makara Dwaja*, which recombines Makara Dwaja with Kapura (camphor), Jatiphala (nutmeg), Samudra Phena (Pisti of cuttlefish shell) and Kastūrī (deer musk).

Right:

1) venting fumes from a kūpī jar; 2) Vālukā Yantra/sand bath; 3) mercury vapour and deposits of mercuric sulphide in the neck of the kūpī; 4) protective layering of cloth and clay; 5) mercury/ gold/sulphur; 6) heat source

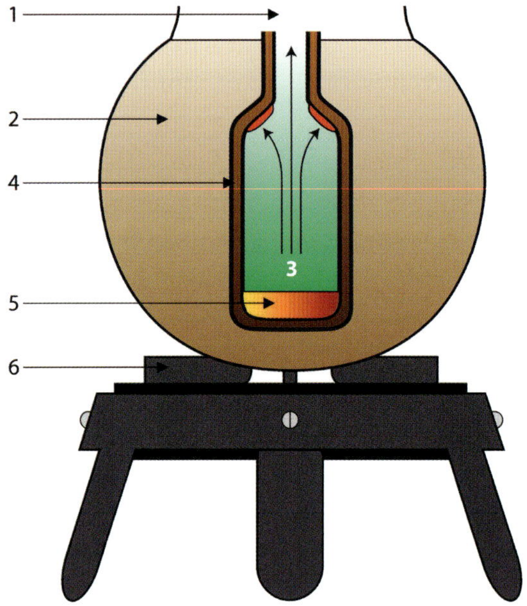

Makara Dwaja preparation method

Ingredients

1 part Swarna Patra (24 carat-gold sheet)[136]

8 parts Pārada (purified mercury)

16 parts Gandhaka (purified sulphur)

Kumārī (*Aloe indica/barbadensis*)

1 The ratio of mixture for Makara Dwaja is 1pt–8pt–16pt. In this instance, we start with 5g of 24-carat gold sheet to which 40g of purified mercury is added and triturated until all vestiges of gold are amalgamated. Periodically during the amalgam process, the mercury's shine diminishes and appears to crust. This change in the surface of the liquid metal was thought to indicate that the mercury was 'feeding' – feasting upon the gold. Further trituration sees the mercury appear to brighten as the liquid metal 'digests' the gold.
2 After amalgamation, 80g of purified sulphur is added and triturated into a smooth Kajjali (black sulphide form of mercury). The aloe juice is then added

136 Gold is heated until glowing, then quenched (nirvapana) into a variety of herbal mediums. A decoction of Kanchanara (*Bauhinia variegata*) is very popular for this preparation.

to the Kajjali, and the material is pasted and inserted into a pre-prepared kūpī.[134]

3 The kūpī is then submerged neck-deep in a sand bath. During this first stage of heating (three to four hours), the heat intensity is kept low until visible sulphur fumes emerge from the bottle neck.

4 During the second stage of heating (five to six hours), the heat is increased, and the kūpī is continually studied for signs of sulphur fume cessation. The absence of sulphur fumes usually requires re-opening the bottle neck, which generally becomes blocked by heavy sulphur deposits. A narrow iron rod is inserted into it and lowered until reaching the base of the kūpī to open the bottle neck. The rod is slowly rotated, clearing the neck while agitating the contents of the bottle. This action is essential in preparation, ensuring all liquefied contents are thoroughly mixed and interacting.

Left, preparing gold amalgam

5 During the final stage (seven to twelve hours), the intensity of heat is increased to its highest level. The kūpī still shows no signs of fuming until a plume of bluish-coloured smoke appears, followed by an orange flame at the opening of the kūpī. At this point, the heating is stopped. The opening of the bottle is then closed with a ceramic plug.

6 The kūpī and apparatus are left to cool over twenty-four hours.

7 Next day the kūpī is retrieved, and its outer protective cloth and mud layers are removed. A piece of twine is then tied about the waist of the bottle, soaked in a flammable liquid and set alight. Rapidly cooling the burning

Left, and far left, red mercuric sulphide crystals accumulated about the neck of the kūpī

twine, using a wet cloth, cracks the glass and exposes the bottle neck. Just above the neck line, the concentration of fumes will have deposited silvery/ red mercuric sulphide crystals. These crystals are carefully dislodged and finely ground to produce a deep red-coloured powder. This powder is then stored in an airtight glass bottle.

Makara Dwaja – XRD/EDX

XRD/EDX analysis was conducted on three distinct batches of Makara Dwaja, revealing the presence of mercury and carbon in each sample. The first sample, which exhibited significant attenuation, displayed indications of the transition metal tantalum, along with polonium and sulphur. However, determining the significance of their concentrations is challenging as they coincide with the spike associated with mercury. Given that tantalum and polonium are rare elements, it is probable that their concentrations in this sample are relatively low. Furthermore, there is a possibility of other trace elements being present, but the level of attenuation hinders detection.

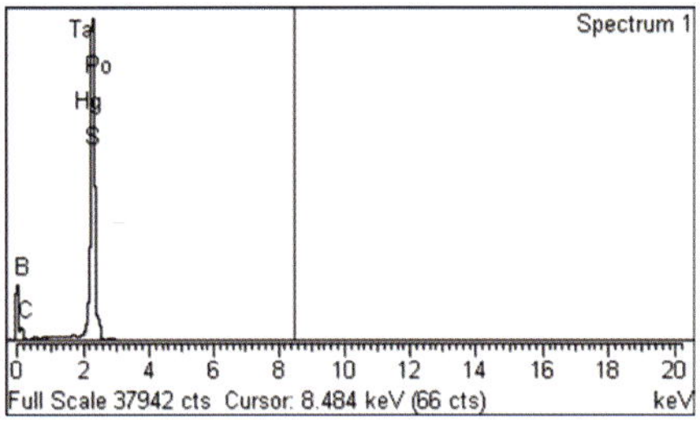

Right, Makara Dwaja sample 1, Sri Lanka 2005

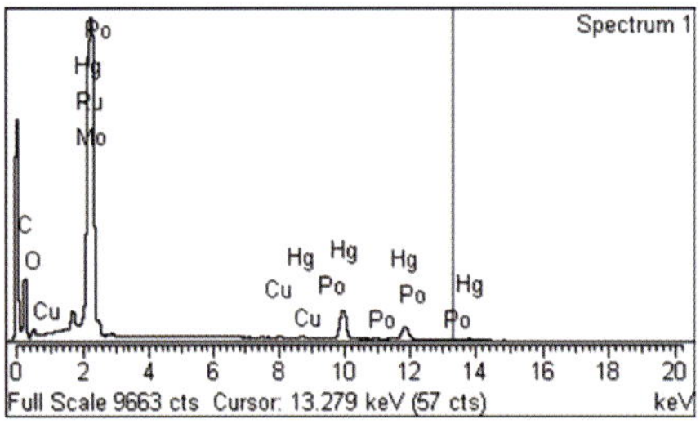

Right, Makara Dwaja sample 2, Sri Lanka 2005

Samples 2 and 3 exhibit varying degrees of attenuation, indicating lower concentrations of the predominant elements, mercury and carbon. Moreover, these samples reveal the presence of other elements, albeit in low or trace amounts. In Sample 2, the presence of copper is evident, as well as the exceptionally rare transition metals ruthenium and polonium. Sample 3, on the other hand, demonstrates the presence of trace quantities of molybdenum. Interestingly, neither sample

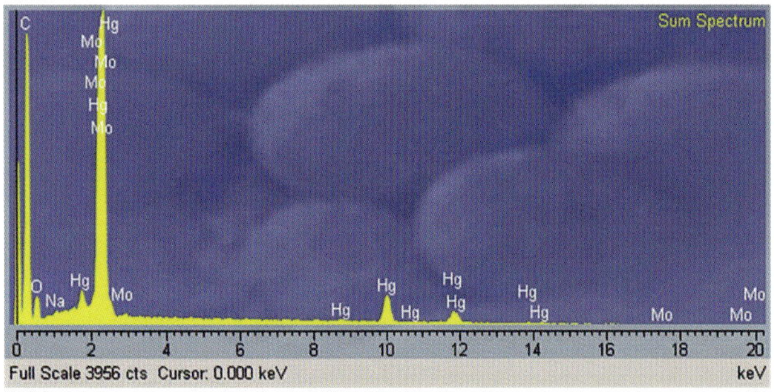

Above, Makara Dwaja sample 3, India 2012

2 nor sample 3 shows any traces of sulphur, which suggests that the original material may have been exposed to higher temperatures or for longer durations, leading to the complete conversion of any sulphur content into sulphur dioxide before being vented.

All three samples exhibit similar arrays of detected elements, with variations observed in the trace elements. The occurrence of these rare elements in the samples offers insights into the source of the mercury ore, potentially originating from placer deposits.[137]

Benefits of Makara Dwaja

Makara Dwaja offers a variety of benefits. Along with its strong Rasāyana and aphrodisiac properties it enhances the immune system, improves complexion, strengthens respiratory function, promotes intelligence, and enhances Agni (metabolic functioning) and overall metabolic balance.

For the administration of Makara Dwaja, the most suitable Anupāna includes honey, betel leaf, warm milk, and saffron or date palm jaggery. It is recommended to administer this remedy during the cooler winter months, as the heating nature of mercury and similar mercurial remedies align well with the colder climate.

7.10/ Extraction of mercury from cinnabar

> 'Hiṅgula is to be rubbed for one day with lime or lemon juice and then subjected to the process of Urdhvapatāna, upward sublimation by means of Vidyādhara Yantra.[138] Mercury thus obtained is pure and free from all the blemishes and coverings. It may be used in everything without being subjected to the eight indispensable operations of mercury (Ashta-saṁskāra).'
>
> *Rasa Jala Nidhi*

137 Placer deposits are a geological term used to describe the separation of minerals by gravity during the process of sedimentation.

138 Also, a Damaru Yantra

Rasa Shāstra states that mercury derived from cinnabar through direct sublimation should be used in the production of all medicines, including Rasāyana, without requiring Ashtasaṁskāra. However, this assertion can be perplexing since those seeking to purify mercury would likely opt for this method to eliminate the complexities of Ashtasaṁskāra and minimise material loss.

Considering the relative simplicity of sublimation, it is surprising that access to cinnabar did not render other mercury purification methods obsolete. One possible explanation for this could be limited availability. Even today, a significant portion of cinnabar sold in the market is artificial and prepared by cooking mercury and sulphur in sealed iron crucibles. The look of the artificial material is very different to the naturally occurring ore (see images overpage).

7.11/ Hiṅgula (cinnabar)

> 'Hiṅgula destroys disorders created by all three humours. It fuels digestive fire, it is a strong rejuvenator and it cures all disease. It is an aphrodisiac and has praiseworthy attributes in Mārana procedures. Pārada extracted from Hiṅgula attains a quality similar to Pārada which has undergone Mārana with Gandhaka.'
>
> *Rasa Ratna Samuccaya, Srī Vāgbhatāchārya*

> 'Dan is the name of a stone, its character in Chinese symbolises a piece of cinnabar in a well, the character also means red. Taoist alchemists processing cinnabar would release its mercury content and grind it with the fat of pig, sheep and bull. This mixture was then pounded with Tong Cao[139] and using Guanmu Tong[140] as a wick, candles were fashioned and used to locate treasures such as gold, silver and copper. The light emitted from these candles was said to also reveal the hiding places of devils and snakes.'
>
> *Bencao Gangmu, Li Shizhen*

139 Tong Cao/*Medulla tetrapanacis* (rice paper pith)

140 Guanmu Tong/*Aristolochiae manshuriensis*

Cinnabar had diverse applications in ancient times, spanning medical, artistic, and ceremonial practices, contributing to its widespread presence across various cultures. As a red pigment,[141] it was commonly used to treat wooden structures, protecting against insects and the elements. Additionally, vermilion-dyed fabrics adorned both priests and their assistants during ceremonies. The alchemists of antiquity were familiar with cinnabar's sublimation process, understanding that the amalgamation of mercury and sulphur formed the basis of all metals. Mercury symbolised permanence, while sulfur represented impermanence. When cinnabar was heated, its unique silvery essence would mysteriously appear in small droplets. Combining it with sulfur and reheating it would transform it again into an enigmatic vermilion-coloured powder.[142]

Cinnabar acquired a legendary reputation and came to be referred to as Dragon's or Phoenix Blood[143] due to its perceived supernatural powers. In ancient Japan, vermilion paint adorned both warships and warriors. At the same time, the country's royalty were laid to rest in

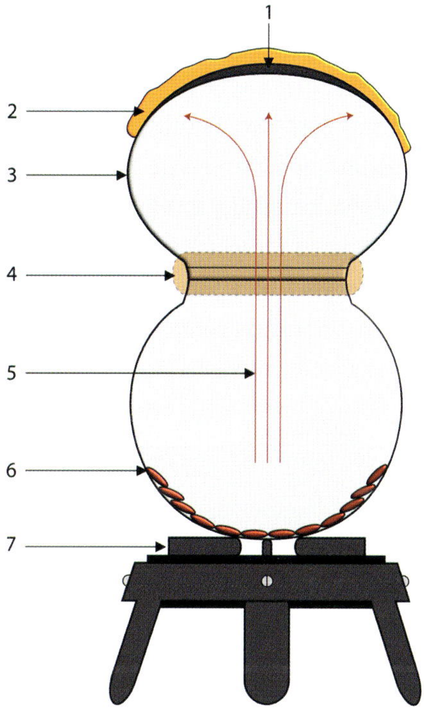

Left, Damaru Yantra:

1) sublimated mercury;
2) damp cloth;
3) earthen pots;
4) joints between pots sealed with fire clay;
5) mercury vapour;
6) Hiṅgula cakrika;
7) heat source

141 Cinnabar (vermilion) is also a highly effective pesticide and fungicide. Prolonged exposure and oxidation intensifies its reddish hue.

142 *Vahni Mṛtsnā*, a mixture of three parts chalk, one part iron oxide, and one part rock salt and fresh milk.

143 Cinnabar was believed to be the solidified blood of dragons or phoenix birds, both creatures associated with longevity. Interestingly, the dragon and phoenix are both important symbols in Feng Shui.

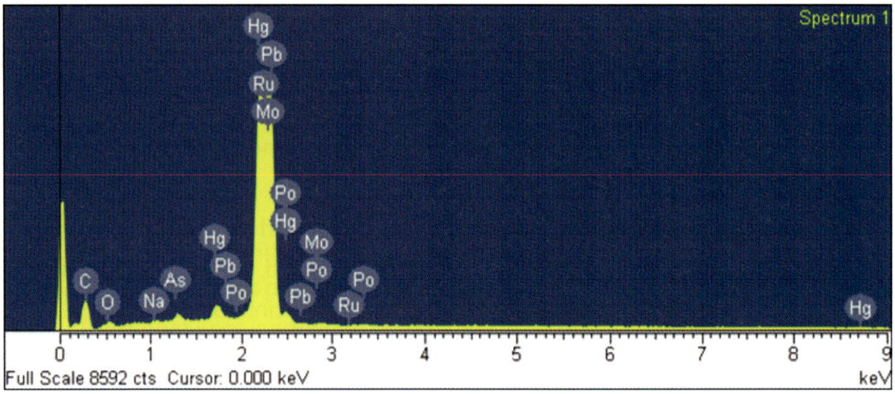

Right, Hiṅgula sample 1 XRD), Pakistan 2008

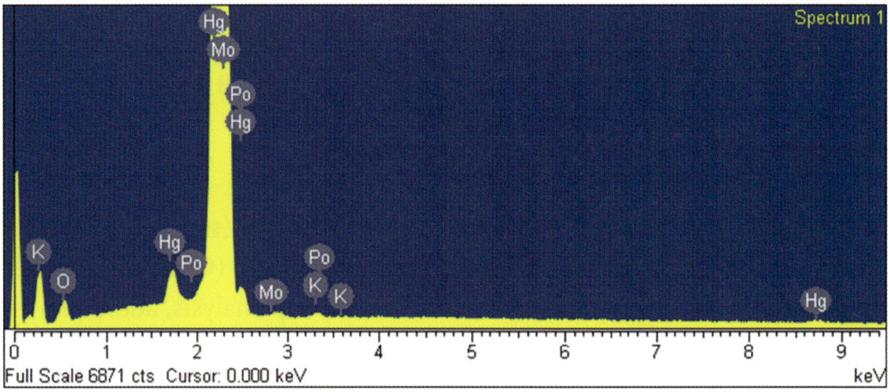

Right, Hiṅgula sample 2 (XRD), author's sample, Sri Lanka 2005

Benefits of Hiṅgula

Hiṅgula offers several benefits, including its strong Rasāyana (rejuvenating) and aphrodisiac properties. It has been found to improve complexion, enhance intelligence, promote healthy Agni (digestive fire), and balance metabolism.

When it comes to choosing an Anupāna (vehicle) for consuming Hiṅgula, milk, honey, or betel leaf juice are the most suitable options. These substances can complement the properties of Hiṅgula and aid in its effective absorption and assimilation within the body.

Preparation of Kushta Sangraf

I have included Kushta Sangraf (Unani tradition) to demonstrate the effects of higher temperatures on this substance. It's important to note that Rasa Shāstra and classical Chinese medicine do not recommend heating Hiṅgula. However, the Unani Tibb tradition favours this method for preparing certain types of Kushta, their equivalent of Bhasma, which is derived from the sulphide form of mercury. The method is given below.

The introduction of nested Sharaava in the Arabic tradition makes this approach to cinnabar interesting. The external sharavaa is filled with plant ash, creating an effective filtration system which dampens the ejection of mercury vapour. Kushta Sangraf is considered a highly beneficial medicine for older and debilitated people. It possesses strong Rasāyana properties and is known to empower digestion, aid assimilation, and restore nervous system functionality.

Left, Kustha Sangraf (finished Bhasma)

Ingredients

120g Sangraf rumi (cinnabar)
Q.S. Latex of Arka (crown flower) *(Calotropis gigantean)*
Q.S. Nimbuka *(Citrus limon)*
Q.S. Priyangu (*Callicarpa arborea)*
Q.S. Ash of Apamarga *(Achyranthes aspera)*

1 The cinnabar is ground with Arka latex until a fine paste is achieved. This paste is then dried under sunlight.
2 When dry, the cinnabar is re-ground with lemon juice or the juice of priyangu.
3 Cakrika are formed from the paste and dried in the sun. These are then sealed in a Sharaava using the clay and cloth method. The crucible is then dried in the sun.
4 The Sharaava is placed inside a second larger Sharaava, and packed with the ash of Apamarga. This second Sharaava is sealed using the clay and cloth method and dried under the sun.
5 The Sharaava is given puṭa, using approximately 5kg of cow dung cakes.

The heating process takes two to three hours with a slow diminution of heat until retrieval twenty-four hours later.

6 Upon opening the inner crucible, its contents should have converted to a dull reddish-brown colour.

7 The cakrika are ground into a fine Kushta, sieved and stored in a glass bottle.

Kushta Sangraf – EDX

EDX analysis was performed on a sample of Kushta Sangraf. It was observed that mercury (Hg) was absent from the prepared sample, which can be attributed to the high temperatures applied during its preparation, leading to the vaporisation of mercury. Additionally, due to the heating process, a significant portion of sulphur is likely to have been oxidised to sulphur dioxide (SO2).

Right, nested crucibles are used in the preparation of Kustha Sangraf

The composition of this particular sample primarily consists of silicon (Si), calcium (Ca), and iron (Fe). Furthermore, trace amounts of sulphur (S), aluminium (Al), phosphorus (P), potassium (K), and sodium (Na) are present, which are typical of rock-forming minerals commonly found in such samples.

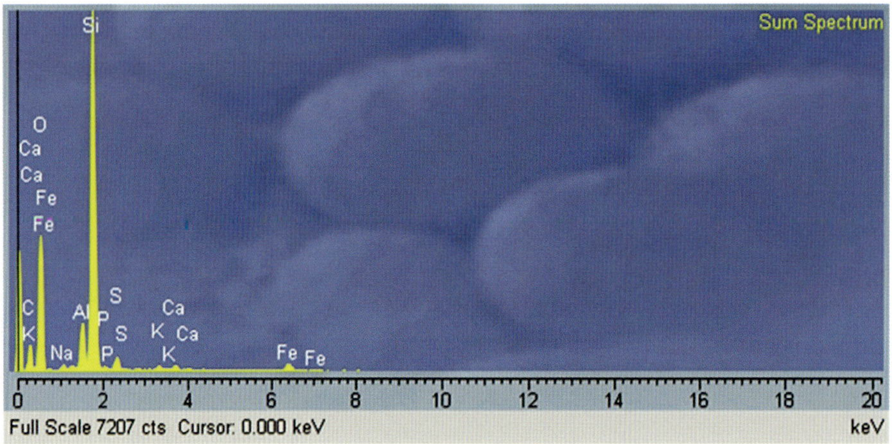

Right, Kushta Sangraf (EDX) sample, 2012

Benefits of Kushta Sangraf

Kushta Sangraf offers several benefits, including removing catarrh (excessive mucus) and relief from indigestion. It serves as a potent rejuvenator, particularly beneficial for older people, and can be used effectively in cases of sexual debility.

When selecting an Anupāna (vehicle) for consuming Kushta Sangraf, butter, cream, or honey are the most suitable options. These substances can enhance the properties of Kushta Sangraf and aid in its absorption and assimilation within the body.

7.12 / Summing up and dangers of mercury

Mercury, a highly toxic element, has a complex past as both a medicinal substance and a poison. In the Almadén cinnabar mine in Spain, Roman slaves devised crude gas masks in a desperate attempt to avoid inhalation of the toxic dust. Working in Almadén was seen as a virtual death sentence.

In more recent history, mercury has left its mark on countless products, ranging from batteries, fluorescent tubes, and anti-fouling paints to its inclusion as an antibacterial agent in vaccines. Its remarkable adaptability and versatility have rendered it indispensable across the ages. However, in recent times, the demand for mercury has waned due to the emergence of safer and environmentally friendly alternatives. Advancements in reclamation techniques have facilitated the revival of a significant portion of the mercury already in commercial circulation. Furthermore, global awareness regarding the severe health risks posed by this element has been steadily growing. Remarkably, China, which boasts ample reserves of mercury and a long history of its utilisation, now stands as both the largest producer and consumer of this metal in the last two decades.

When humans are exposed to mercury vapour, approximately 80% of this toxic gas is absorbed through their respiratory system. In contrast, mercury ingestion or skin absorption remains relatively low, accounting for less than 2%. Once inside the body, mercury tends to accumulate primarily in the brain, spine, kidneys, and liver. As a known neurotoxin, it targets both the autonomic and sympathetic nervous systems, exerting detrimental effects throughout the entire body. The consequences of high levels of mercury exposure encompass a range of symptoms, such as degeneration of mucous membranes, excessive salivation, tremors, receding gum lines, tooth loss, tachycardia, kidney damage, epileptic seizures, muscular spasms, abdominal pain, impaired vision and hearing, vomiting, dementia, and, in cases of prolonged exposure, death.

Furthermore, from a psychological perspective, mercury exerts measurable effects on one's personality, giving rise to various mental states, including depression, irritability, shyness, attention deficit disorder (ADD), religious euphoria, obsessive-compulsive disorder, rage, psychosis, and isolationism. These traits collectively contributed to the origin of the phrase 'mad as a hatter' during the 19th century. Hat-makers of that era, who utilised mercurous nitrate in the carroting process (which involved matting together the pelts of deceased animals), were particularly prone to developing these abnormalities due to prolonged exposure to the effects of mercury.

In the past decade, the potential health risks associated with mercury have gained significant attention in mainstream news. One concern revolves around the possibility of protracted toxicological effects from exposure to amalgam fillings. It is believed that absorption into the bloodstream can occur by leaching material from the fillings into saliva or the direct inhalation of mercury vapour emitted from decaying dental amalgam.

Extensive research has been conducted on this subject, and readers should explore the claims made against amalgam fillings and the counterarguments put forth in support of their continued usage. For further information, the website http://iaomt.org/mercury is recommended as a resource to delve deeper into this topic.

Detoxification and chelating mercuric deposits in the body

The human body possesses a specific capacity to eliminate heavy metals; however, the challenge lies in the level of exposure, dietary choices, and lifestyle factors. Sweating is considered one of the effective methods for mercury removal, alongside consuming foods rich in sulphur and selenium. Mercury has a strong affinity with sulphur, leading it to compete with molecular structures in the body and deplete their sulphur supply. This indiscriminate harvesting of sulphur can result in the dysfunction of certain structures. Sulphur plays a crucial role as an intracellular messenger in the human immune system. Elevated levels of mercury in the body can hinder the immune system's responsiveness, making the body more susceptible to internal and external pathogens.

Furthermore, several modern chelating agents are available that can aid in the removal of mercury and other heavy metals. These agents are designed to bind with heavy metals and facilitate their excretion from the body.

Rasa Shāstra provides treatment protocols for removing improperly treated or excess mercury from the body. The underlying principle is to utilise the same technology employed in the purification of mercury. Several well-known agents in Āyurveda for reducing mercury toxicity include Himalayan garlic,[149] piper betel leaf, lemon juice, Kanji (fermented rice water) and goat's milk. Purified sulphur, combined with ghee or fresh milk and sweetened with honey, is particularly useful in this regard.

Citrus lemon, vinegar, milk, and ghee are all potent antioxidants and excellent Anupānas, aiding in the body's internal cleansing and providing protection against pathogens.

Left, mercury (Hg)

149 Kashmiri 'snow mountain' garlic pearls

Mercury has the unique ability to penetrate all tissues, even crossing cellular membranes. Fortunately, ghee and milk both possess a similar quality. So, when combined with specific herbs and minerals, particularly sulphur, they can neutralise mercury, nourish damaged tissues, and facilitate the elimination of toxins through the body's central digestive pathway. Āyurveda recognises that something as simple as consuming high-quality dairy products and certain minerals can contribute to the long-term removal of deep-seated toxins like mercury.

Long-term use of honey water is also prescribed as a detoxification method. It involves diluting fresh honey with warm water and consuming it on an empty stomach each morning. This practice is typically recommended for a period of two to four months. It is considered a *Pūrvakarma*, which refers to pre-treatments conducted prior to the main therapeutic interventions. This detoxification method is particularly utilised in cases of extreme debility, where the body has already experienced significant weakening due to an overload of āma or toxicity. By incorporating honey water into the regimen, Āyurvedic practitioners aim to support the body's natural detoxification processes and gradually alleviate the burden of toxins.

Shilajit (bitumen) may also be used in cases of mercurial poisoning. It is believed to possess a unique ability to absorb and metabolise mercury, aiding in its elimination from the body. Shilajit is particularly known for its affinity with the kidneys and urinary system, which are often affected by mercury accumulation. A strict diet is advised in combination with the treatments already mentioned. This diet should be rich in foods with two of the six tastes recognised in Āyurveda – bitter, associated with the elements of æther and air, and astringent, which is associated with the elements of air and earth. Both of these tastes are considered effective in deep tissue cleansing. By following a diet abundant in bitter[150] and astringent tastes, one can support the body's natural detoxification processes and promote the elimination of toxins, including mercury.

150 Karavella (bitter gourd) is frequently used in cases of mercury toxicity; its ability to remove Visha extends even to the king of poisons (arsenic).

Section 8

Minerals

8.1/ Use of mineral-based medicines

This section examines three specific materials Rasa Shāstra acknowledges for their potent medicinal properties. Furthermore, we explore the diverse techniques utilised by Rasa Shāstra for the processing and purification of these materials:

1. Gandhaka/ sulphur
2. Shilajit/ bitumen
3. Haritāla/ orpiment

Within subsequent subsections, we explore a range of processing methods for minerals, encompassing grinding techniques and implementing advanced procedures such as puṭa and different types of yantra, which enable more intricate forms of processing. While specific guidelines exist for purifying and preparing each material, whether metal, mineral, or gemstone, it is essential to recognise that resource availability plays a crucial role in comprehending the diverse manufacturing techniques. Moreover, local and regional disparities contribute to the utilisation of materials at hand.

In addition, I have included analysis graphs for each mineral discussed in this section that depict the chemical composition of each after purification. These samples represent completed medicinal grades typically employed independently or as components of more intricate formulations. Furthermore, the final medicinal preparations are enhanced by incorporating distinct Anupānas, which induce diverse healing effects.

Far left, preparation of Rasa Maanikya/ arsenic trisulphide (As_2S_3)

8.2/ Sulphur

> 'There was a powerful demon with long arms named Lelihāna whose mass covered 264 miles of the Himalayas; his body lay there after being killed by Vishnu's cakrika (discus). As he decomposed upon the mountains his medas (muscle fat) became known as Lelīhtaka (sulphur).'
>
> *Origins of Sulphur, Caraka Saṃhitā*

A brief exploration of Vedic alchemy textbooks reveals a multitude of formulations that rely on the elemental union of mercury and sulphur, establishing the foundation of their potency and therapeutic value. Beyond serving as essential components of these formulas, these two elements can also be considered as Anupāna, facilitating the effective delivery of the complete formulation, ensuring it reaches its intended destination within the body.

It is crucial to note that even in its purified state, mercury is not recommended for direct introduction into the bodily system without first being bound to sulphur. This combination is indispensable, as it enables the primary means of conversion for mercury, enhancing its compatibility with the energetic dynamics of a living organic system. Once it gains access to this system, mercury exhibits unparalleled adeptness in traversing through the various tissue layers.

Above, Gandhaka (sulphur)

Mercury possesses a distinctive property known as all-pervasiveness, which elucidates its capacity to inflict considerable harm to bodily tissues in an unpurified state. Whether in its impure form or as a vapour, it swiftly targets the core operational systems of the body, specifically Majjā Dhātu, or the nervous system. Majjā tissue is located deep within bone cavities, nerve channels, and the cerebrospinal fluid; gaining access to the most sensitive components of this tissue pathway can lead to irreversible damage.

Conversely, once properly purified and processed, mercury, upon gaining access to Majjā Dhātu, obtains a favourable passage through this delicate network of refined tissue. It can transport its payload of beneficial elements directly to the source, ushering in a rapid change to the deepest bodily tissues.

Sulphur as a curative agent

While the close relationship between sulphur and mercury in their synergistic interplay forms a fundamental cornerstone of Rasa Shāstra, sulphur has also been acknowledged for its potent healing effects. In a brief passage from the *Caraka Saṃhitā*, which laments the prevalence of various skin diseases, we encounter the following statement:

> 'The administration of Lelīhtaka (sulphur) in combination with the juice of jati (Āmalakī) and honey is considered the unparalleled remedy for treating seventeen types of kustha (persistent skin diseases), including leprosy.'
>
> *Use of Sulphur, Caraka Saṃhitā, Vol. III*

This short excerpt emphasises the remarkable therapeutic effectiveness associated with the combination of sulphur, jati (Āmalakī) juice, and honey. It highlights their collective potential in addressing a diverse range of obstinate skin ailments. Furthermore, the *Caraka Saṃhitā* notes that this remarkable capacity to combat skin diseases is rivalled only by Maksika Dhātu (copper pyrite[151]) when consumed along with cow's urine.

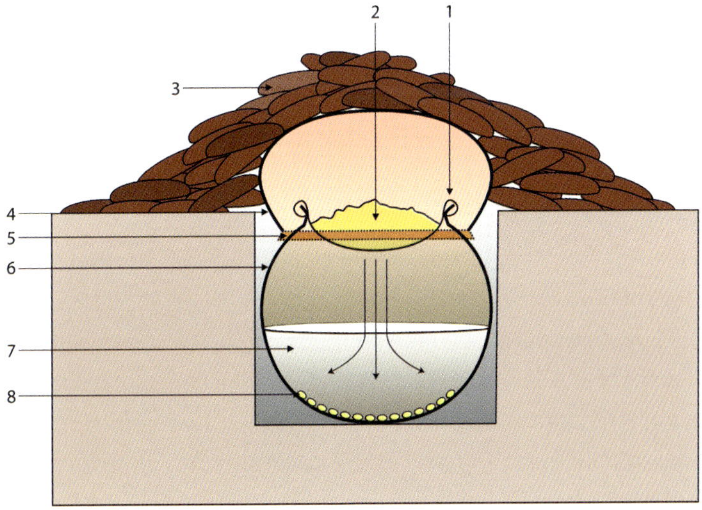

Left, Adhahpatana Yantra:

1) cotton cloth suspended; 2) sulphur powder; 3) cow dung cakes; 4) upper vessel; 5) joint between vessels; 6) lower vessel; 7) milk; 8) purified sulphur

151 For more information about Maksika see Appendix 2, Catalogue of Materials and Their Use.

Sulphur's significant reputation for its healing prowess has led to the development of numerous purification and processing techniques, far exceeding what can be detailed here. Nevertheless, there are several methods that consistently emerge in the literature of this science, and it is these methods that are addressed in the following discussion.

Sulphur purification – method 1 (used for quantities over 1kg)

Processing large quantities of sulphur can lead to messy, more time-consuming procedures. While this method of purification is preferred for large-scale operations, it may not be practical for quantities below 1kg.

Ingredients

1kg Gandhaka (powdered)
3–4 litres of raw milk
Copper wire
Two large earthenware vessels
Fine weave cotton cloth

1. The sulphur crystals are finely powdered.
2. To apply temperature to the sulphur powder without it burning, a variation of a yantra known as *Adhahpatana Yantra* is employed, involving digging an earthen pit and burying your vessel up to its neck.
3. The lower vessel is half filled with fresh milk. A fine cotton cloth holding the sulphur powder is positioned securely suspended above the milk bath.
4. A secondary smaller upturned vessel is placed over the first vessel; this will act as both lid and heat shield.
5. A sufficient quantity of cow dung cakes are placed around and on the outer surface of the exposed vessel and ignited.
6. Cow dung cakes produce a high temperature, but the open space between them and the sulphur/cloth is enough to melt the sulphur (119°c) while ensuring the cloth remains intact.
7. As the sulphur melts, it slowly drips out through the cloth and drops into the milk bath below.
8. The quantity of cow dung cakes depends on the size of vessel. For the purposes of this example, a working figure of about thirty cakes is used, taking about one and a half hours to be reduced to ash.
9. After cooling, the lower vessel is excavated, and its contents retrieved.

Water is added to the milk, mixed and slowly poured off. As this process is repeated, small sulphur pills will be seen at the bottom of the vessel.

10 The sulphur recovered is spread out on paper and dried under sunlight. When all the moisture has evaporated, the pills are ground in a ceramic mortar, returning the sulphur to a fine yellow powder, smelling slightly sweet and milky.

This purification process is usually repeated three times. In practice, most modern pharmacies perform it once.

Sulphur purification – method 2 (Used for quantities less than 1kg)

Ingredients

500g Gandhaka (powdered)
2–3 l Raw milk
Q.S. Ghee (clarified butter)
Fine weave cotton cloth

Method 2 offers a practical and efficient purification approach specifically designed for small quantities of sulphur. In this method, ghee serves as a medium to dissolve the sulphur, followed by solidifying it through immersion in cow's milk. This process ensures swift purification with optimal results.

1 An iron pan is heated, and a small quantity of ghee is added. This is usually about one-third of the weight of the sulphur powder. An additional pot, half filled with fresh room temperature milk, is covered with a fine cotton cloth and secured in position. This cloth must not sag into the milk.

2 Sulphur powder is added to the ghee and mixed thoroughly until a creamy paste is formed.

3 As the sulphur liquefies, it will dissolve into the ghee, and as heat is continually given, the liquid will become slightly orange, like the colour of egg yolk. At this point, the heat is discontinued, and the

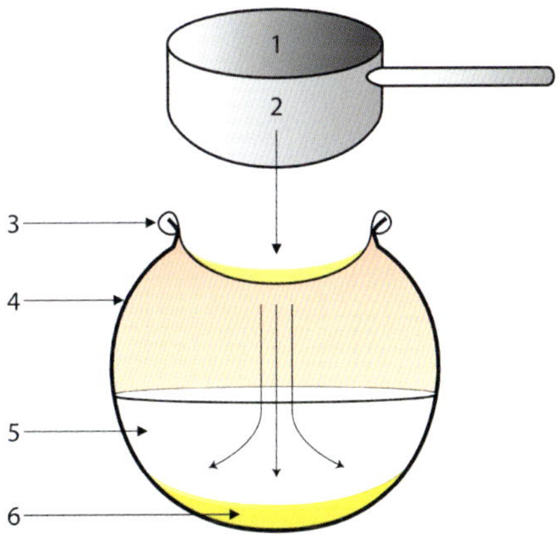

Left, sulphur processing method 2:
1) iron pan;
2) pouring liquid sulphur dissolved in ghee;
3) cotton cloth;
4) vessel;
5) milk;
6) solidified sulphur

pan is removed from the heat. Note: do not overcook sulphur; if the liquid starts to look red, it is overcooking.

4 Assuming the liquid is now yolk coloured, it can be poured through the cotton cloth and into the milk. Note: pouring slowly is best, as it allows the liquid to pass through the cloth membrane without backing up and overflowing.

5 Once the sulphur has solidified, it can be retrieved. The ghee and milk can be removed by washing the solidified sulphur in hot water with some lime juice added. The sulphur is then dried and powdered.

It is recommended that this process is repeated six times for minimum purification, with some texts recommending 100 times or until the material no longer smells like sulphur. However, most modern pharmacies feel three times is adequate. The use of a cotton cloth is also dispensed with, as modern sulphur supplies have no impurities.

Some variations on method 2 substitute milk with Bhringaraj decoction, the juice of Dhattura leaves, goat's milk, onion juice or the gel of *Aloe vera*. Indian beech oil and eranda (castor oil) can be used in place of ghee.

In a fascinating paper published in *Āyurveda: Science of Life*, Dr Damodar Joshi conducted a series of four distinct experiments utilising the aforementioned purification methods. The objective was to determine the medium with the most significant impact on the sulphur sample. Subsequently, the purified samples underwent XRD analysis, and the findings are presented in the table opposite.

Above, sulphur undergoing purification

Gandhaka purification – comparative methods			
Method		Material loss	XRD
100g powdered sulphur, mixed with 25g ghee	Heated and poured into Bhringaraj juice (×6)	15g	All samples showed the removal of trace elements Cr, Mn, and Zn after purification
100g powdered sulphur, mixed with 25ml Indian beech oil and 25ml castor oil	Heated and poured into goat's milk (×3) and Dhattura juice (×3)	15g	
100g powdered sulphur, mixed with 25g ghee	Heated and poured into aloe gel (×5) and onion juice (×5)	25g	
100g powdered sulphur, mixed with 25g ghee	Heated and poured into cow's milk (×6)	15g	

Sulphur purification – method 3 (Used for quantities under 250g)

Ingredients

250g Gandhaka (powdered)

Raw milk or Bhringaraj decoction

This method is suitable for very small quantities of sulphur and relies upon the use of milk or Bhringaraj decoction. It also involves lots of grinding with a pestle and mortar. Method 3 is popular when there is ample labour and less access to materials such as ghee or oil.

1 The sulphur is ground in a ceramic mortar, and a small quantity of fresh milk or Bhringaraj decoction is added and triturated for about one hour.
2 The mortar is left under sunlight until the milk has evaporated and the mixture dries.
3 A second amount of raw milk or Bhringaraj decoction is added, and the previous step is repeated.
4 This process is repeated seven times in total.
5 Upon the last drying of sulphur, the remaining powder is washed in warm water and lemon juice to remove any remaining milk solids. Usually, four to five washes would be performed to clear any remaining deposits. Failure to remove milk solids can lead to mould forming on the powder.

Naturally occurring sulphur

Many of the world's leading sulphur-producing countries obtain this element by extracting or recovering it from the by-products of natural gas and petroleum. Native sulphur samples found in hot water springs, acid crater lakes, and thermal volcanic vents are often rich in minerals, metals, and trace elements, with notable examples including arsenic, antimony, copper, nickel, selenium, tin, lead, and molybdenum.

Recognising the potential for contamination in minerals like cinnabar, orpiment, and realgar, it is likely that practitioners of Rasa Shāstra developed suitable extraction methods to eliminate impurities, as contamination was perceived as a significant concern. Given the inherent Visha/poisonous qualities associated with some of these elements, it is no surprise that purifying naturally acquired sulphur samples underwent rigorous Shodhana practices.

Until recently, the Frasch process served as the predominant method for sulphur extraction. This technique involves injecting super-heated water at high pressure into abundant underground sulphur deposits, resulting in the liquefaction of the material and subsequent displacement of the molten sulphur back up to the surface.

Right, sulphur XRD sample 1

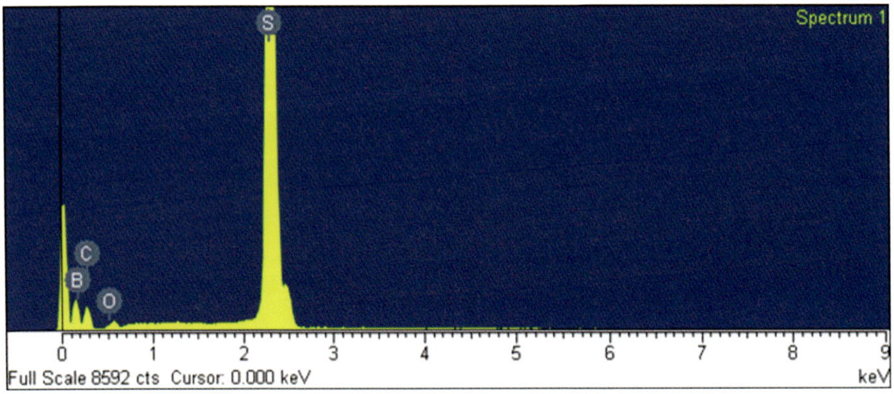

Sulphur XRD sample 1 was purified in the United Kingdom in July 2009 using method 2, as described earlier, which involved immersing the sulphur repeatedly into fresh milk six times. The accompanying graph demonstrates the absence of any toxic heavy metal contaminants in the sample. Sulphur XRD sample 2 was purified in the United Kingdom that same year using method 3.

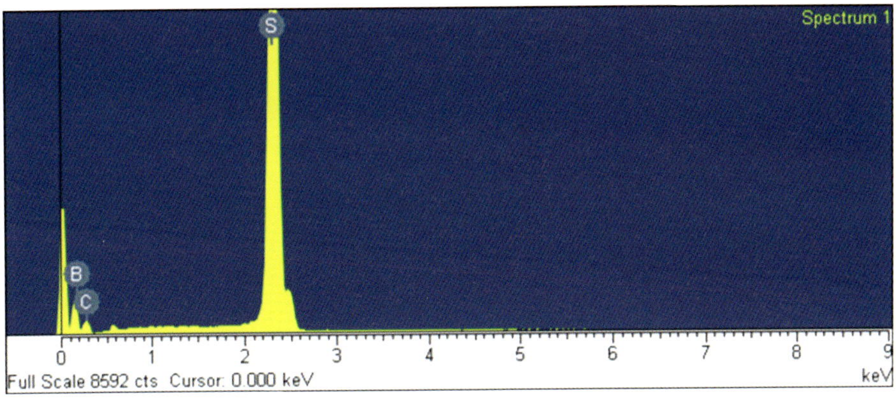

Left, sulphur XRD sample 2

X-ray diffraction (XRD) Analysis indicates both samples are similar, i.e., sulphur with minor traces of carbon, boron, and oxygen. Notably, boron, commonly found in tourmaline, is frequently associated with regions of active volcanoes. Similarly, sulphur is often obtained from areas exhibiting geothermal activity, such as those volcanic regions.

Sulphur is an essential element within the human body, contributing to the structure of fats, bones, and muscles. Additionally, sulphur plays a crucial role as an intracellular messenger in the functioning of the immune system. The health benefits associated with sulphur are wide-ranging, with particular relevance to conditions affecting the skin, liver, and digestive processes. Some common ailments where sulphur exhibits its therapeutic affinity include itching, leprosy, digestive disorders, excessive mucus production, high levels of āma (toxic accumulation), intestinal parasites, fevers, and mercury poisoning.

8.3/ Bitumen

'This is an elixir for long life and happiness. It prevents ageing and diseases. It is an excellent drug for producing sturdiness of the body. It also promotes Medhā (intellect), Smṛti (memory) and Dhana (wealth). While taking this recipe a person should live on milk.'

Shilajit Rasāyana, Caraka Saṃhitā

'A kind of gelatinous substance secreted from the sides of the mountains when they have become heated by the rays of the sun in the months of Jyestha and Áshádha. This substance is what is known as Shilajit and it cures all distempers of the body.'

Shilajit: Its Origin and Properties, Susrutha Saṃhitā

'In the days of yore, when the ocean was being churned with the mountain Mandara, the sweat of this mountain came in contact with Somā (nectar). This substance with innumerable properties called Shilajit should then be useful to the creatures of the earth; this was the desire of the gods. Because of the heat of the summer sun, it exudes from the mountains.'

Iatro-Chemistry of Āyurveda/Rasa Shāstra

While mercury holds a prominent position in Rasa Shāstra, Shilajit[152] (*Asphaltum punjabianum*) has garnered significant acclaim throughout Āyurvedic and alchemical literature. It is hailed as a panacea for a wide range of ailments. The use of Shilajit as a medicinal substance traces back to ancient times. Notably, Caraka dedicates an entire section in his *Saṃhitā* to its healing potency, and Susrutha does the same in his treatise on diabetes.

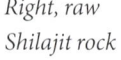

Right, raw Shilajit rock

Caraka and Susrutha discuss different grades of Shilajit, attributing its potency and taste (Rasa) to the metals present in its surroundings. They mention six metals in connection with

152 Also known as *Adrija*.

Shilajit: tin, lead, copper, silver, gold, and black iron.[153] The effectiveness of the metal and Shilajit combination follows a similar pattern, with tin having relatively lower potency and iron offering more significant medicinal benefits. Specific criteria are used for identification to ensure high purity. Pure Shilajit should exhibit a black, glossy appearance, possess substantial weight, and be free of sandy particles. Additionally, it should emit an odour resembling that of cow urine.

Modern supplies of Shilajit commonly undergo a purification process involving dissolution in water and filtration. This procedure effectively removes any small stones or sand particles present in the material. Shilajit is typically available in three forms for purchase: hard, brittle blocks (suitable for powdering), the dark brown malleable variety, and sticky black resin (ideal for decoction with other drugs). Scientific analysis[154] has led some researchers to classify these grades as follows: hard/brittle (high exposure), brown/malleable (medium exposure), and black/sticky (low exposure). The term 'exposure' refers to environmental factors that influence the formation and maturity of Shilajit, such as the humus reserve, the absorption rates of plant exudation, the decomposition rate of the material, and the formation rate of fresh humus.

Above, purification of Shilajit

In scientific circles, there are different theories regarding the true origin of Shilajit. However, many agree that it stems from an organic source. Possible candidates include latex from *Euphorbia royleana Boiss* and *Trifolium repens Linn*. Others propose a more primordial origin, considering it a fossilised plant mineral exposed by the ongoing elevation of the Himalayas. Various tests have been conducted over the years to shed light on this question, with one of the more recent studies concluding that its origins are organic. Due to its abundance of bioactive compounds, Shilajit is believed to be a composite mass of humus (organic carbon), fulvic acid, xylose (wood

153 Black-iron is normally associated with Kanta Pashana (magnetic iron ore).

154 Ghosal, S. (1990) 'Chemistry of Shilajit, an Immunomodulatory Āyurvedic Rasāyana', Varanasi: Dept. of Pharmaceutics, Banaras Hindu University.

sugars), plant latex, glucose, arabinose (monosaccharide), antioxidant lipids, trace minerals, and antibacterial tannoids suspended in a resinous, water-soluble form.

Bitumen purification – method 1

The purification process of Shilajit is relatively straightforward, although, like other Rasa procedures, it can be influenced by weather conditions, specifically favouring warm, bright, and dry weather. Whenever feasible, it is recommended to utilise the indigenous rock for purification. It is always wise to purify this substance, regardless of its origin.

Ingredients

1–2kg Shilajit (raw rock)
Q.S. Triphala decoction
Q.S. litre of milk
Earthenware vessels
Cotton cloth
Baking tray

1. The larger pieces of Shilajit rock are broken up and wrapped in cotton cloth. A 'pottali' or cloth bag is formed and immersed in water and heated for one to two hours using the Dolā-yantra method.
2. During cooking, the Shilajit within the rock dissolves into the water leaving only the rock's stone, sand and grit. Often, this will be a mixture of micaceous limestone and various silicates. It is not unusual for a foamy material to appear on the water's surface; this can be scraped off and discarded. After heating, the pottali is removed and squeezed to recover any remaining Shilajit.
3. The remaining contents are then gently heated to evaporate more water. As the last of the water is removed, a black, tarry material will appear, this will be the yield from the original rock.
4. The Triphala decoction is prepared, added to the remaining material, stirred well and cooked again. An equal quantity of milk is sometimes added at this point and stirred into the mixture, although this is optional. In some instances (in the absence of Triphala), milk alone can be used. However, this is usually watered down a little.
5. When the mixture is hot, it can be filtered through cotton cloth into a

second vessel, allowing any extraneous material, such as stones and fine sand, to be removed. It might be necessary to filter a few times to remove any contaminants.

6 After filtration, the remaining liquid is poured into a flat-bottomed baking sheet and exposed to sunlight over a period of days. Dependent upon the weather, this can be a slower evaporation process. Still, a crust will appear on the surface during this period. This crust is skimmed off as it forms and placed on another flat baking sheet to dry. This process is repeated until all the contents of the first baking sheet are dried up and removed.

7 The dried pieces of Shilajit are separated from the sheet with a spatula and broken into smaller pieces. With a greater surface area exposed, the smaller pieces are exposed to sunlight until brittle, upon which they can be further reduced to a course powder using a stone mortar. Shilajit can stored in a glass jar when fully dry.

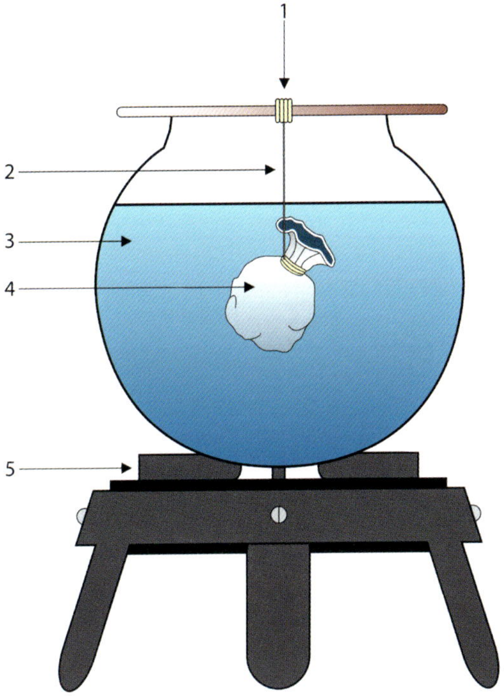

Above, Dolā-yantra/ swing device: 1) wooden support; 2) string or copper wire; 3) medium; 4) pottali; 5) heat source

Bitumen purification – method 2

In addition to the method described in *Caraka Saṃhitā*, some opt for an alternative approach to extract Bitumen from Shilajit that does not use heat. This alternative method involves soaking raw Shilajit rock in Triphala decoction until it dissolves. The dissolved mixture is then filtered and left to dry, carefully removing it from the surface as it forms a crust. From the author's perspective, both methods are valid, each with its own advantages and disadvantages.

Either method may sometimes result in a resinous grade of Shilajit that remains soft and pliable, making it impossible to powder. This particular grade of Shilajit is occasionally combined with resins like Guggulu (*Commiphora mukul*), honey, or jaggery, among other substances.

Analysis of Bitumen

Fortuitously, I managed to acquire unrefined rock sourced from Pakistan, from which I successfully extracted unrefined Shilajit. I employed traditional purification methods to refine the extracted resin, which involved boiling it in Triphala decoction. It was astonishing to discover that the rocks contained a remarkably abundant amount of resin, surpassing initial expectations.

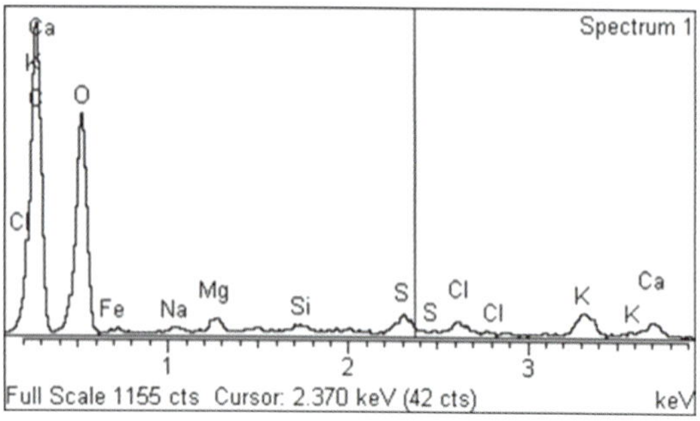

Above, bitumen (XRD) sample 1 Sri Lanka

When subjected to hot water, the Shilajit rock swiftly disintegrated into a soft resin alongside micaceous limestone, sand, and other impurities, separated and removed during the purification process.

The graphs on this page present an analysis of two samples of Shilajit, revealing their significant contents of carbon, calcium, and potassium, along with trace amounts of sodium, iron, silicon, magnesium, and sulphur. Sample 1 was obtained from Sri Lanka from a company based in Punjab, India, while sample 2 was directly acquired from India (with an unknown source). Both samples exhibited similar properties, although sample 2 exhibited slightly higher levels of carbon and iron.

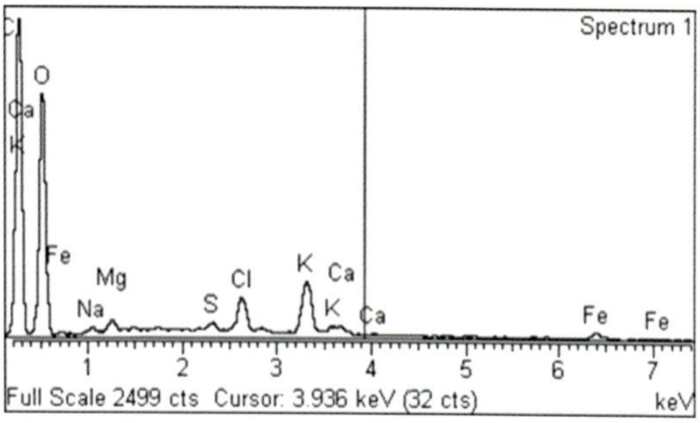

Above, bitumen (XRD) sample 2, India

The presence of carbon in Shilajit is not surprising, considering its catalytic properties. This substance is well-known for its antioxidant qualities and effectiveness in scavenging free radicals. Throughout the centuries, Shilajit has proven to be a valuable combination of water-soluble minerals, renowned for its nourishing properties. Additionally, carbon, iron, and calcium are essential elements that play a crucial role in constructing, supporting, and maintaining the integrity of bodily tissues.

Benefits of Bitumen

Shilajit offers several notable health benefits, particularly in affinity with the kidney, skeletal, and urinary systems.

This remarkable substance is beneficial in various conditions, such as:

1. **Kidney stones** Shilajit is often utilised to support the treatment and prevention of kidney stones, aiding in their dissolution and removal.
2. **High Vāta** Shilajit has properties that help balance Vāta Dosha, which is associated with nervous system disorders, joint pain, and digestive issues.
3. **High blood sugar or diabetes** Shilajit has traditionally been used to regulate blood sugar levels and support individuals with diabetes.
4. **Asthma** The anti-inflammatory properties of Shilajit make it valuable in managing asthma symptoms, helping to alleviate bronchial inflammation and improve breathing.
5. **Piles** Shilajit is believed to assist in reducing the symptoms and discomfort associated with piles, also known as haemorrhoids.
6. **Nausea** Shilajit has been utilised to alleviate nausea and vomiting, relieving gastrointestinal discomfort.
7. **Leprosy** In Āyurvedic practice, Shilajit has been historically employed in treating leprosy, a chronic infectious skin disease.
8. **Excessive tissue growth** Shilajit regulates tissue growth and prevents abnormal or excessive cell proliferation.
9. **Cough** Shilajit is often used to soothe coughs and respiratory ailments, providing relief and promoting respiratory health.
10. **Oedema** Shilajit has diuretic properties that may aid in reducing oedema, the accumulation of fluid in body tissues.

8.4/ Arsenic trisulphide

> 'Hiranya-Kaśipu, the demon adorned in golden garments, met his demise at the hands of Lord Narasimha as the sun dipped below the horizon. From the demon's vomit came Haritāla (orpiment), while Manah śhilā (realgar) materialised from his armpits. Hiranya-Kaśipu was notorious for his excessive indulgence in alcohol, and as he breathed his last breath, *Manohvā* sprang forth from his armpits. The name Manohvā itself signifies "that which brings delight to the mind".'
>
> *Mythological Origin of Haritāla and Manah śhilā*

With its reputation as 'the king of poisons', arsenic has a rich and complex history, serving as both a deadly weapon and a potential remedy. One of its forms, known as orpiment or *Haritāla* in Rasa Shāstra, is termed 'yellow-leafed'. This aptly captures the essence of high-quality Haritāla, which exhibits a leaf-like texture, a radiant golden hue, substantial weight, and a faintly oily or greasy quality. Arsenic, due to its highly toxic nature, was frequently employed with lethal efficiency in regicide, hence earning the moniker 'king's yellow'. Its brilliant yellow colouration made it a beloved choice among artists seeking vibrant pigments until it was later substituted with cadmium yellow in the early 1900s.

Above, Haritāla (orpiment)

Caution: orpiment and realgar are highly toxic materials.

Arsenic purification – method 1

Given the toxic nature of arsenic, its purification methods are diverse and quite intricate. The most common methods employed require the following materials:

Ingredients

- Haritāla (orpiment/yellow arsenic)
- Kushmanda (*Benincasa hispida Linn*), also known as ash pumpkin
- Lime water (calcium carbonate)

Sesame seed oil
Triphala decoction
Kanji (rice vinegar)
Ash of Tila (*Sesamum indicum*)
Borax (sodium borate)

1 A Kushmanda is selected and halved, then, using a coconut scraper, its contents are removed and pulped. The pulp is then placed into a clean cloth and squeezed, collecting its juice. Enough juice should be collected to fill a small earthenware pot, with extra for top-up later on.
2 The Haritāla is broken into manageable pieces but not powdered; an iron mortar is the most practical for the reduction of this material. The broken pieces are then placed into a pottali, and a Dolā-yantra is constructed.
3 The pottali, containing the Haritāla, is cooked in Kushmanda juice, simmering for about six hours and topping up as necessary.
4 After this process, the pottali is removed, and the contents washed in warm water infused with some borax (sodium borate), then dried.
5 The Haritāla, now purified, is ground into a fine powder and stored in a glass jar. If purified Haritāla is to be used to produce Rasa Maanikya (see Section 8.5/ Rasa Maanikya), it is not necessary to powder this material.

Left, top to bottom, Haritāla undergoing purification in kushmanda juice

Arsenic purification – method 2

1. The juice of Kushmanda is collected and filtered.
2. The Haritāla is ground to a powder.
3. The Kushmanda juice is triturated with the Haritāla powder until a yellow, creamy paste is formed. This paste is continually ground for one hour or until the liquid content has evaporated.
4. The above process is repeated seven times in total.
5. The Haritāla, now purified, is ground into a fine powder and stored in a glass jar.

Arsenic purification – method 3

1. Pieces of Haritāla are placed into a pottali and given Dolā-yantra for three hours. This process is repeated four times in four different solutions: Kushmanda juice, lime water, a mixture of Triphala decoction and sesame seed oil (ratio: one part oil to two parts parts Triphala decoction), and finally kanji.
2. Between each Dolā-yantra process, the Haritāla is removed from the pottali and washed in a solution of warm water and borax.
3. The Haritāla, now purified, is ground into a fine powder and stored in a glass jar.

Arsenic purification – method 4

1. Pieces of Haritāla are placed into a pottali. They are then soaked in the following liquids for seven days each: Kushmanda juice, lime water, sesame seed oil, Triphala decoction and Kanji. No heating is required.
2. Each liquid is changed daily.
3. Between the changes of liquid, the Haritāla is removed from the pottali and washed in a solution of warm water and borax.
4. The Haritāla, now purified, is ground into a fine powder and stored in a glass jar.

Note: Instead of using Kanji, you can use alkaline ash (Ksara) made from incinerated sesame seeds mixed with water as a substitute. It has also been mentioned that equal quantities of lime water and kanji can be used as a substitute for other liquids. However, the author has not personally tried this.

Two purified samples of arsenic trisulphide were subjected to XRD analysis, revealing expected concentrations of arsenic (As) and sulphur (S). Additionally, other detectable elements were identified, including bromine (Br), stibnite (Sb), silicon (Si), carbon (C), oxygen (O), potassium (K), and polonium (Po).

Both samples share similar properties, but sample B lacks bromine and instead exhibits traces of polonium. The primary constituents of both samples are arsenic, sulphur, stibnite, and minor amounts of bromine. These additional elements are commonly found in minerals associated with hot springs (geothermal activity), and other components, such as cadmium and iron, not detected here. The presence of carbon may be a result of processing, possibly involving heating, while silicon is naturally present in the material's matrix. It is worth noting that polonium, as previously discussed, is a rare element and unlikely to be highly concentrated. Arsenic, stibnite, and polonium are all metalloids with metal and mineral properties. This particular batch of arsenic trisulphide was obtained from China.

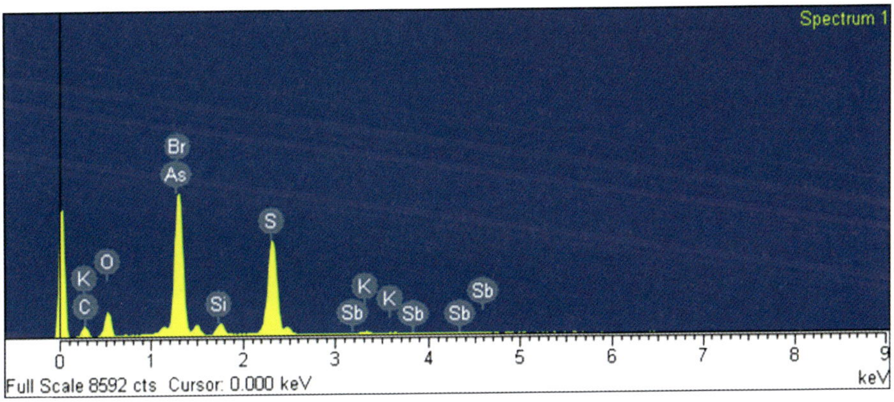

Left, Haritāla sample A was prepared using method 1

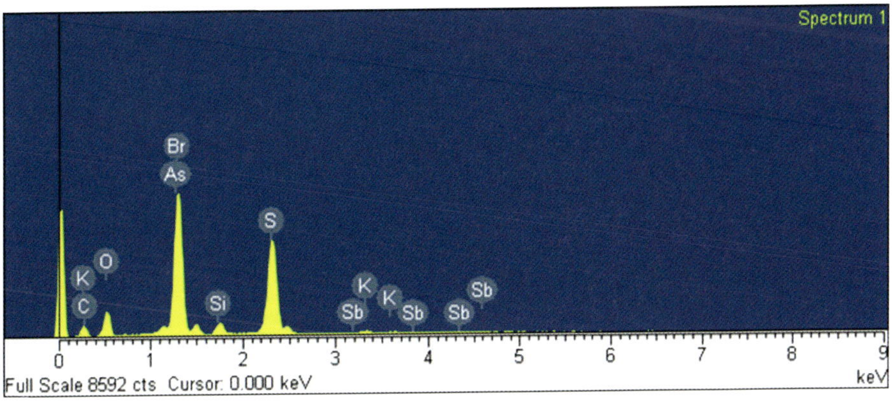

Left, Haritāla sample B was prepared using method 3

Benefits of Haritāla

Haritāla is believed to possess several health benefits, particularly for addressing skin-related issues and certain autoimmune disorders, including specific types of cancer. It has a long-standing traditional association with the treatment of various conditions, including eczema, psoriasis, urticaria, syphilis, gout, anal fistula, high fever, and haemophilia.

8.5/ Rasa Maanikya

Right, Rasa Maanikya crystals

Maanikya means 'ruby', and its use here refers to the crystallised form of Haritāla after exposure to heat, either directly upon an open flame, in a crucible, or using Kūpīpākwa. Rasa Maanikya is a commonly prepared Rasa Shāstra drug used to treat skin diseases, high fever, asthma, persistent cough, piles, fistula and the general rejuvenation of bodily tissues.

Rasa Maanikya preparation – method 1

1. Purified pieces of Haritāla are selected (ideally rice-grain size).
2. Small sections of mica are cut (single sheet), and a few grains of Haritāla are then sandwiched between two sheets (allowing a good margin around these). Typically, sheets could then be drawn together using copper wire, however this tradition has been replaced by the modern use of staples (see illustration opposite).
3. Once secured between the thin mica sheets, the Haritāla is exposed to an open flame and heated until red-hot. As the heat builds, the material will liquidise into a dark red mass, spreading outward from the centre of the sheets.
4. When fully dissolved, hold the sheet up to the light; any dark areas show pieces not fully heated.
5. If the mass between the mica sheets seems evenly spread, leave the contents to cool and crystallise.

6 When the sheets are parted again, any flat polished surfaces that look like ruby can be ground to into a bright vermilion-coloured powder.

Above, mica sheet with small pieces of purified arsenic sandwiched in between

Rasa Maanikya preparation – method 2

1 A flat-bottomed Sharaava is lined with sheets of mica, and small pieces of purified Haritāla are laid upon its surface. To ensure the pieces do not move around, the sheets can be stapled at the corners to pinch the material within.

2 The Sharaava is then sealed using the clay and cloth technique and allowed to dry. Note: it is not advised to use an open crucible as this gives rise to direct contact with arsenic vapour.[155]

3 The sealed Sharaava is then heated slowly. Heating is maintained for one to one-and-a-half hours, slowly increasing the temperature at regular intervals to avoid thermal shock to the Sharaava.

4 There is no exact timing for the heating period, but usually, when the sulphury smell has gone, the material is ready. When cool, upon opening, dark red crystals like Maanikya (rubies) will be seen inside the mica sheet. These crystals are easily ground to a fine red-orange powder and stored in a glass jar.

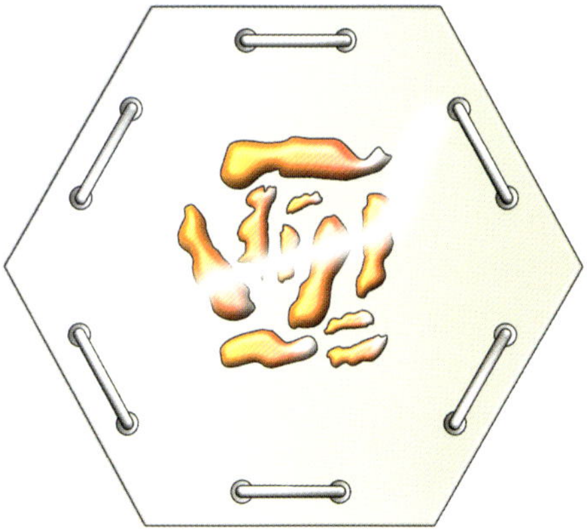

Above, Rasa Maanikya method 1 (small quantities)

Note: Alternatively, a secondary method of preparation can be utilised to minimise direct heat on the Sharaava. This method involves using a Vālukā Yantra, or sand bath. It reduces fume exposure and ensures more even heating of the crucible.

155 *The Materia Medica of the Hindus* states in its section on Haritāla that 'the man who roasts orpiment, dies very soon'.

Rasa Maanikya preparation – method 3 (Kūpīpākwa)

As with all Rasa Maanikya preparations, Haritāla is first purified prior to its use.

1 A suitable glass kūpī is wrapped with seven layers of cloth and clay.
2 Purified Haritāla is then placed into the kūpī jar.
3 Vālukā Yantra is heated slowly, periodically checking for sulphur fumes. When heated, this material produces toxic arsenic vapour as well as sulphur dioxide.
4 The Haritāla begins to melt at 350°c and boils at 550°c.
5 The heating duration depends on a number of factors, including the size of the sand bath, the heating capacity of the burner (gas or wood), and the distance between the heat source and the bottom of the buried kūpī. As the heat slowly percolates up through the sand, the temperature is elevated in stages and with uniformity, protecting the integrity of the glass.

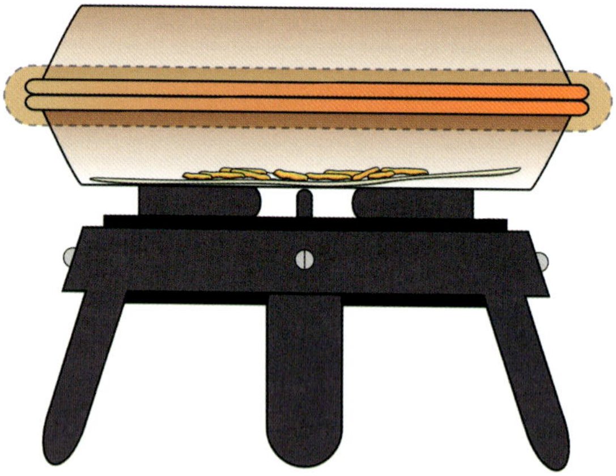

Right, Rasa Maanikya method 2 (larger quantities)

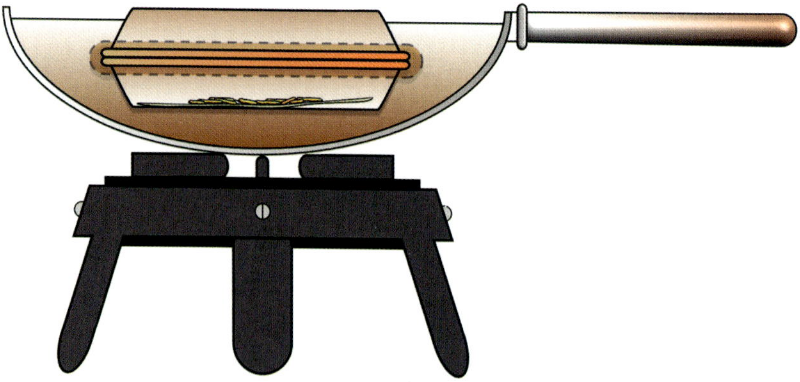

Right, Rasa Maanikya method 2 (alternate) using sand bath

6 The conclusion of the process is indicated by the cessation of fumes; this can typically be expected after five to six hours of gradual but continuous heating. The ideal temperature of the sand bath should be somewhere between 450 and 500°c.

7 After the heating stage, the apparatus is allowed to stand for twenty-four hours before retrieval of the finished medicine. For the opening of the kūpī, see Makara Dwaja preparation. Unlike the aforementioned, Rasa Maanikya is harvested from the bottom of the jar.

8 The maanikya crystals are easily ground to a fine red/orange powder and stored in a glass jar.

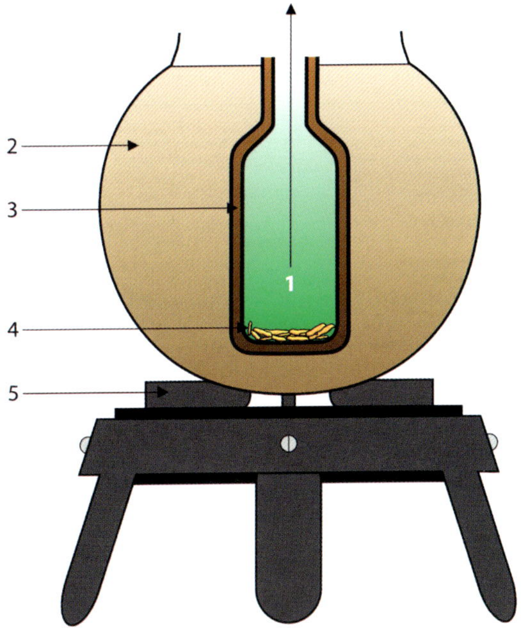

Above, Rasa Maanikya method 3, using kūpī:

1) arsine gas;
2) sand bath;
3) kūpī, wrapped with cloth and mud;
4) Haritāla;
5) heat source

Rasa Maanikya – XRD

XRD analysis of Rasa Maanikya reveals that the sample primarily consists of (As) arsenic and (S) sulphur, with traces of (C) carbon and (B) boron, the latter commonly found in regions with geothermal activity. In addition, small amounts of (Po) polonium and (Sb) stibnite are again detected. The presence of polonium, a rare element associated with bismuth, was observed in the same peak as sulphur, indicating that it is present in very small quantities due to its rarity. Similarly, the traces of stibnite found in the sample are minimal, which is not unexpected since arsenic deposits are often associated with iron, silver, and antimony ores.

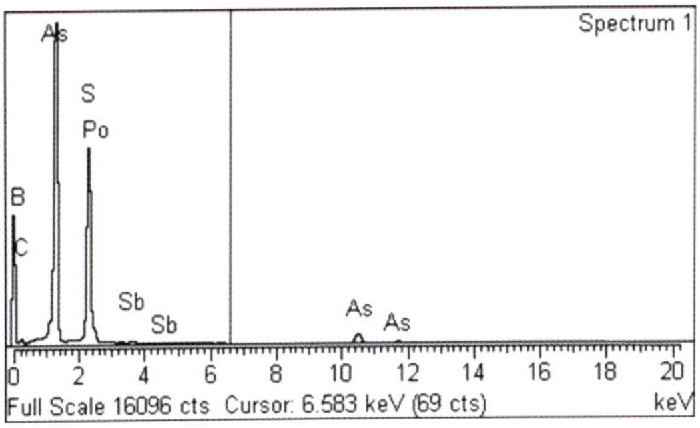

Above, Rasa maanikya XRD

8.6/ Lavaṇavarga (types of salt)

> 'Salt is a purifier, it increases one's interest in food, and promotes digestion. It is an increaser of Pitta and Kapha, yet reduces Vāta. Salt is a destroyer of manliness, it slakens and softens the body, impairs strength and induces the shedding of water through the mouth, causing inflammation to the cheeks and neck.'
>
> *Rasa Jala Nidhi*

In ancient times, salt was highly valued as both a condiment and a medicinal substance. Āyurveda recognises salt as *Sooksmagami*, meaning it possesses the remarkable ability to swiftly permeate all tissues and channels in the body. This makes it an incredibly useful medicine.

Salt exhibits a strong affinity with the stomach, which explains its use in saltwater emetics. However, the salty taste has been associated with Somā (the Moon), a celestial body strongly linked to sensitivity and emotional disturbances that can disrupt the digestive process. According to ancient teachings, *Lavaṇa*, or salt, is regarded as the destroyer of manliness as it softens bodily tissues and causes cheek inflammation, neck stiffness, and joint looseness. Excessive consumption of salt was observed to increase desire and possessiveness, while also contributing to conditions such as intestinal ulcers, ophthalmic disorders, tooth loss, premature greying of hair, and excessive thirst.

On a more positive note, salt was also found to have beneficial effects. It clears and irrigates the body's channels, particularly liquefying Kapha in the lungs, and helps in the reduction of obstinate masses like benign tumours, polyps, lipomas, or narrowed arteries.

The addition of different types of salts in varying quantities has been observed to pacify Vāta, reduce the viscosity of Kapha, induce a laxative effect, and enhance digestion, albeit with the potential to aggravate Pitta. By stimulating digestion, salt promotes the secretion of digestive enzymes, cleanses the taste buds, and, in small amounts, aids in the initial stages of digestion by providing lubrication and emulsification.

Āyurvedic recommendations on salt

The recommended order of preference for Lavaṇa (salts) according to Āyurveda is listed below.

1. **Saindhava (rock salt)** This is the most favoured type of salt, known for its therapeutic properties. It is considered to be beneficial for digestion and is often used in Āyurvedic preparations.
2. **Viḍa (black salt)** Also known as *Kala Namak*, it has a distinct sulphuric flavour and is used for its digestive stimulant properties. Black salt is believed to pacify Vāta and increase appetite.
3. **Viḍa (black salt) artificial** This is an artificially prepared black salt, which is a substitute for natural black salt. It is used in a corresponding manner and has similar effects.
4. **Samudra (sea salt)** Sea salt is obtained from seawater and is widely used in cooking. It is considered to be cooling in nature and is used to balance Pitta Dosha.
5. **Sauvarchala (Saltpetre)** This salt is naturally occurring, but can be also manufactured. Its use is less common in Āyurveda but popular in Rasa Shāstra, along with other medicines that treat the urinary tract.
6. **Narasara (Sal-ammoniac)** Narasara, also known as Sal-ammoniac, is a mineral salt with specific uses in Āyurveda. It is used in certain formulations to reduce Kapha, cleanse the eyes and counteract the effects of poison.
7. **Romaka (Sambar)** Romaka, also referred to as sambar, is a type of salt that is less commonly used in Āyurvedic preparations. Its use and specific properties are not extensively mentioned in Āyurvedic texts.

Note: It is important to note that the selection and usage of salts may vary based on specific Āyurvedic formulations. For example, Tripatu,[156] a standard Āyurvedic herbal blend, typically includes the first three salt varieties, while pancalavaṇa[157] may involve the first five. Each salt has its own unique qualities and medicinal benefits, contributing to the overall therapeutic effects in Āyurvedic treatments.

Saindhava (rock salt)

Pink Himalayan rock salt, a naturally occurring salt primarily found in the Sindh and Punjab regions, earns its name from the unique reddish or pinkish hue it exhibits due to its rich assortment of trace minerals.

In the realm of Rasa Shāstra, rock salt takes precedence as the most commonly recommended salt. It is known for its sweet taste, lightness, and ability to generate minimal heat within the body. Its overall effect is to enhance Agni, the digestive fire, without causing imbalance. Additionally, rock salt is revered as an aphrodisiac, a cardiotonic, and beneficial for ocular health and vision.

Above, Saindhava (rock salt)

By default, rock salt always occupies the top position in Lavaṇavarga, this is because it is less likely to provoke Dosha aggravation when used appropriately. Whenever salt is recommended without specifying a particular variety, Saindhava, or rock salt, is implied as the correct choice.

Viḍa (black salt)

Viḍa is a naturally occurring land salt which, due to its composition, possesses distinct properties – it contains small amounts of sulphur and iron, as well as numerous trace minerals found in Saindhava. These combined elements lend Viḍa its characteristic black colour. When ground, it transforms into a deep

156 Three preferred salts

157 Five preferred salts.

reddish powder, which can be distinguished from pink rock salt not only visually but also by its recognisable sulphurous aroma.

In terms of its qualities, Viḍa is initially warming, light, and carries a pungent taste after digestion. It primarily acts as a cardio-tonic, is calmative, a laxative, and increases both appetite and Agni, the digestive fire. Its therapeutic applications are extensive, particularly in the treatment of conditions such as colic, constipation, flatulence, intestinal obstruction, abdominal tumours, and heart ailments.

Viḍa (black salt) artificial

It is worth noting that some commercially available black salt may be artificially produced. Viḍa (artificial) is a black salt that bears striking resemblance to naturally occuring Viḍa. Despite being labelled as such, Viḍa (artificial) demonstrates excellent medicinal properties when prepared correctly. Its significance can be seen from the fact that it typically occupies the third position in the Lavaṇavarga hierarchy.

The process of creating artificial Viḍa involves combining 95% rock salt with 5% ground, dried amla fruit. This mixture is then heated to approximately 800°c until it liquefies. The molten salt is then poured onto a metal sheet and allowed to cool before being broken into smaller pieces.

Various alchemical texts present alternative recipes for black salt preparations. The primary motivation behind producing Viḍa artificially lies in its availability and abundance, as naturally occurring black salt

Left, ground amla to make Viḍa (artificial)

Left, rock salt and amla mixture

Left, Viḍa (artificial)

requires not only the presence of land salt but also proximity to natural reserves of magnetite iron ore.

Although this type of Viḍa is artificially produced, it possesses virtually identical properties to the naturally occuring variety. Its qualities are, therefore, the same, and there is no need to repeat them.

Samudra Lavaṇa (sea salt)

Sodii muras, commonly known as sea salt, has a chemical formula of NaCl (sodium chloride). In addition to sodium chloride, sea salt contains several trace minerals, including magnesium, potassium, calcium, and various sulphates. The process of obtaining sea salt involves filtration followed by evaporation, typically carried out in hot, arid climates with minimal rainfall. As the saltwater evaporates, sodium crystals form and are collected from the surface through raking. Due to its abundance, there is no need for artificial sea salt production.

Sea salt is characterised as heavy and pungent, yet it exhibits a sweet taste after digestion. However, it has a tendency to create stickiness in the body, affecting bodily fluids and electrolyte balance. As a result, sea salt is not extensively recommended for medicinal purposes. It has a propensity to aggravate Kapha Dosha and, to a lesser extent, Pitta Dosha. On the other hand, sea salt pacifies Vāta Dosha.

Souvarcala (saltpetre)[158]

Souvarcala, known by its chemical formula KNO3 and commonly referred to as saltpetre, possesses distinct characteristics. Its crystals are highly unique, growing in long, thin, and transparent shards. This remarkable appearance led the ancients to associate it with intelligence, as the crystals resembled a scribe's pen. It was believed that this salt not only promoted, but also possessed intelligence, inherently knowing where it would best serve in the body without prior instruction.

Saltpetre tends to naturally form in areas where animals defecate and urinate in large quantities, often appearing as sharp yellowish crystals atop dung. Alchemists sometimes prepare this salt by collecting layers of manure interspersed with potash (potassium carbonate). Over time, the mound is

158 Also know as Sūrya Kṣara

turned and aerated, with fresh urine poured on top. Ideally, the mounds should be protected from rain and allowed to mature. After a certain period, crystals may start to form. To increase crystal density, the fermented mass is soaked in water and decanted to remove solids. This water is then left to stand for a few weeks before undergoing filtering. The crystals can be obtained and collected by heating and cooling the remaining water.

Today, saltpetre finds application in various industries, including the production of matches, fireworks, explosives, fertilisers, and glass.

Souvarcala plays a significant role in several Āyurvedic preparations, most notably in Sorakaa, Swetha, and Sheetal Parpati. Its analgesic and antimicrobial effects make it highly useful in eye and dental diseases. The salt exhibits excellent anti-fungal and antibacterial properties but also has a tendency to increase Kapha Dosha.

Souvarcala (saltpetre) artificial

In instances where saltpetre is unavailable, alchemical literature in India provides a recipe for an artificial Sauvarcala mixture. This preparation involves grinding equal quantities of natron (potash) and rock salt together and then dissolving the mixture in water. The solution is then heated to remove the water content. Once the mixture is dry, the remaining powder is ground further and stored for future use.

Narasara (sal-ammoniac)

Narasara, also known as NH4Cl or ammonium chloride, occurs naturally in caves near volcanic regions. However, it is believed that ancient civilisations artificially prepared it by subliming the soot of animal dung to collect the salt.[159] Today, Narasara is 100% artificially produced in laboratory settings.

Narasara possesses the ability to pacify all three Doshas (Vāta, Pitta, and Kapha). It is characterised as subtle, light, and lubricating, aiding in digestion and promoting appetite. It has a liquefying effect on Kapha Dosha and helps counteract the effects of poisons. In alchemical operations, Narasara facilitates the liquefaction of both metals and minerals. Indian alchemical

---

159 The collection of this salt was famous at the temple of Zeus-Ammon in Siwa Oasis, hence the name sal-ammoniac, i.e. ammon's salt.

literature recognises the significance of Narasara as an essential ingredient in various formulas, particularly those aimed at treating heart ailments, eye diseases, and specific types of skin conditions.

It is worth noting that some Indian alchemical texts provide methods for artificially preparing sal ammoniac, although these recipes actually produce sodium chloride rather than ammonium chloride.

Sambar (Romaka)

Sambar, also known as Romaka, is an earthen salt that derives its name from the lake where it is collected, specifically Lake Shakambari near Jaipur in Rajasthan. The water of this lake has a saline composition similar to that of seawater, and it was once the source of the Romavati River, which is now extinct.

The chemical composition of sambar salt is diverse and includes sodium chloride, sodium carbonate, sodium bicarbonate, and sodium sulphate. Its overall coloration is a pale reddish-brown.

According to Āyurveda, Sambar salt is described as light and corrosive, with astringent (Kshar-like) properties. It strongly aggravates Pitta Dosha, leading to inflammation and suppuration of the skin. However, it is beneficial for eliminating phlegm through the nostrils and eyes, and it is renowned for its ability to penetrate even the smallest bodily channels. It also pacifies Vāta Dosha.

Section 9

Metals

9.1/ Use of metal-based medicines

'Very thin copper sheets are to be coated with the milk of Vajrī (snuhi) and salt. Place these upon fire and, when fully heated, dip into the juice of Nirgundi. Repeat this process of immersing the heated copper sheets seven times.'

Rasendra Maṅgalam

Given the significance of metallurgy and the progress in metal-working technology, it is logical to infer that the early use of herbo-mineral-metallic medicines can be traced back to revolutionary advancements in metal-working equipment. These innovations enabled the maintenance of high heat levels over prolonged periods. The synergy between ancient metal-working and ceramic technology supplied the necessary means to explore various combinations of Dhātu.

Given the inherent durability of metals, extensive processing is required to transform them into a suitable form for administration within bodily tissues. This section clarifies the core principles of metal conversion, which likely emerged from the experimentation conducted by ancient metal workers who discovered the effects of combining various solutions and minerals with different metals during their daily endeavours. Recognising that pure metals and alloys exhibited resistance to environmental factors and the ravages of time, it is unsurprising that efforts were made to harness and concentrate these timeless attributes into a form of elixir, capable of imparting the same enduring properties to human bodily tissues.

In the subsequent subsections focusing on Dhātu, I have chosen three commonly prepared materials. These materials differ significantly in their purification methodologies, providing the reader with a solid foundation in the techniques employed to purify various grades or castes of metal. Copper, tin, and zinc, all highly valuable and esteemed in their own right, serve as

Far left, purification of copper sheet using rock salt and lemon juice

essential medicines and frequently feature in numerous popular Rasa remedies. Each metal possesses a distinctive character and exhibits unique behaviour. When working with any Dhātu, it becomes evident that the masters of this science developed an intimate relationship with metals or, at the very least, cultivated a deep appreciation for these metals among those who regularly worked with them.

Like all the primary materials discussed in this section of the book, each item undergoes multiple purification and formulation methods. Furthermore, each finished medicine can be enhanced using different Anupāna, thereby inducing diverse healing effects.

9.2/ Copper

> 'Copper is bitter and astringent in taste; its effective taste after digestion is sweet. It is hot by nature, useful in treating disorders of Pitta and Kapha humours with āma, stomach disorders and dermatosis, as well as microbial disorders. It purifies both upper and lower body (by inducing vomiting and loose motions), it is useful in obesity, it increases appetite and cures difficult-to-treat forms of tuberculosis and anaemia. It is good for ophthalmic health and has scraping properties.'
>
> *Rasa Ratna Samucchya*

> 'Visha is not the only poison, Tamra is Visha. Visha has only one Dosha, but Tamra has eight, hence Tamra should be subjected to proper Shodhana before its administration as a medicine.'
>
> *Rasendra Sāra Sangraha*

Right, copper is heated until red hot and quenched in Kāñjī (vinegar); far right, purification of copper using butter

Copper was highly valued for its exceptional curative properties. It was considered a semi-precious metal, surpassed only by gold and silver. The finest quality of copper, known as *Nepalaka*, meaning 'Nepalese in origin', was particularly esteemed for its remarkable resistance to high temperature and visually appealing deep red colour. However, it was essential to handle copper with caution due to its toxic effects if processed incorrectly. This precautionary approach extended even to the treatment of minerals containing copper, such as copper sulphate, bornite, and copper pyrite, necessitating additional purification measures to ensure safety.

Top and above, purification of copper sheets before and after puṭa

The scraping effect of copper makes it an excellent agent for cleansing the body's subtle channels. Moreover, water stored in copper vessels acquires enhanced energising properties.

Commonly, copper is alloyed with tin and zinc to create bronze and brass, which possess their own therapeutic qualities. Interestingly, copper is among the essential metals the body requires to maintain optimal health. Its presence acts as a catalyst for several vital bodily functions, including enzyme production, skin pigmentation, preserving connective tissue elasticity, and facilitating the production of anti-inflammatory agents. However, excessive levels of copper can disrupt liver function, leading to digestive disorders and muscle fatigue. To restore balance, consuming zinc-rich foods or zinc supplements to counteract elevated copper levels is recommended.

Copper purification (method for copper sheets)

The quantities mentioned in this section of the book likely represent traditional preparation amounts. It is important to note that 200g of copper may appear small compared to the larger batches processed by modern pharmacies, which often handle several kilos at once.

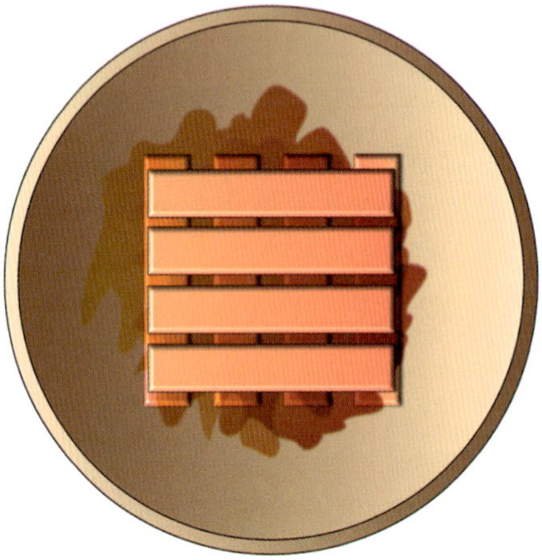

Far right, copper sheet is staggered inside Sharaava, interspersed with a mixture of ground Gandhaka (sulphur) and Hiṅgula (cinnabar)

Ingredients

200g Tamra (copper) sheet
50g Purified Gandhaka
50g Purified Hiṅgula
1 Suran root (*Amorphophallus paeoniifolius*)
Rock salt
Lemon juice
Pol vinakiri (Coconut vinegar)

1. The juice of several lemons is extracted and mixed with rock salt, then ground in a mortar until pasted.
2. 200g of Tamra (thin copper sheet) is selected and washed to remove any surface contamination or grease.
3. The pasted lemon juice and salt is brushed upon the sheets and allowed to stand until verdigris (cupric salt) forms.
4. Using tongs, each sheet is heated until red-hot and quenched in *Pol Vinakiri* (coconut vinegar). This procedure is repeated seven times for every sheet.
5. The copper sheet is washed and allowed to dry. At this stage, the copper will appear slightly withered and, hopefully, reddish in colour.
6. The sheets are regimentally laid down inside a Sharaava (see illustration).
7. Equal quantities of purified Gandhaka and Hiṅgula are ground together and lightly sprinkled over the sheets of copper. Fresh sheets are alternately laid over the top. This stacking and powdering is continued until the Sharaava is full.
8. The Sharaava has to be sealed very carefully, using clay and cloth, to avoid disturbing the stacked sheets inside. When dry, the Sharaava is placed into a Varaha Puṭa, and cooked at around 800–900°c.
9. After twenty-four hours, the Sharaava is retrieved and its contents removed. If

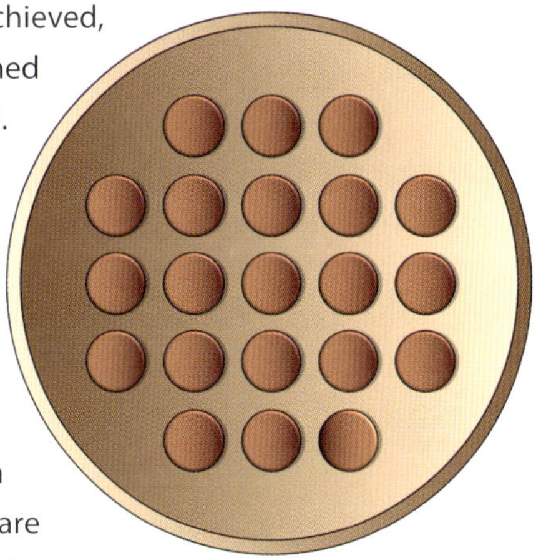

Left, copper cakrika positioning inside Sharaava

the correct temperature has been achieved, the contents should yield a darkened brown powder – copper oxide (CuO).

10 The contents are then ground with lemon juice until well pasted. This paste is then allowed to dry until stable enough to form cakrika. These are then placed into a new Sharaava.

11 The cakrika are then again heated in a puṭa. Upon opening, the cakrika are again retrieved and ground with lemon juice. This process is repeated twelve to fifteen times in total until a brown/reddish powder is achieved.

12 The final purification stage, Amṛītakarana, sees the remaining copper ash mixed with lemon juice and moulded into a ball. A Suran root[160] is then hollowed out to accommodate the copper. This last processing is an extra opportunity to cleanse the copper deeply and nectarise it, making it easier to digest and assimilate.

13 The Suran root is then wrapped using the cloth and clay method; usually, four to five layers are sufficient. When complete, the bound root is dried under sunlight.

14 Using Kukkuta Puṭa, the root is cooked. During this heating process, the root softens and exudates its juice, which then surrounds the copper. As the heating continues, the temperature starts to drive off the root's moisture, drawing toxins out from the copper and dispersing them into the flesh of the root.

15 When cool, the copper is retrieved and allowed to fully dry.

16 The remaining copper Bhasma can be powdered and stored in an amber glass bottle.

160 Suran, or elephant yam (*Amorphophallus campanulatus*)

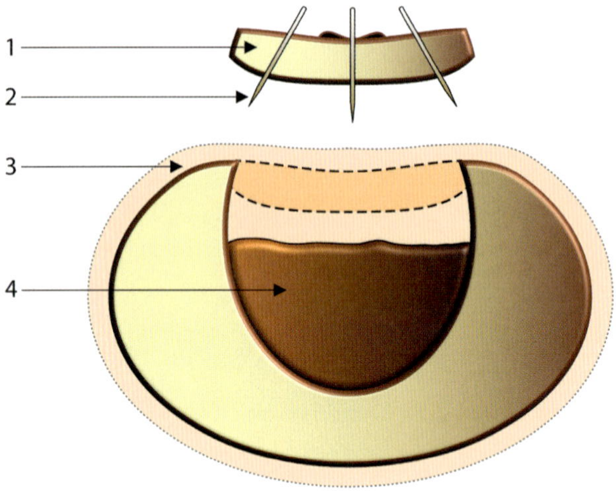

Right, Suran root is used in a process known as Amṛītakarana. Whilst cooking inside this root, its juices permeate into copper placed inside and draw out its Visha (poison):

1) lid cut from the top of Suran;
2) bamboo staves pin the lid in place;
3) cloth and mud are used to seal the root;
4) copper paste inside Suran root

Benefits of copper

Copper offers several notable health benefits, particularly with diseases affecting the blood, liver, spleen, and skin. It has been found to be beneficial in addressing various conditions such as anaemia, abdominal disease, piles, fever, leprosy, cough, asthma, oedema, hyperacidity, consumption, catarrh, parasitic infections (worms), colic, obesity, gastritis, colic, poisoning, and eye diseases.

It is recommended to consume copper with suitable Anupānas, such as honey or ghee, to enhance its effectiveness, since these substances can complement the positive effects of copper on health.

9.3/ Tin

'Tin has a bitter taste, is hot in potency. It is drying and hence increases Vāta Dosha. It cures urinary disorders, Kapha disorders, acts on adipose tissue and is anthelmintic.'

Rasa Ratna Samucchya

'Vanga (tin) which is soft, unctuous and melts quickly, is heavy and emits no sound upon beating. Khuraka variety is best.'

Rasendra Sāra Saṅgraha

According to Rasa Shāstra, tin is classified as a Puti (impure) metal. It undergoes a purification process that differs slightly from copper. Like other metals, tin is subjected to the Ḍhālana method, which involves heating and pouring the molten metal into various liquid mediums.

One of the common applications of tin is in the production of bronze when it is alloyed with copper. Āyurvedic texts suggest that tin has the potential to address twenty different types of urinary diseases. It is also recommended for ailments related to the inner ear, especially tinnitus.

However, it is important to note that excessive use of tin can aggravate Vāta Dosha, as it reduces the Medas (fat) tissue, automatically decreasing Kapha. Even trace amounts of tin have been observed to impact the body's growth rate. Insufficient tin levels can hinder physical and mental development and contribute to adrenal debilitation. On the other hand, maintaining healthy tin levels is said to help alleviate anxiety, depression, fatigue, and low vitality while supporting the functionality of the thyroid and adrenal glands.

Tin purification – method 1

Ingredients

200g Vanga

150g Apamarga (*Achyranthes aspera*)

Nirgundi leaf juice (*Vitex negundo*)

Fresh Haridra (turmeric)

Godugdha (raw milk)

Above, tin ingot

1 200g of tin is washed and dried.
2 Pithara Yantra (an iron vessel with a hole of approximately 2cm in diamenter in the lid) is then prepared (see illustration). Nirgundi leaf juice and freshly juiced Haridra are mixed and poured into the yantra using the general principle of Ḍhālana.
3 The tin is melted in an iron ladle, a process which only requires a very mild melting temperature (232°c).
4 When liquid, the tin is poured into the yantra. Note: eye protection and gloves are essential, as tin is volatile upon contact with some liquids.
5 Cooling is instantaneous, and the tin can be retrieved and washed in warm water. This process is repeated seven times in total.
6 Using a kadai (iron pan), the tin is re-heated and liquefied.

7 Dried Apamarga (*Achyranthes aspera*) is slowly added to this liquid metal. As the herb ignites, it burns down to ash and is stirred into the tin. When enough Apamarga has been added (usually this is when it is sufficient to obscure the tin), the heat is stopped, and the contents of the kadai are vigorously stirred. As the contents cool, the tin will miraculously vanish into the ash.

8 The remaining material is called tin ash. It is collected and washed in cool water, and the sediment is allowed to settle. When the liquid starts to clear, the excess water is poured off. This is repeated until only a greyish paste remains, which is then collected and dried.

9 This dried tin ash is then ground in a stone mortar with milk.[161] When ground sufficiently, it should be possible to form crude cakrika from the paste.

10 The cakrika are then sealed inside a Sharaava and dried, using the clay and cloth method. The Sharaava is then cooked in a Laghu Puṭa.

11 After twenty-four hours, the Sharaava is retrieved and the contents removed. The recovered cakrika are again triturated with milk until a paste is created that can then be reformed into new cakrika. These cakrika are re-sealed into a new Sharaava.

12 This Sharaava is again cooked in a Laghu Puṭa, then left to cool for twenty-four hours. After this period, the contents are again removed and reground using milk. This process is repeated seven times in total.

13 After the final Laghu Puṭa, tin Bhasma is obtained and subsequently ground into a fine, light-grey-coloured powder. This resulting material is then stored in a glass jar for further use.

It is worth mentioning here that the process for purifying lead follows an identical procedure to this one.

161 Note: some texts advise the addition of purified Haritāla and the Arka latex.

Tin purification – method 2

Ingredients

200g Vanga
150g Apamarga (*Achyranthes aspera*)
1 litre Taila (sesame seed oil)
1 litre Takra (curd/buttermilk)
1 litre Gomutra (cow's urine)
1 litre Kanji (rice vinegar)
1 litre Kulatha (horse gram)
Godugdha (raw milk)

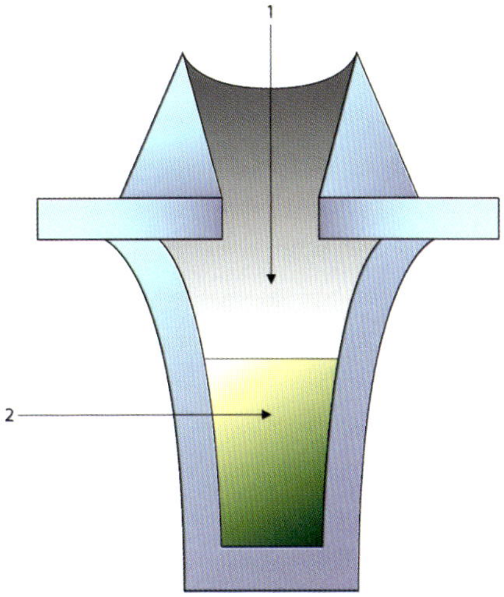

Left, cross-section of Pithara Yantra, its tapered aperture helps reduce any volatility of tin upon contact with liquids inside:

1) Liquid tin;
2) Nirgundhi leaf juice and turmeric

Method 2 is essentially the same as method 1; the only difference is that Nirgundi leaf juice and turmeric replaces the more traditional five liquids method during Ḍhālana. This substitution slightly varies the process while preserving the core principles established in method 1.

Benefits of tin

Tin is believed to offer several health benefits, particularly concerning the urinary tract, thyroid gland, and pancreas. It has been traditionally associated with treating various conditions, including diabetes, anaemia, swellings, urinary disorders, kidney stones, abdominal bloating, piles, eczema, and psoriasis.

Suitable Anupāna or adjunctive substances, such as honey or ghee, are often recommended when consuming tin. These can enhance the therapeutic effects and aid in the assimilation of tin in the body.

Above, Vanga (tin) undergoing purification

9.4/ Zinc

Zinc, like tin, is classified as a Puti (impure) metal in Rasa Shāstra. It is often used in conjunction with copper to manufacture brass. Primarily obtained from zinc carbonate ($ZnCO_3$), known as calamine or smithsonite, it appears in its native form as *Rasaka* in Maha Rasa. Zinc holds a significant position in Rasa Shāstra due to its medicinal power and importance as a complementary metal in mercurial works.

One of the key uses of zinc is to rebalance an excessive amount of copper in the body. As a trace mineral, zinc plays a crucial role in various enzymatic functions, including DNA/RNA formation, white blood cell (T-cell) production, and free radical removal. The body typically contains around 2–3g of zinc, with significant deposits found in the eyes, liver, kidneys, and bones. Zinc levels in the body have been shown to have a significant impact on stomach acid production (HCl) and the body's ability to utilise essential fatty acids. These indirect effects can contribute to hormonal imbalances and deficiencies in the nervous system, leading to feelings of anxiety and depression.

Above, Yasada (zinc) undergoing purification

In Rasa Shāstra, zinc is also regarded as a panacea for infertility and low libido. Its oxide form is incorporated into various formulations aimed at improving the strength of Shukra (reproductive fluids) and ultimately enhancing Ojas (supporting the immune system and promoting rejuvenation). It is interesting to note the biochemical similarities between modern research and ancient wisdom regarding zinc's effects on the body.

Note: An appropriate respirator mask is advised when preparing zinc as its fumes are toxic.

Zinc purification – method 1 – ingot or wire (method for smaller quantities of zinc)

Ingredients

200g Yasada (wire or ingot)

Q.S. Apamarga (*Achyranthes aspera*)/ dried opium poppy *(Papaver somniferum)*

Godugdha (raw milk)

Kumārī (*Aloe vera*)

1 The zinc wire or ingots are rinsed in water and dried.
2 The zinc is then heated in an iron ladle. When liquefied, it is poured into milk. This process is repeated twenty-one times, replacing the milk after each quenching.
3 After processing, the remaining material is collected, washed, and dried.
4 Next, the zinc is heated in a kadai until liquefied.
5 A herbal component is then added to the liquid metal, either Apamarga, Kaffir (dried lime leaves) or Khaskhas (dried poppy heads). All are suitable for zinc processing, but Apamarga is more commonly available. As the herbal content is added, the heat quickly reduces it to ash, therefore more material is added until the zinc becomes literally covered by ash. At this point the heat can be stopped, and the contents of the pan vigorously stirred. The zinc will reduce to small pieces and become mixed into the ash as both cool. This mixture is known as zinc ash.
6 When sufficently cool, the zinc ash is washed in water and then allowed to settle out before pouring the excess water off. This process is then repeated until the water remains clear. After washing, the remaining material is allowed to dry.
7 The zinc ash is now ground using a granite mortar, adding Kumārī (*Aloe vera*) gel or milk. This forms a rough paste material which is then formed into cakrika, dried, and sealed in a Sharaava using the cloth and clay method.
8 The Sharaava is then cooked in Gaja Puṭa and allowed to cool for twenty-four hours. The melting point of zinc is 416°c, however the puṭa will reach over 900°c, converting the silver-grey cakrika into a white oxide upon opening.
9 This white oxide is then removed and reground and mixed with more fresh

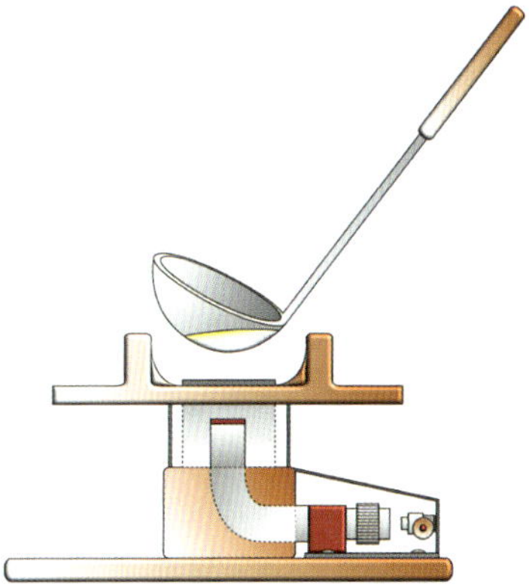

Above, iron ladle (Palika Yantra) on a gas burner, is used to ignite powdered zinc

aloe gel or milk to form a second batch of cakrika. This second batch will be much easier to form as the powdered zinc oxide and aloe gel will form a nice thick paste.

10 Once the new cakrika are prepared and dried, they are again sealed into a Sharaava and heated in the same manner. Upon cooling and opening the Sharaava, the cakrika are again ground. The rotation of grinding, forming cakrika and re-heating is repeated seven times in total, the end result being a very fine white powder.

11 This powder is stored in a glass jar for further use.

Zinc purification – method 2 (powdered zinc)

Due to the high loss of material during zinc Shodhana, an alternate method is to use zinc powder. Although this can be a viable substitute, in the author's opinion the end result is less preferable.

Ingredients

200g Yasada (powdered)
Q.S. Kaffir (*Citrus hystrix*)
Q.S. Apamarga (*Achyranthes aspera*)
Q.S. Dried opium poppy *(Papaver somniferum)*
Godugdha (raw milk)
Kumārī (*Aloe vera*)

1 200g of zinc powder is washed with water and dried.

2 The powder is ground with milk for one hour, after which it is allowed to air dry.

3 Fresh milk is again added, and the process of grinding is continued for another hour. Steps 2–3 are repeated seven times in total.

4 The ground zinc is now washed in water and allowed to settle out. The excess water is poured off, and the process repeated until the water remains clear.

5 The remaining material is air-dried.

6 Using an iron ladle or Kadai, this material is then heated until a slight colour change occurs upon its surface, this is usually a white patch, with a yellow border. Once this patch is seen, it should be allowed to burn, no further heating is needed. The powder can be agitated a little to produce a

greenish-yellow flame in order to aid oxidation. When all of the powder has been engulfed and becomes white, the burning process will naturally subside. Keep in mind: the overall content of the ladle or Kadai will swell, so it is vital not to overfill either to allow for the expansion of the material.

Note: Do not undertake this preparation in a confined space. An appropriate respirator mask is also advised to be worn. Zinc fumes are toxic.

7 The now oxidised zinc is collected and reground with either milk or aloe gel. The resulting paste is formed into cakrika, dried and sealed in a Sharaava, using the cloth and clay method.

8 For the remaining procedure, follow steps 8–11 in method 1.

Benefits of zinc

Above, ignition and oxidation of zinc powder

Zinc is widely recognised for its numerous health benefits, specifically in promoting the well-being of the eyes, digestive tract, and reproductive system. Throughout history, it has been associated with treating various conditions like abdominal distension, diabetes, urinary disorders, skin diseases, consumption, loss of appetite, and impaired vision.

To maximise the effectiveness of zinc consumption, it is often advised to include suitable Anupāna or complementary substances such as honey, milk, butter, or cream. These substances enhance the therapeutic effects of zinc and facilitate its absorption within the body.

Section 10

Gemstones

10.1/ The origins of gemstones

In the distant realms of antiquity, a formidable demon named Bala, whose very name evoked strength, ruled with terror. His power grew to such immense proportions that he managed to defeat the god of lightning, Indra, together with the rest of the gods, and the celestial armies. However, despite his demonic nature, Bala allowed them to perform sacred Yagyas,[162] even solemnly pledging his own participation in these rituals.

Little did Bala know that his seemingly benevolent gesture would sow the seeds of his ultimate downfall. The cunning gods devised a plan, conspiring to orchestrate a Yagya that would serve as their means to reclaim their lost dominion.

Finally, as the stars and planets aligned in a propitious configuration, and the entire heavenly realm gathered, a grand Yagya commenced in honour of Bala and his formidable deeds – or so it appeared. However, when the time came to offer the sacrifice, nothing materialised, and none among the assembled dared to step forward and offer their own life.

The Yagya, lacking the necessary sacrifice, teetered on the brink of ruin, inviting a curse upon its cause. After careful deliberation, Bala himself, true to his promise to the gods and fearing the dire consequences of a spoiled Yagya, took a courageous step forward and offered his own life.

His sacrifice was met with gleeful acceptance. Yet, at the very instant of his demise, his lifeless body underwent a remarkable transformation, shimmering like coloured glass and emerging as the future seed of all gemstones.

Witnessing this noble metamorphosis of Bala's body, the gods, moved by a profound sense of awe, ventured forth to seize this extraordinary prize. However, their efforts to claim the glass-like corpse descended into fierce combat, resulting in its shattering into innumerable fragments that scattered in all directions.

Far left, Navaratna – the nine gemstones, arranged in their traditional Yantra

162 *Yagya* was a fire ceremony, undertaken in order to undo the effects of negative karma.

Subsequently, the remnants of the demon rained down from the celestial abode upon the vast oceans, majestic rivers, towering mountains, and untamed wilderness. These scattered pieces of Bala would one day be rediscovered in the deepest mines, as fabulous gemtsones.

Among the precious gemstones, some were infused with the power to absolve all sins, while others possessed the ability to counteract the most potent poisons. Rubies, pearls, emeralds, bloodstones, topaz, diamonds, sapphires, garnets, and cat's eyes acquired sacred significance, becoming associated with the nine planets and worn as protective talismans, shielding the body from the ravages of time and disease.

The curative power of gemstones has been understood for millennia, from the simple act of wearing a stone ornamentally to the extraction of their essences in liquids such as milk and alcohol, and their reduction in fire to produce fine Bhasmas.

Note: As with all materials covered in this part of the book, all have multiple methods of purification and ultimate formulation. Each finished medicine will be augmented with certain Anupāna to induce different healing effects in different parts of the body.

10.2/ Red agate (Akika)

This subgroup of chalcedony is available in a wide variety of colours; it is an amorphous silicate mineral enriched with a ferrous content. The red variety of this material is thought to have the more significant medicinal effect, whether prepared as Bhasma or Pisti. Agates often display a distinctive colour banding when cut and polished; it is a popular mineral preparation in Unani Tibb.

Right, coarsely ground agate

Red agate purification – method 1

Ingredients

200g Akika (red agate)
Gulāba Jala (rosewater)

Above, mixing ground agate with Kumārī (Aloe vera) gel

1. 200g of agate material is broken into small pieces and soaked in a solution of warm salt water for twenty-four hours.
2. Once removed, the pieces are dried and powdered with an iron mortar. Distilled organic rosewater is then added to the powder and triturated into a smooth paste. Ideally, fifteen to twenty hours is an acceptable amount of grinding to meet the requirements of a Pisti. Towards the end of this period an iron mortar should be replaced by a ceramic one.
3. A small amount of ground Pisti is placed upon the tongue and rolled about the mouth and teeth to feel for grittiness. The Pisti should not feel scratchy; its particle size should be small enough to fill the indentations of the fingerprints when rolled between the fingertips.
4. When complete, the Pisti should be stored in an amber-coloured glass jar.

Red agate purification – method 2

Ingredients

200g Akika (red agate)
Gulāba Jala (rosewater)
25 Iklil-u-malik (white lotus) seeds

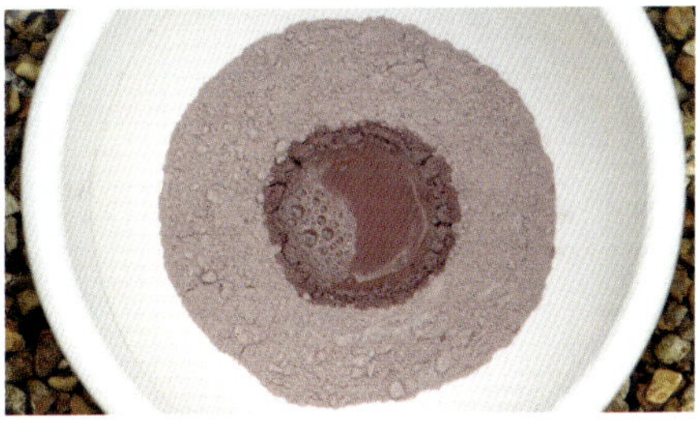

Left, agate after multiple puṭa

1. 200g of agate material is broken into small pieces and soaked in a solution of warm salt water for twenty-four hours.
2. Once removed, the pieces are dried and powdered with an iron mortar. Distilled organic rosewater is then added to the powder and triturated into a smooth paste for approximately one hour. The paste is then allowed to dry until cakrika can be formed.
3. The cakrika are then sealed into a Sharaava and heated in a puṭa. After twenty-four hours, the crucible is retrieved and opened.
4. One handful (about twenty) white lotus seeds are lightly ground and heated in water to prepare an approximately 200ml decoction. This is then cooled and filtered.

5 The agate cakrika are ground in a mortar, adding the lotus seed decoction. The mixture is then triturated for about one hour, after which its liquid content is allowed to evaporate, and cakrika prepared. These are then re-sealed into a Sharaava and heated in a puṭa.

6 This process is continued until a whitish-pink Bhasma is formed. The material is then sieved and stored in an amber-coloured glass jar.

Red agate – EDX

An analysis of red agate Bhasma revealed its content to be (Si) silicon, (Fe) iron, (C) carbon and (K) potassium. The chemical composition of agate is SiO_2 (almost identical to quartz) but of a different molecular structure. Various impurities within the crystal matrix of agate produce different intensities and variations of colour. This Bhasma was produced using the red variety and, as such, contained higher amounts of iron.

Right, an agate Bhasma sample using XRD analysis

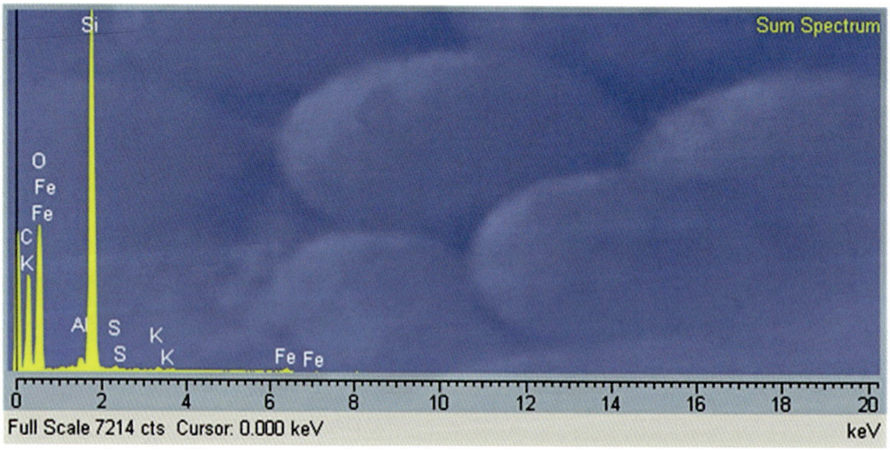

Benefits of red agate

Benefits of agate include the treatment of heart disease, internal bleeding, infertility, insanity, menorrhoea, general debility, urinary calculi, and weakness of gums, teeth or jaw. Agate is seen to have a special affinity with the heart and brain; it is considered to have a nutritive/Rasāyana effect upon both.

Suitable Anupāna for red agate include honey, milk and butter.

10.3/ Blue sapphire (Nilama)

This gemstone is traditionally prepared as either Bhasma or Pisti.

Blue sapphire purification – method 1

Ingredients

100g of Nilama (blue sapphire)

Q.S. Gulāba Jala (rosewater)

1. 100g of blue sapphire is broken into smaller, manageable pieces, which are then soaked in a solution of warm water and rock salt for twenty-four hours.
2. After removal, the pieces are dried and reduced to a fine powder using an iron mortar. Rosewater is then added to the ground powder and triturated for approximately fifteen to twenty hours, replenishing the water when necessary. The Nilama Pisti eventually reduces to a bluish-white powder.
3. The Pisti is usually placed upon the tongue and rolled around the mouth to check its composition. If the powder feels gritty or scratchy, it requires additional grinding. If the material feels smooth in the mouth and fills one's fingerprint indentations after being rolled between the forefinger and thumb, the Pisti is sufficiently ground.
4. The completed Pisti is then stored in an amber-coloured glass bottle.

Left, Sri Lankan blue star sapphire

Left, sapphire being ground with rosewater

Blue sapphire purification – method 2

Ingredients

100g of Nilama (blue sapphire)

100g Purified Haritāla (orpiment)

100g Purified Manah śhilā (realgar)

100g Purified sulphur

Kulatha (horse gram, *Macrotyloma uniflorum*) decoction
Lemon juice

1 100g of blue sapphire is broken into smaller, manageable pieces, which are then soaked in a solution of warm water and rock salt for twenty-four hours.
2 The pieces of sapphire are collected and suspended in the juice of lemon or lime and boiled for three hours, using the Dolā-yantra method.
3 After removal, the sapphire pieces are dried and reduced to a fine powder using an iron mortar.
4 A decoction of horse gram is prepared and ground with powdered sapphire. Grinding is continued until the liquid has evaporated and only a paste remains, usually after about one hour. This paste is then allowed dry under sunlight, at which point horse gram decoction is again added to the dried material. This whole process is repeated seven times in total. After this processing is complete, the final material is reground.
5 Equal quantities of orpiment, realgar and sulphur (each purified according to its own method) are then added to the powdered gemstone and triturated with lemon juice for one hour or until the material is smooth.
6 This mixture is then formed into cakrika and sealed into a Sharaava using the clay and cloth method. When dry, the Sharaava is heated in a puṭa and then allowed to cool over twenty-four hours. Upon cooling, the contents are removed and reground into a fine powder.
7 Lemon juice is once more added to this powder which is again triturated for one hour, and the cakrika remade. Steps 5–6 of this process are repeated eight to ten times or until a fine whitish-pink ash is achieved.
8 Upon completion, the Nilama Bhasma is stored in an amber-coloured glass jar.

Right, Sri Lankan blue star sapphire; far right, finished Nilama (sapphire) Bhasma

Blue sapphire – EDX

The chemical formula of sapphire is Al_2O_3. Sapphire is one of the corundum gemstones; others include ruby, topaz, amethyst, and emerald.

In this particular analysis, the material shows itself to be predominantly an aluminium silicate. (Al) Aluminium, a very common element representing about eight percent of the earth's crust, is the major component of this gemstone. The higher levels of iron (Fe) might well be indicative of this gem's particular appearance, i.e. Sri Lanka blue star sapphire. This gemstone is well known for its purple/blue smokey colouration.

Trace elements in this scan include (C) carbon, (Na) sodium and (Yb) ytterbium. Ytterbium is a lanthanoid, a rare-earth element, reducing the likelihood of any concentration in this sample. Ytterbium is one of a number of rare-earth elements frequently recovered from monazite.

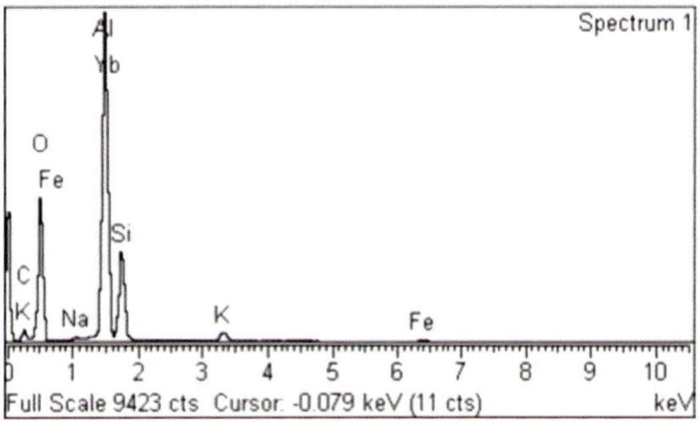

Above, sapphire Bhasma sample using EDX analysis

Benefits of blue sapphire

The benefits of sapphire include the treatment of tuberculosis, arthritis, poisoning, brain dysfunction, central nervous system diseases, infertility, low Agni, skin diseases, general debility and weakness. Nilama reduces Vāta and Pitta types of diseases.

Blue sapphire was seen to represent Shani, the planet Saturn. As such, it was thought to remove or calm ailments presided over by this feared planet. Saturn is traditionally the planet of old age and sickness; he was thought to be highly inauspicious, bringing dryness, debility and pain to the tissues of the body. Saturn is Vāta increasing by nature. The use of blue sapphire was thought to lessen his grip upon the body while enhancing the more desirable qualities, such as the fruition of good karmas, reliability and endurance. Most importantly, this gemstone is said to extend longevity. Suitable Anupāna for blue sapphire include honey and milk.

10.4/ Diamond (Hiraka)

> 'Diamond is a bestower of long life, a tonic, an allayer of the three derangements (namely Vāta, Pitta and Kapha), a killer of all ailments, a fixer of mercury, a subduer of death – in short it is like nectar . . . Vajra Bhasma increases life span, induces wellbeing, enhances strength, beauty and eradicates disease and untimely death. Vajra Bhasma has been attributed to many more therapeutic qualities.'
>
> *Rasendra Sāra Saṅgraha*

> 'Diamond is to be roasted over fire until it becomes as lustrous as the fire. Afterwards this roasted diamond is to be immersed in the oil of *Kankalakhecari* (one of five precious plants) . . . The diamond, on being immersed ten times in the oil of Kankalakhecari, is reduced to ashes (calyx). This method is also to be applied for reducing the ripe seed of gold (*Hemapakvabija*) ripened by its heating to ash (calyx).'
>
> *Rasanavakalpa*

During Sagar Manthan, as the gods secured Amṛīta from the milky oceans, a few drops of that nectar fell upon the earth. These drops miraculously transformed into diamonds, one of nature's hardest substances. Hiriaka (diamond) is considered the most potent of all gemstones and finds its use in a number of formulations that attempt to tackle the most debilitating diseases. Diamond is associated with Shukra (reproductive tissues) and the planet Venus, both of which support Ojas (immunity).

Above, Shiva Ratna (quartz) is often prepared as a diamond substitute

In general, gemstone processing is divided into two categories: Pisti (finely ground powder), and Bhasma (alchemical ash). The methodology to reduce gemstones into Pisti remains quite consistent across all types of gemstones except diamond, which has its own unique purification methods. Diamonds are considered to be the apex of all gems, so it is not surprising that this particular stone has been singled out for special treatment.

As with all alchemical preparations, there are a number of processing variations, and the ones explained here are by no means exclusive – they are simply the ones I have experimented with and found to give good results. Prepared correctly, Hiraka Bhasma is used to treat all illnesses in the body.

Diamond purification – method 1

Ingredients

75g Hiraka (diamond)
75g Purified Haritāla (orpiment)
75g Purified Manah śhilā (realgar)
75g Purified sulphur
Kulatha (*Macrotyloma uniflorum*) decoction
Snuhi (*Euphorbia ligularis*) latex

1. 75g of clear diamonds are collected and soaked in a solution of ground rock salt and warm water for twenty-four hours, then removed and dried.
2. The pieces of diamond are placed into a crucible and heated until red hot. The contents of the crucible are then emptied using iron tongs into a decoction of Kulatha. This process is repeated twenty to thirty times. In practice, the material quickly becomes brittle after being quenched in Kulatha decoction. This first part of the processing weakens the structure of the diamond as well as purifying it.
3. The purified diamond is powdered in an iron mortar and ground; it is then added to equal quantities of orpiment, realgar and sulphur (all having previously undergone their own prescribed purification methods). When mixed, Kulatha decoction or snuhi latex is added and triturated until a smooth paste is achieved. The paste is then formed into cakrika and air-dried. These cakrika are then sealed into a Sharaava, using the clay and cloth method, and heated in a puṭa.
4. After twenty-four hours, the Sharaava is removed and its contents removed. Usually, the crucible will have become blackened with sulphur deposits. Orangey red orpiment crystals may also be evident.
5. Having removed the cakrika, more Kulatha decoction is added, and the cakrika reground for about an hour. New cakrika are prepared. When dry, these cakrika are again sealed in a crucible and heated in a puṭa. This cyclical process is repeated twelve to fourteen times or until a fine white Bhasma is

achieved. After the final cooking, the cakrika are finely sieved and ground using a ceramic mortar. The completed Bhasma is stored in an amber-coloured glass jar.

Diamond purification – method 2

Ingredients

75g Hiraka (diamond)
75g Purified Haritāla (orpiment)
75g Purified Manah śhilā (realgar)
Kulatha (horse gram, *Macrotyloma uniflorum*) decoction
Sudhā (lime water)

1 75g of clear diamonds are collected and soaked in a solution of ground rock salt and warm water for twenty-four hours, then removed and dried.
2 The diamond pieces are steamed in a decoction of Kulatha for three hours using the Dolā-yantra method, or soaked in lime water for forty-eight hours.
3 After purification, the diamond pieces are washed, dried and heated until red hot using a crucible. The material is then quenched in Kulatha decoction or lime water twenty to thirty times.
4 The resultant material is pulverised and triturated with equal amounts of orpiment and realgar, along with a Kulatha decoction. Typically, the mixture is ground for one hour and formed into cakrika. Upon drying, the cakrika are then sealed into a Sharaava (using clay and cloth) and heated in a puṭa.
5 To complete this preparation, follow step 5 of the previous method.

Diamond – XRD

Analysis of diamond Bhasma confirmed it to be overwhelmingly (C) carbon and (O) oxygen with trace amounts of (Mg) magnesium and (Si) silicon. This XRD analysis shows no surprises or contamination in this batch of Hiraka Bhasma.

Benefits of diamond

Carbon is an essential element for all living organisms, forming part of the DNA molecule. Diamonds represent one of nature's purest forms of carbon.

Traditionally, diamonds were considered the most potent of all gemstones, hence their association with Venus, the planet of potency and physical beauty. Diamond Bhasma finds good use in cases of extreme debilitation where the body's immune system has begun to collapse.

The use of diamond is indicated in cases of cancer, tumours, immunodeficiency diseases (for example HIV), consumption, diabetes, anaemia, impotence, diminished eyesight, urinary tract infections, skin diseases, oedema and learning disabilities. Suitable Anupāna for diamond includes ghee, milk, cream or jaggery.

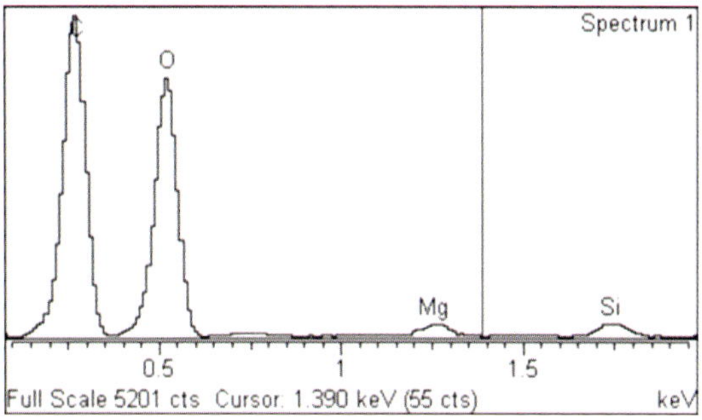

Above, diamond Bhasma sample using XRD analysis

Section 11

Animal products

11.1/ Use of animal products as medicines

> 'Mṛgavarga (deer) are categorised as Hariṇa (antelope), Kurañga (a type of deer), Ṛksa (white footed antelope), Gokarṇa (deer/antelope), Mṛgamatrika (red coloured deer), Śambara (deer with branched horns) and Caruṣka (gazelle), all are known as Mṛga.'
>
> *Aṣṭāñga Hṛdayam*

Drugs of animal origin are popular in Rasa Shāstra and include such diverse materials as conch shell, cowrie shells, abalone shell, pearls, mother-of-pearl, red coral, cuttlefish bone, deer musk, and ambergris. Animal horns from deer, buffalo, antelope, and cow are also popular. Each has a slightly different chemical signature and energetic, but ($CaCO_3$) calcium carbonate constitutes the greater part of their makeup so they are easier for the human body to assimilate.

Many of these materials are readily available and can be collected from forests, beaches, rivers and seas. However, I should stress here that the following descriptions and preparations are not to be taken as an endorsement for removing these items from nature. The following information is solely for reference purposes and is considered by this author to be purely academic. That being said, most of these materials hold a time-honoured position in Rasa Shāstra as potent life-extending medicines, so their inclusion in this book is essential.

As before, all materials covered in this part of the book have multiple purification methods and resulting formulations. Each finished medicine will be augmented with the use of certain Anupāna to induce different healing effects in different parts of the body.

Far left, Mrga Śrnga (fallow deer horns), prior to processing

11.2/ Deer horn

Śrnga means horn and was used to refer to a number of different varieties used for medical purposes. Horns could sometimes be burnt, and their fumes inhaled, in cases of lung disease, or sharpened and used in processes such as bloodletting.

Above, deer horn after a single puṭa

The Bhasma obtained from horns was highly prized for their potency, both as aphrodisiacs and circulatory stimulants. *Mrga Śrnga*, or deer horn, is an excellent heart tonic, simultaneously rejuvenating and improving circulatory functionality. Calcined deer horn also has a strong affinity with the respiratory system, helping to improve the overall strength and capacity of the lungs as well as having strong demulcent properties, as in the case of high Kapha, i.e. phlegmatic congestion.

Deer horn purification – method 1 (brown Bhasma)

Ingredients

- 1.5kg Mrga Śrnga (deer horn)
- 4 litres of Pol Vinakiri (coconut vinegar)
- Bhringaraj (*Eclipta alba*) decoction
- Arka (*Calotropis gigantea*)
- Kumārī (*Aloe vera*)
- Godugdha (raw milk)

1. The cut horn is first boiled in vinegar for one to two hours and then washed thoroughly with clean water. After drying, the reduced horn pieces are placed into a Sharaava and sealed using clay and cloth.
2. Once its contents are dried, the Sharaava is then heated in a puṭa. It is then allowed to cool for twenty-four hours before opening.
3. Upon removal, the incinerated horn is ground and sieved. The powdered horn is then ground with milk for one hour and then re-formed into cakrika. In some texts the use of plant latex such Arka (crown flower) or Kumārī (*Aloe vera*) gel is advised.
4. The cakrika are dried and resealed in a Sharaava and heated a second time,

using the same method previously mentioned. Upon removal, the cakrika are ground with Bhringaraj decoction for one hour and formed into new cakrika. This process of trituration, reforming cakrika and heating is traditionally performed a total of three times. In practice, however, the material usually requires a fourth puṭa. The final Bhasma is a brownish black.

5 When complete, the Bhasma is then sieved and stored in an amber-coloured glass jar.

Deer horn purification – method 2 (white Bhasma)

White Bhasma has similar effects to its regular dark brown counterpart; however, its potency in treating heart conditions is generally accepted to be greater.

Ingredients

1.5kg Mrga Śrnga (deer/antelope horn)
Katura Murunga (*Sesbania grandiflora*)
Godugdha (raw milk)

1. Lengths of deer horn are cut and then soaked in Katura Murunga leaf juice for three days.
2. The soaked horn is then removed and allowed to air-dry.
3. The lengths of deer horn are placed into a Sharaava and sealed using the cloth and clay method. This is then allowed to dry and given one puṭa.
4. Upon opening the Sharaava, the pieces of deer horn are removed and powdered. The powdered material is then mixed with milk and triturated for one hour, then formed into cakrika.
5. The dried cakrika are again sealed into Sharaava and heated in a puṭa. Steps 4–5 are repeated three times in total, eventually producing a whitish light-grey Bhasma.
6. When complete, the Bhasma is stored in an amber-coloured glass jar.

Deer horn – XRD

Analysis of the deer horn Bhasma shows it to be comprised mainly of (C) calcium and (P) phosphorus, with trace elements (K) potassium and (C) carbon. The presence of carbon is not unexpected due to the process of heating organic material in a low-oxygen environment, restricting the amount of carbon converted into carbon dioxide (CO_2). After the incineration of the bone mass, the only remaining elements are potassium, phosphorus and calcium. All three are essential elements for life and metabolism.

Potassium salts are thought to lower blood pressure; phosphorus and calcium combine to form calcium phosphate, essential for the maintenance of bones and the transportation of lipids (fats) in all the areas of the body closely associated with good cardiovascular function.

Benefits of deer horn

The benefits of deer horn include the treatment of heart disease, pleurisy, pain in the sides of the chest, eye diseases, sinus and migraine conditions, cough, and chronic hiccups.

Deer horn is highly efficacious in cases of heart disease. Applied topically after being boiled in milk, it helps reduce swelling. Some reports indicate that inhaling burning Śrnga fumes has a beneficial effect in cases of hiccups and asthma. Suitable Anupāna for deer horn includes butter, milk and ghee.

Right, deer horn Bhasma sample using EDX analysis

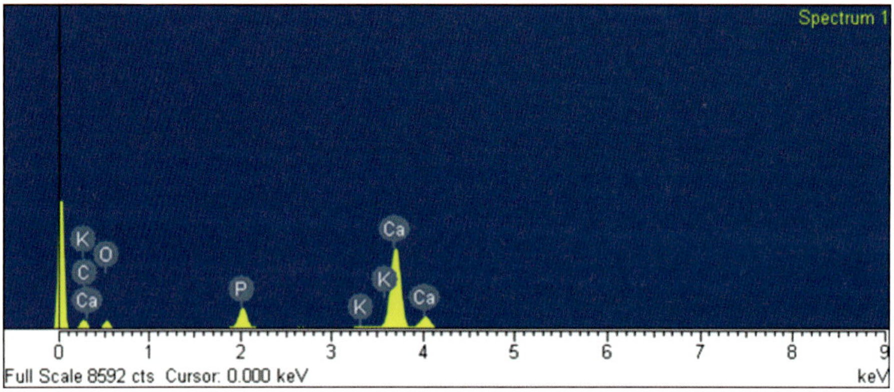

11.3/ Pearl

> 'Pearls are easily digestible, cold in potency, sweet in effective taste. They enhance lustre of the skin, vision and digestive power. They counter the effects of toxins; they have a mild purgative action.'
>
> *Rasa Ratna Samuccaya, Srī Vāgbhatāchārya*

In Rasa Shāstra, pearls hold great significance as a precious gemstone and are highly valued for their medicinal properties. Several criteria are used to select suitable pearls, including a large and lustrous appearance, substantial weight, white colour, spherical shape, cool touch, and freedom from all flaws.

While the following section mainly focuses on the processing of pearls, it is important to note that the methods described are equally applicable to oyster shells (*Ostrea concha*), abalone shells (*Haliotidis concha*), and mother-of-pearl (*Margaritiferae concha*). These alternatives were considered acceptable substitutes when pearls were not available. However, they were deemed to be less potent therapeutically, although administering higher doses or taking them for longer periods helps compensate for this.

Above, freshwater biwa pearls (Shiga, Japan)

In ancient times, procuring oceanic pearls was labour-intensive and perilous. Thankfully, modern pearl farming has now eliminated the need for such hazardous endeavours. Japan has emerged as a primary source of top-quality cultured pearls, closely followed by China, Indonesia, Thailand, and, more recently, Tahiti. Chinese companies dominate the production of freshwater pearls, also known as sweetwater pearls.

Modern pearl farming was developed by Kokichi Mikimoto (1858–1954), a Japanese pioneer who helped perfect the Akoya method, a method that uses the mollusc *Pinctada fucata martensii*. With the advent of cultured marine pearls,[163] obtaining good-quality specimens became far less problematic.

Today, many countries around the world have adopted similar methods

163 The Chinese were known to have been harvesting freshwater mabé pearls (semi-spherical pearls grown against the mollusc's shell) as early as the thirteenth century.

of intensive farming, offering a bewildering array of pearls in every size, shape and colour. Although Japan continues to hold a prestigious reputation for large marine pearls, the Chinese have rapidly become the world leader in freshwater pearls, cultured from the freshwater mussels *Hyriopsis cumingi* and *Hyriopsis schlegeli*.

From a Āyurvedic perspective, marine cultured pearls are considered to be of inferior quality compared to naturally occurring pearls. This is because the process of creating cultured pearls involves the insertion of a pre-engineered bead nucleus[164] into the living tissue of a mollusc, which stimulates the formation of a nacre.

The process of inserting the bead nucleus itself is extremely challenging and likely to cause death to the mollusc unless performed with a high level of skill and practice. Once the bead and mantle tissue are successfully placed, a pearl sac is formed, and the mollusc begins depositing nacre over the bead in successive layers to alleviate the irritation caused by its presence. The entire process of pearl formation can take anywhere from one to two years. When the pearl sac is finally harvested, it signifies the end of the mollusc's life as its shell is forcibly opened and its contents are removed. This harvesting method is viewed as an act of forceful extraction, further contributing to the unfavourable perception of cultured pearls in Āyurveda.

In more recent times, manufacturers have capitalised on mass production and advanced techniques, leading to the refinement of their products and a substantial increase in their pearl inventory. This has resulted in a significant expansion of available pearl varieties, quadrupling the options for buyers. Manufacturers have also ventured into experimentation with various methods to subtly alter the appearance of pearls. Techniques such as bleaching agents, dyes, silver nitrate, and even gamma radiation have been utilised to achieve desired visual effects in modern pearls.

164 Most pearl-bearing saltwater molluscs are inserted with a bead nucleus. Freshwater varieties originally did not require a nucleus; however, this practice is now being adopted to grow larger 'perfect' specimens. Typically a bead nucleus is about 8mm in diameter and cut and polished from organic material such as freshwater mussel or mother-of-pearl shell.

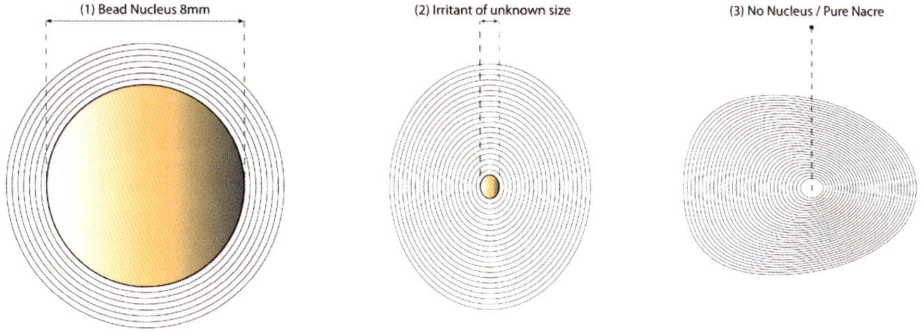

Left, simplified guide to nacresation as imaged by X-ray:

1) multiple spherical layers penetrating to a large bead nucleus indicate a cultured marine pearl;

2) successive asymmetrical/ ovular layers penetrating to a smaller nucleus indicate a wild marine pearl;

3) asymmetrical/ ovular layers with no nucleus indicate freshwater pearls

The introduction of faux-pearls,[165] imitation pearls that closely resemble natural pearls, has influenced buyer preferences towards cultured pearls with an untreated, natural look. Fortunately, this trend has had a positive impact on the availability of medicinal-grade pearls, subsequently enhancing the quality of Bhasma (ashes) or Pisti derived from their nacre.

In recent years, the global demand for pearls has declined somewhat. While partly due to economic factors, the chief cause of the decline is over-production. To revive interest in the pearl trade, some farmers have shifted their focus towards specialised markets, introducing exciting new ranges and varieties of pearls. This focus includes exploring the potential of conch and abalone pearls, which offer unique and distinct characteristics.

With the ongoing advancements in production methods, distinguishing lower-grade and faux pearls has become a challenging task, requiring a keen and trained eye to identify these undesirable imitations. In the past, one method to detect imitations was to lightly rub the pearl's nacre against one's own teeth. Genuine pearls tend to feel gritty against tooth enamel, unlike imitations. Another aspect to examine closely is the drilled hole. Fakes often have larger holes, as genuine pearls are valued for their overall weight. When inspecting large batches of pearls, major buyers often employ X-ray technology. Under magnification, X-ray images swiftly reveal the layered structure of a nacre, resembling rings on a cross-section of a tree trunk, extending from the pearl's core or bead nucleus. These studies provide insights into the formation of the pearl and its nacre density.

---

165 Faux-pearls are mostly hollow silica glass or plastic beads with a pearlescent coating, sometimes made from finely ground pearl powder suspended in a lacquer.

Many commercially produced pearls now adhere to the AAA grading system to help standardise their overall aesthetics; see the pearl grading system table below.

Pearl grading system table	
AAA	Metallic high lustre; almost perfectly spherical. Approximately (95%) blemish-free surfaces
AA+	Some irregularity, slightly ovular in shape, largely blemish-free (90%) with surface imperfections only detectable upon close inspection
AA	Ovular, uneven when rolled. Its appearance can have up to 20% of surface blemishes. Average lustre
A	Low jewellery grade, ovular in profile, low lustre. Appearance can have 25% of surface imperfections, detectable from a distance

Nacre

The captivating characteristic found in certain molluscs is created through the secretion of semi-opaque calcium carbonate variants, including conchiolin. This organic matrix provides stability and serves as a scaffold onto which mineral aggregates adhere. Subsequently, these aggregates crystallise in successive layers of nacre, consisting of hexagonal aragonite platelets. Over time, these deposits develop into a remarkably robust and free-from-cracks exterior.

The translucent nature of this material allows light to interact with its surface, causing refraction at various wavelengths. This phenomenon gives rise to the classic iridescent metallic sheen commonly observed on pearls and the inner surfaces of mollusc shells. The interplay of light, together with the structure of the nacre, are what contribute to the captivating lustre exhibited by these objects.

Marine vs. freshwater pearls

Rasa Shāstra prefers ocean-sourced pearls, but contemporary marine-cultured pearls often incorporate an artificial bead nucleus. Freshwater pearls, on the other hand, do not require a bead to form, resulting in a higher concentration of nacre. Since all pearl-bearing molluscs can be naturally stimulated through

injury or contamination[166] of their mantle tissue, any resulting pearls would consist entirely of nacre. Thus, a trade-off emerges between the ocean-bound mollusc, which relies on an artificially crafted nucleus, and its freshwater counterpart, which develops solid nacre pearls from inserted mantle tissue.

Pearls possess a natural rejuvenating effect on the body, particularly in soothing the mucous membranes. Ancient civilisations recognised the profound influence of the Moon on the oceans and its ability to govern the tides. They collected the precious salt, which was known to kindle the digestive fire and enhance water retention in the tissues, through the evaporation of seawater. Both pearl Bhasma and pearl Pisti were acknowledged for their ability to restore balance to Agni, the digestive fire, thereby aiding in the elimination of toxins from the body's tissues. Moreover, pearls were revered for their cooling properties, which were reminiscent of moonlight, and were utilised to alleviate fevers. Furthermore, they were considered a potent aphrodisiac due to their association with the god of the Moon, who was traditionally linked to fertility.

Above, Purification of pearls; soaking in yoghurt for three days

In today's world, finding naturally occurring sea pearls is arduous and often unattainable for most people. Consequently, most modern pearl preparations utilise cultured pearls. During the purification process, the bead nucleus is typically extracted before the pearls are powdered, leaving only the nacreous shell behind. Based on my experience, well-prepared cultured freshwater pearls can yield outstanding outcomes that often rival those of their oceanic counterparts.

Above, reducing pearls to powder in an iron mortar

166 There is now some debate about the inception of natural pearl formation. It was thought to be initiated by the 'encapsulated irritant theory' whereby a confined irritant was subjected to the natural formation of nacre. More recent research favours the formation of pearls due to mantle tissue injury.

Pearl purification – method 1 (Pisti)

Ingredients

200g Mukta (cultured saltwater or freshwater pearls)
Q.S. Gulāba Jala (rosewater)
Q.S. Fresh live yoghurt

1. The selected pearls are washed to remove any contaminants.
2. The pearls are then soaked in a solution of coconut vinegar[167] for three days, changing the solution daily. Alternatively, they can also be immersed into fresh yoghurt for three days, changing the yoghurt each day.
3. The pearls are washed and dried, then powdered using an iron mortar. Freshly distilled rosewater is added to the pearl powder, which is ground until a smooth paste is achieved. As the rosewater evaporates, a fresh amount is added and the Pisti reground.
4. Ideally, all Pisti should be ground for about twenty hours, first in a stone mortar before swapping to a ceramic mortar. The powder can be sieved at regular intervals to help remove larger pieces.
5. The final Pisti powder is stored in a glass jar.

Above, pearl powder formed into cakrika

167 Grape vinegar is more commonly available in Europe; its Asian counterpart is coconut vinegar.

Pearl purification – method 2 (Bhasma)

Ingredients

200g Mukta (cultured saltwater or freshwater pearls)
Q.S. Coconut vinegar
Q.S. Fresh live yoghurt
Godugdha (raw milk)
Q.S. Kumārī (*Aloe vera*) gel

1. The selected pearls are washed to remove any contaminants.
2. The pearls are then soaked in fresh yoghurt for three days, changing the yoghurt each day. At the end of this period, the pearls are removed, washed in warm water, and then dried.
3. After drying, the pearls can be broken up coarsely using an iron mortar or placed directly into a Sharaava.
4. Using the clay and cloth method, the Sharaava is sealed, dried and heated in Laghu Puṭa and allowed to cool for twenty-four hours. Upon removal, the pearls can be easily powdered.
5. The powdered pearl is then triturated in raw milk or Kumārī (*Aloe vera*) gel for one hour. The ground material is then formed into cakrika and dried. The cakrika are then sealed into a Sharaava and heated in Laghu Puṭa.
6. Repeating steps 4–5, pearl powder is processed four times in total, using either milk or *Aloe vera*.
7. The final Bhasma achieved at the end of the processing is stored in a glass jar.

Note: lemon juice can be substituted for yoghurt if the latter is unavailable.

Pearl – XRD

The analysis of pearl Bhasma reveals the presence of calcium, potassium, carbon, and chlorine, along with trace elements such as iron, sodium, phosphorus, and sulphur. Calcium and carbon, which form calcium carbonate ($CaCO_3$), are the primary elements associated with pearls, and this is consistent with XRD findings, including nacre (the iridescent inner layer of a pearl). These elements play crucial roles in biological processes and are vital for sustaining life. They serve as caretakers and maintainers by bonding with undesired substances within the body and facilitating their elimination while also ensuring the structural integrity of tissues. The remaining trace elements detected in the sample are indicative of those commonly found in a marine environment.

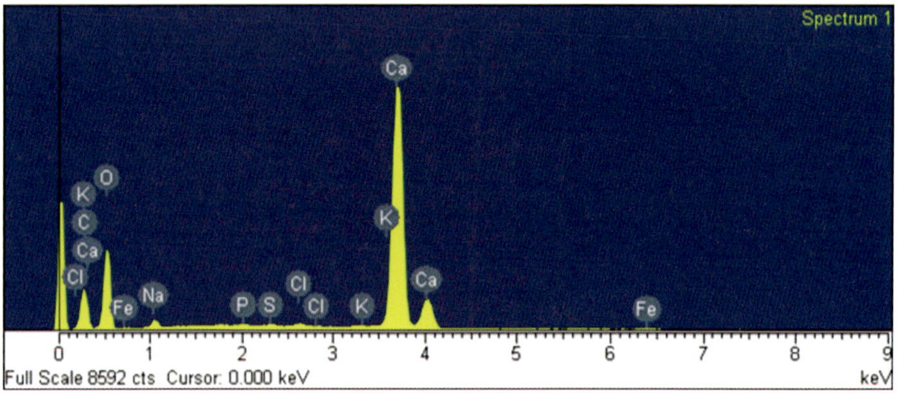

Right, pearl Bhasma sample using XRD analysis shows it to be a concentration of calcium, sodium and trace elements of sulphur and iron

Benefits of pearl

Pearl, particularly in the form of Pisti, possesses unique and remarkable efficacy. Ground into a fine powder and mixed with organic rosewater, then dried under the cooling beams of the Moon, referred to as Chandra Puṭa, it becomes incredibly beneficial for various ailments. Pearl Pisti is known to be highly effective in treating conditions such as high fevers, impaired eyesight, persistent cough, poisoning, digestive disorders, inflammation, asthma, general weakness, stomach ailments, bone disorders, heart disease, excessive sweating, asthma, and bronchitis. Pearl Pisti and Bhasma are considered especially valuable as an anti-Visha (counteracting poison), particularly in cases of lead poisoning. Pearl is also believed to have aphrodisiac properties.

When consuming pearl, it is recommended to take it with suitable Anupāna. These may include milk, butter, cream, or jaggery, all of which enhance the effectiveness of pearl preparations.

The eight sacred types of Moti (pearl)

> 'Pearls are found in the temples of elephants and wild boars, in conch-shells, in oysters, in the hoods of cobras and in the hollow stems of bamboo. The origin of a species of pearls is ascribed to thunder. Pearls found in oyster shells abound in numbers and are usually included within the category of gems.'
>
> *Garuda Purana*

The ancients classified pearls into eight distinct types, each associated with its own unique value. Although some of these pearls still exist today, they are primarily utilised for ceremonial or decorative purposes. These precious objects are highly esteemed and often find their place in protective talismans and royal jewellery.

According to Rasa Shāstra, pearls obtained from oysters, conches, and fish stimulate the appetite and enhance digestion, like pearls found in the ocean. Elephant pearls were believed to grant immunity from all diseases, and, along with cobra and boar pearls, were thought to increase wealth and social status. Sky pearls were regarded as a symbol of divine intervention, showing one favoured by the gods and destined for greatness.

The information overleaf has been given in the *Garuda Purana*.[168]

Modern pearl experts maintain a sense of scepticism regarding the existence and authenticity of pearls numbered 3–8. There are ongoing questions and doubts about these pearls' validity. Some of the pearls referred to in this part of the table might actually be *Mustika* or bezoar stones. Bezoar stones are believed to originate from the gallbladders, stomachs, and intestines of mammals, potentially forming like ambergris in sperm whales. Bezoar stones, also known as gallstones of cattle or *Calculus bovis*, have also found significant use in Āyurvedic as well as Unani Tibb formulations.

168 One of thirty-six religious texts written in Sanskrit and in story form, cataloguing the history of the universe (cosmology), genealogies of gods, demigods and kings, and the cycle of world ages. There is no agreement as to the *Puranas* true age; they were an oral tradition long before being committed to writings.

Different sources of pearl	
Type	**Description**
Oyster (Chandra Moti)	Genuine marine pearl, capable of being pierced from end to end. Lustrous, large, white, and heavy. Auspicious when using in Pisti and Bhasma.
Conch (Shankha Moti)	Found in the entrails of the conch shell. These pearls are seldom spherical and often devoid of lustre. The colouration of the conch pearl often matches the interior of its host.
Cobra (Nāga Moti)	Recovered from the hoods of the King Cobra, Nāga Moti are said to emit effulgence. Upon repeated washing these pearls obtain the lustre of a well-polished sword.
Boar (Varaha Moti)	Found in the temples of wild boar, these pearls are marble-sized and of dark colouration. Typically, Varaha Moti are said to be devoid of lustre.
Elephant (Gaja Moti)	Large, brown, and heavy, often seen with a distinctive hemispheric division upon its surface. Also devoid of lustre.
Bamboo (Venu Moti)	Found in the stems of bamboo, Venu Moti are ovular and rough in appearance (it is thought that they look like petrified wood). These pearls are without definite colour and are without lustre.
Fish (Matsya Moti)	Recovered from the mouths of whales, or the eyes of certain fish. Matsya Moti are spherical and of a whitish-yellow hue.
Sky (Akash Moti)	Formed in celestial realms during thunderstorms. These are the most illustrious of pearls, outshining all others. They resemble asymmetrical coloured rock crystal.

11.4/ Peacock feather

This unique form of Bhasma is derived from the exquisite tail feathers of the Indian peafowl or peacock (*Pavo cristatus*). The stunning array of colours exhibited through the refraction of light on their feathers has captivated generations, making this bird a prominent figure in myths and legends throughout Asia. Peacocks hold a revered status and are often depicted accompanying various deities. Maha-Mayuri (the Peacock King) is the most notable of these, and is depicted seated atop a resplendent golden

Right, eye tail feather from the male peacock

peacock. The feathers of these majestic birds are often placed on altars or displayed in shrines during *Homa* (fire ceremonies), further accentuating their significance.

Peacocks have gained a reputation for their unique ability to neutralise toxins by consuming venomous snakes, spiders, and insects. It was believed that this ability transferred to their bodily tissues and, subsequently, to their feathers. This belief extended to other materials displaying peacock-like colours, such as Sasyaka (bornite), which was used in Rasa Shāstra to detoxify the body of poison. Additionally, the juice extracted from peacock meat was utilised for bathing the eyes, since it was believed it aided in the restoration of vision. This practice is similar to the medieval doctrine of signatures, where the presence of 'eyes' on the peacock's tail feathers was seen to indicate their potential medicinal use for the human eye.

Above, prior to use, tail feathers are washed in warm water and dried, and their spines removed

Peacock feather purification

Ingredients

- Mayūr piccha (peacock feathers)
- Q.S. Ghee (clarified butter)
- Q.S. Madhu (honey)
- Kumārī (*Aloe vera*) gel

1. The peacock feathers are selected for radiance of colour. They should also be free from blemishes, insects and chemical lacquer.
2. The feathers are clipped to remove the eye; the spine is not used in this preparation. They are then washed in hot water and air-dried.
3. The peacock feathers are cut down into smaller pieces.
4. Ghee is added to an iron kadai and melted on medium heat. The cut eye-feathers are added to the ghee and stirred until they start to become 'sticky' and agglutinate.
5. Honey is added, usually a little less than the quantity of ghee used, and stirred quickly to avoid the honey burning. When absorbed, the heating is discontinued.
6. The finished calcined material is allowed to cool and mixed with a small

quantity of aloe gel. This is then triturated until a paste is formed, and cakrika can be prepared.

7 Upon drying, the cakrika are placed in a Sharaava, and this is then sealed using the clay and cloth method. When dry, the Sharaava is seated atop twenty to thirty cow dung cakes and incinerated.

8 The cooking period is relatively short due to the smaller number of cow dung cakes. The Sharaava is opened when cool, and the cakrika ground into a fine matt black powder.

9 The finished Bhasma is then stored in a glass jar.

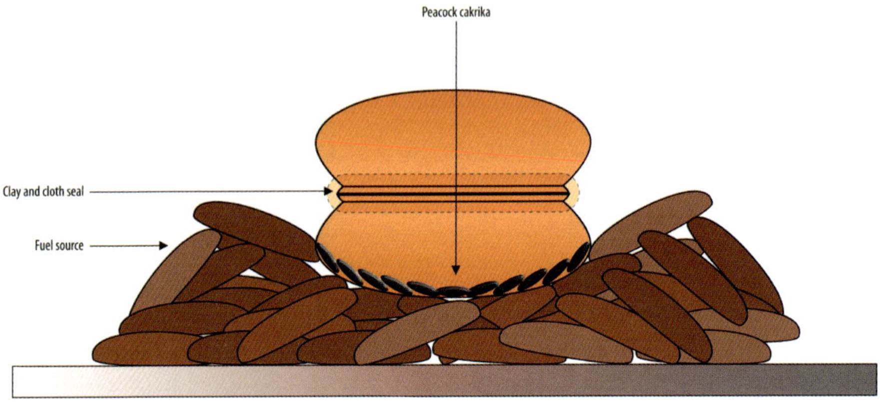

Right, puṭa for peacock feather Bhasma

Peacock feather – XRD

The XRD analysis of peacock feather Bhasma reveals that it consists primarily of calcium and potassium, along with small amounts of sodium and sulphur as trace elements. The presence of carbon depends on the extent of incineration during its preparation. The Bhasma predominantly comprises calcium, suggesting it possesses similar health benefits to pearl and deer horn. The trace amount of sulphur detected in the analysis likely arises from the natural sulphur content in the feather's amino acids, which are similar to those found in human hair. Additionally, aloe gel, which is used in the preparation of Bhasma, could potentially serve as another source of this trace element.

In 1996, gamma-ray spectrometry (GRS) conducted in India detected elevated levels of copper, manganese, zinc, and mercury in Mayūr Piccha Bhasma. However, none of these elements were found when a sample was

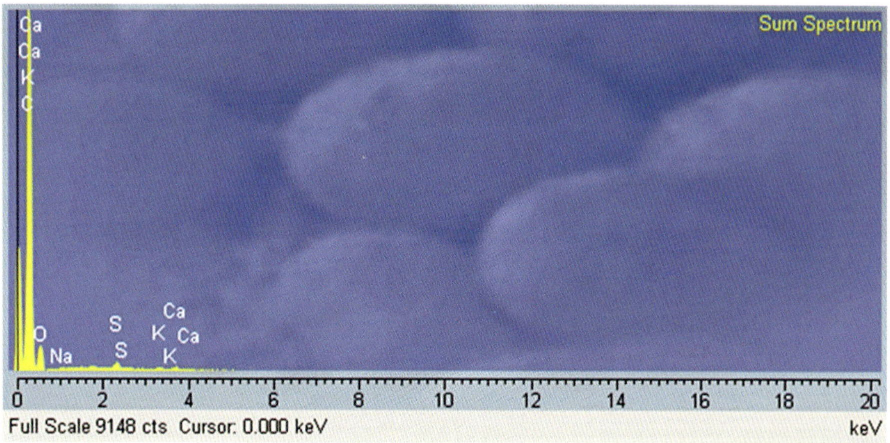

Left, Peacock feather XRD analysis

submitted for XRD analysis in 2010. The reason mercury was detected in the original 1996 experiment remains unclear. It is possible that the medicine became contaminated during processing, or there could have been issues with the calibration or attenuation of the spectrometry equipment used. Another plausible explanation could be environmental sources. In recent years, there have been concerns about peacock poisoning caused by contaminated crops or seeds treated with mercury-based pesticides. This could potentially account for the 1996 results. Detectable levels of manganese, zinc, and copper may not be unusual as these elements are essential, and their presence would be expected at some level of attenuation.

Mayūr piccha: ancient and modern thoughts

The vibrant colours observed in the plumage of peacocks led to the belief that copper concentrations were present in their feathers, as copper is known to oxidise and produce vivid blue and green hues.

A recipe found in *Rasa Jala Nidhi* suggests a method for extracting copper from peacock feathers. It advises combining the ashes of the feathers, obtained through their incineration, with ghee and honey, along with Guggulu, borax, natron, honey and Gunja seed powder.[169] This mixture, known as *Pancamitra*, is then subjected to a high temperature in a sealed crucible, resulting in a material resembling copper. Indeed, research conducted in 2003 supports the notion that

169 Gunja (*Abrus precatorius*) is also known as Indian liquorice.

Top, and above, possible copper-like extraction from peacock feathers

the iridescent characteristics of peacock feathers are primarily attributed to their reflective properties rather than inherent pigmentation. Scientists discovered that the vibrant colours of the feathers are not generated by pigments but instead by complex two-dimensional crystal-like structures present within them. These intricate microscopic structures interact with light in a manner that selectively filters and reflects different wavelengths, resulting in the diverse range of iridescent hues observed in the feathers. The slight variations in the spacing of these microscopic structures contribute to the unique play of colours, creating the stunning visual effect that is characteristic of peacock feathers. This research[170] shed light on the underlying mechanisms responsible for the mesmerising iridescence of the feathers, revealing the crucial role of structural properties rather than pigmentation.

Inspired by this report, I became intrigued by the original recipe found in *Rasa Jala Nidhi* and embarked on recreating it, which yielded fascinating outcomes. To replicate the process, I collected a substantial quantity of eye feathers, approximately 100 specimens of high quality. I transformed them into a paste by combining them with ghee and honey. This resulting mixture was then blended with an equal amount of Guggulu, borax, natron, and Gunja seed powder. Finally, a sealed crucible subjected the combined material to high temperatures. The culmination of this experiment was the formation of a small, carbonated mass. The top section, which was exposed during the process, displayed a matte black appearance. However, the underside of the mass exhibited an intriguing and unexpectedly shiny, coppery colouration, as can be seen in the images.

While these observations are undoubtedly intriguing, it is important to

170 Peacock Plumage Secrets Uncovered, National Geographic News, 17 October 2003

note that without further scientific analysis and investigation, it would be premature to definitively attribute this coppery appearance to the actual extraction of copper through this procedure. Nevertheless, this experiment provides a captivating exploration into a historical recipe, and the obtained results, along with the accompanying image, offer a valuable insight into the outcome of this recreation.

Benefits of peacock feather

Mayūr Piccha Bhasma holds significant therapeutic value, particularly in treating bronchial debility and hiccups. Additionally, it is indicated for conditions such as accidental poisoning, elimination of toxins, asthma, chronic colitis, and general weakness of the lungs. It is regarded as an excellent Rasāyana (rejuvenating tonic) specifically beneficial for lung health.

To enhance its efficacy, the most suitable Anupāna for administering Mayūr Piccha is a combination of ghee, long pepper and honey. These complementary ingredients help optimise the therapeutic effects of the Bhasma, providing a well-rounded approach to respiratory wellness and overall health.

11.5/ Ghee (clarified butter)

Over the past few decades, the use of ghee in Western cuisine has gained popularity among those who have developed a taste for South Asian foods. Many people are now exploring and experimenting with this ancient and flavourful oil.

Ghee is essentially clarified butter. In simple terms, it involves gently heating butter over a low flame to remove its water content and separate the heavier milk solids. The result is a golden-coloured oil that retains a semi-solid consistency at room temperature. Ghee exhibits excellent stability when heated at high temperatures and can be stored in a cool environment away from direct sunlight for extended periods.

Ghee holds a special place in Āyurvedic medicine, where it is valued for its ability to preserve and stabilise herbal infusions. Medicated ghee, known as *Gṛtha*, plays a significant role in the Āyurvedic pharmacy. Ghee's inherent stability and longevity were found to impart these qualities to the herbal contents it was infused with. This symbiotic relationship between herbs and oil

Right, preparing ghee

enhances the therapeutic effects of both. It also improves the oil's ability to penetrate deeper into the body's tissues.

Traditionally, ghee was considered to embody the unique qualities of cows, which held a revered status. Their abundant and unconditional milk, given with love, promoted rapid growth and strength in their offspring. The same milk, when churned into butter and clarified to make ghee, was believed to enhance the power of the digestive fire, Agni, which when balanced and efficient becomes the foundation of good health. Undigested or unassimilated food is believed to be the root cause of internal degeneration. A strong digestive power enables the easy assimilation of food, together with burning and eliminating toxins through various bodily wastes, which are known as *malas*.

According to Āyurveda, all cow products, excluding their flesh, are revered for their healing properties. This reverence extends not only to fermented dairy products but also to the dung and urine of the animal. Interestingly, the dung is commonly used as the primary fuel source for cooking fires in traditional households, while the urine is utilised for purification processes and as an antibacterial wash.

Ghee holds a special significance during *Puja*, an Indian prayer ritual. It is employed to enhance the strength of the flame and facilitate the conversion of other food offerings during the ceremony. The inclusion of ghee is believed to please Agni, the fire god, aiding in his digestion of the offerings and enabling their conveyance to heaven and the gods.

Ghee as Rasāyana

In Āyurveda, ghee is regarded as a potent Rasāyana – a substance that can restore youthfulness and vitality. Ghee, when infused with herbal ingredients, becomes a highly recommended component in the Āyurvedic pharmacy, particularly for the formulation of rejuvenation, longevity, and convalescence remedies. These formulations aim to promote overall well-being, restore vitality, and support the healing process during recovery. Ghee serves as a carrier for the beneficial properties of herbs, enhancing their effectiveness in these specialised formulas.

Preparing ghee

For those who are unfamiliar with the process of making ghee, here is a step-by-step method:

1. Slowly melt the butter in a large stainless-steel vessel, preferably on a gas stove at a low temperature. It's important not to rush the process of making ghee.
2. Once the butter has melted, stir it well to ensure thorough mixing.
3. Allow the butter to cook for approximately thirty minutes, stirring occasionally.
4. During the cooking process, a pale foam will appear on the surface. This foam can be skimmed off and discarded. At this stage, the melted butter will appear opaque.
5. As the cooking continues, heavier solids will sink to the bottom of the pan, while the foam formation gradually disperses. Stirring becomes unnecessary at this point.
6. After several hours, the remaining ghee should become transparent and have a slightly caramelised or sweet aroma. It's important not to overcook it.
7. There are a couple of methods to determine the degree of pāka (cooking) of the ghee. One method is to drop a single droplet of water into the oil. If there is an immediate crackling or spitting reaction, it indicates that the ghee is now free of water and can be removed from the heat. Another recommended method is to make a small wick from cotton, dip it into the heated oil, remove it, and ignite it. If there is no crackling or spitting sound, the heating process should be stopped.

8 Allow the ghee to cool before decanting it, preferably into Kilner jars. It's advisable to filter the ghee during decanting to remove any loose debris or remaining milk solids.
9 Let the ghee stand at room temperature for twenty-four hours before using. Ghee should never be refrigerated, as this can dampen or diminish its inherent Agni (digestive fire). Despite the intensive preparation involved, some people may still be tempted to store it in the refrigerator, treating it like regular butter, thus compromising its digestive-stimulating properties.

By following these steps, you can successfully make ghee at home.

Note: Always use unsalted butter.

Benefits of ghee

Ghee is rich in essential vitamins such as A, E, and D, as well as Omega 3 (monounsaturated fats). It also contains beneficial short-chain fatty acids like butyric acid, and essential medium-chain fatty acids. These support liver function without contributing to excess adipose tissue. While fats and oils are often associated with health conditions like high cholesterol, ghee's Omega 3 oils actually help reduce unhealthy cholesterol levels and inflammation, particularly in areas like the stomach, liver, and small intestine. Incorporating ghee into the diets of individuals with ulcerative colitis has become a recommended dietary practice.

Ghee is recognised for its excellent antioxidant properties, acting as a potent scavenger of free radicals. The combined effects of its conjugated acids (linoleic and butyric acid) and carotenoids stimulate the immune system, reducing oxidative stress and enhancing T-cell production. This boosts the immune system's strength and overall vitality.

Some additional points about ghee:

1 It is reputed that some reserves of medicinal ghee in India are aged for over 100 years, and aged ghee is believed to exhibit greater potency and medicinal benefits.
2 In hot climates like India, ghee is traditionally stored in earthen pits lined with granite to maintain its quality.
3 Some medical texts even suggest using ghee obtained from sources such as buffalo, elephants, and even humans.

These provide further insights into the storied history and diverse traditions surrounding the production and usage of ghee.

Bṛāhmi Gṛtha (medicated ghee)

Bṛāhmi Gṛtha is a significant formulation known for promoting general mind health and enhancing brain functionality. It is also beneficial in addressing various conditions, including epilepsy, mania/psychosis, infertility, skin diseases, toxic blood, voice and memory loss, learning difficulties, and low self-esteem. This time-honoured formulation has a historical association with teachers prescribing it to students to help them optimise their learning potential, and in recent years, Bṛāhmi Gṛtha has gained popularity among students in India who are preparing for exams. Its use among students has become a common practice as they strive to enhance their cognitive abilities and academic performance.

Above, early stages of Bṛāhmi Gṛtha preparation

A popular Gṛtha formula in Āyurveda, Bṛāhmi Gṛtha is essentially ghee infused with herbal ingredients. I have outlined two methods overpage.

Bṛāhmi Gṛtha – method 1

Equal quantities of:

Bṛāhmi (*Bacopa monnieri*)
Śuṇṭhi *(Zingiber officinale)*
Marica (*Piper nigrum*)
Pippali (*Piper longum*)
Trivṛt (*Operculina turpethum*)
Danti (*Baliospermum montanum*)
Śaṇkapuṣphi (*Convolvulus pluricaulis*)
Āragvādha (*Cassia fistula*)
Saptalā (*Acacia sinuate*)
Viḍaṅga (*Embelia ribes*)

The following steps will elevate regular ghee into a medicated ghee:

1 First, all the herbs mentioned above are typically prepared as decoctions, commonly made from fresh or dried whole herbs. When using powdered herbs, preparing smaller batches of three to four herbs is advisable to avoid filtration difficulties. Once the decoction is complete, it should be strained and filtered as thoroughly as possible. The recommended ratio of water to herbs for decoctions is eight-parts water to one-part herbs. The water content is then reduced to approximately one-third and filtered.

2 The ghee should then be prepared as previously described in the Preparing ghee section.

Note: It is crucial not to rush the process of making medicated ghee.

3 Begin by warming the herbal decoction in a large Kadai, preferably on a gas stove. When warmed, add ghee and gently stir until it is melted into the decoction. This mixture is then heated to remove all the water. This step can be pretty time-consuming, and it is essential to maintain a low temperature throughout. On average, it takes around six hours to prepare approximately 1.5 kilos of finished medicated ghee.

4 During the initial stages of cooking, stir the ghee well. However, it is best to allow it to simmer gently after a few hours. Towards the end of the process, while there is still some water in the mixture, you can add more of the

primary herb, in this case, Bṛāhmi. A few extra tablespoons added at this stage will enhance the potency of the medicinal ghee.

5 In the last few hours of cooking, most of the water content will be evaporated, and the remaining oil will become transparent, allowing you to see the bottom of the pan. At this point, all the remaining solids should have formed at the bottom, creating a paste known as *kalka*. This paste is considered a waste product; only the remaining ghee is needed. The ghee should have a slightly sweet and fragrant aroma.

6 In the final stage, you can check the water content of the remaining oil by dipping a small cotton wick into the oil and igniting it. If you hear a crackling sound, it indicates the presence of water. If no sound is heard, the ghee is ready, and you can turn off the heat.

7 Allow the ghee to cool in the final stage, and then filter it using a cotton cloth or coffee filter paper, both of which work well.

8 Store the collected ghee in a glass jar with a hermetic seal and let it cool naturally. Avoid refrigerating the ghee. The end result should be yellow in colour, with a slight hue from the herbs used – in this case, slightly yellowish-green.

Bṛāhmi Gṛtha – method 2

Ingredients (ratio is for powdered herbs, except turmeric, which is whole fresh root):

500g Gṛtha
1.5 l Water
500 ml Milk
30g Bṛāhmi *(Bacopa monnieri)*
20g Liquorice (*Glycyrrhiza glabra*)
20g Vaccha *(Acorus calamus)*
20g Ashwagandha
15g Fresh turmeric

This second recipe is a significantly simplified version of the previous one. The preparation method remains unchanged, with the only difference being that just five herbal ingredients are used, together with milk, which is added along with water. Although this recipe may not be classified as Bṛāhmi Gṛtha itself, it remains a remarkably potent medicated ghee, exhibiting numerous, if not all, of the same medicinal properties.

Benefits of Bṛāhmi Gṛtha

For optimal results, Bṛāhmi Gṛtha is typically consumed with warm milk or warm water two to three times a day, over a period of approximately three months to experience its benefits fully.

Āyurveda acknowledges the importance of incorporating healthy lipids into our regular diet, derived from common sources like milk, seeds, nuts, vegetables, and eggs. These lipids play a vital role in memory retention, circulation, joint lubrication, and cardiovascular strength. In alignment with this principle, several Hindu wisdom deities are symbolised by corpulent figures, such as Brihaspati (instructor to the gods) and Ganesha, who has an elephant head and is the elder son of Lord Shiva.

Medicated Grtha (external application)

Ghee is used topically to cleanse and soothe the skin, particularly the eyes and nose. Liquid ghee, warmed for comfort, is administered to the eyes using an eye bath. Bṛāhmi Gṛtha is commonly applied around the inner lining of each nostril to facilitate the assimilation of Prana (life force).

As mentioned earlier, ghee can significantly ignite the body's digestive fire. Similarly, our eyes are believed to possess a form of digestive fire that empowers us to perceive and decipher visual information. Just as consuming ghee promotes or regulates the digestive fire, the direct application of ghee to the eyes stimulates *Tejas* (the fire of intelligence), which resides partly within the eye.

Above, Bṛāhmi Gṛtha, simplified version

Section 12

Plants

12.1/ Use of plant-based medicines

> "Poison, if taken in the prescribed way, is an increaser of vitality and curer and preventer of diseases and senility. It is an increaser of the properties, good or evil, of a thing which is taken with it. It pacifies all three Dosha, is nutritious, and increases semen. The demerits of salutary poisons are removed by their purification. Poisons are, therefore, to be duly purified, before they are used as medicines."
>
> *Rasa Jala Nidhi*

Having traversed the realms of mercury, minerals, metals, gemstones and animal products, we will now delve into the plant kingdom. Here we explore the categories of *Visha* and *Upavisha*, major and secondary plant poisons. Within Rasa Shāstra, numerous Visha plant materials are utilised, but in this section, we will focus on three specimens of exceptional value.

12.2/ Dhattura

> "Dhattura increases intoxication, complexion, hunger and Vāyu. It cures fever and leprosy. It is astringent, sweet, bitter, warming, and heavy. It destroys lice, boils, phlegm, itching, worms and poison."
>
> *Rasa Jala Nidhi*

Dhattura, revered as a sacred plant, is closely associated with Lord Shiva, the primary deity of Indian alchemy.[171] This divine connection automatically elevates Dhattura to a position of great importance. As a result, Dhattura finds its place in numerous Rasa formulations, either as a direct ingredient or

171 It is considered good practice to encourage the growth of Dhattura plants and auspicious to have them growing in close proximity to the Rasa pharmacy.

Far left, Vatsanābha (aconite)

Right, Datura metel Linn *(Blackcurrant swirl)*

as a binding medium for triturating other ingredients. It is worth noting that the entire Dhattura plant, including its seeds, contains highly toxic properties and thus undergoes a purification process before any internal consumption.

In external applications, the juice extracted from Dhattura leaves can be directly applied to address conditions like alopecia. Additionally, when combined with a small amount of dried turmeric powder, it can be made into a paste and applied directly to the skin to alleviate lesions, boils, or inflammation.

The oil derived from Dhattura seeds possesses analgesic and antispasmodic properties, but due to its potency, it is infused into a carrier oil,[172] typically sesame, before direct application. To prepare this infusion, purified seeds are gently heated in oil for approximately thirty minutes and then allowed to cool. The oil is then strained, separating the seeds, which are subsequently ground.

172 The best carrier oil (sesame) is pre-heated for around one to two hours at a very low temperature to improve their absorption of herbal decoctions. Typically, the temperature is kept just below 100°c.

Above, Datura stramonium Linn *(devil's trumpet)*

The decanted oil and the ground seeds are then recombined and left to stand for a few days. During this period, the colour of the oil will deepen. Finally, all solid particles are again filtered out using a fine cloth. This final oil is applied to the limbs to alleviate arthritic pain and inflammation, with caution taken to avoid application to the head area due to its potency. It's worth mentioning that the oil from Dhattura seeds can also be extracted using the Patala Yantra method, which is elaborated upon later in this section.

In most cases the black variety, *Datura metel*, is generally preferred for its therapeutic attributes. However, in its absence the white variety (*Datura stramonium*) can be utilised. While it is a common inclusion in the pharmacopoeia of Āyurvedic physicians, this plant also holds a metaphysical significance in alchemical pursuits. The juice extracted from its leaves, stem and seed pods is sometimes used in the *Murchana* (fixing) of mercury during some of its transformational processes.

While high alchemical procedures are distinct from the practical applications of Dhattura, the plant offers various benefits in different contexts. The dried leaves of Dhattura are commonly used in medicinal cigarettes, providing relief when inhaled by individuals suffering from severe asthma. Moreover, these very leaves can be steamed inside cloth pottali bags and gently applied to the body during Pañcakarma therapy. This technique aims to alleviate the effects of high Vāta, a Dosha associated with imbalance in Āyurveda. It is also worth noting that Dhattura has been recognised in ancient times for its narcotic properties. Its essences have been extracted by unscrupulous individuals for the purpose of immobilising the unsuspecting victims they were exploiting.

Dhattura is indeed among the various alkaloid-containing plants found in the therapeutic pharmacopoeia of Āyurveda. Several other notable plants also contain alkaloids with medicinal properties, including quaker buttons (*Nux vomica*), and opium poppy (*Papaver somniferum*).

Dhattura contains several tropane alkaloids, including scopolamine, atropine, and hyoscyamine, also known as daturine. Daturine acts as both a stimulant and a parasympathetic inhibitor, suppressing nerve impulses and providing analgesic effects. When consumed in low doses, Dhattura can cause a sensation of heaviness and tiredness in the body and limbs, along with a dry palate, hoarse voice, and a high likelihood of blurred vision. Higher dosages can lead to acute memory loss, muscle spasms, increased body heat, visual hallucinations, and, ultimately, difficulty in respiration. For this reason, it is important to note that Dhattura should be used with extreme caution due to its potent effects and potential side effects.

In Western herbalism, tinctures of *Datura stramonium* were historically utilised to calm those individuals exhibiting signs of mental disturbance, particularly during the full moon period. The name 'stramonium' itself signifies a tendency towards madness. Ironically, the narcotic content of the plant is known to fluctuate significantly over twenty-four hours, often peaking in alkaloid concentration during the full moon. This increase in hallucinogenic compounds coincides with the plant's full flowering stage and the release of its renowned intoxicating aroma. Such observations undoubtedly reinforced the ancient association of the plant with Shiva, the deity commonly depicted with a crescent moon, as well as with Somā, the divine elixir associated with the Moon and immortality.

Above, Datura stramonium *seeds, with husks removed*

Dhattura purification method

Ingredients

Q.S. Dhattura seeds *(Datura stramonium)*
Q.S. Gomutra (cow's urine)
Q.S. Godugdha (raw milk)

1 The Dhattura seeds are collected and placed into a Pottali (cotton cloth bolas) then suspended in fresh cow's urine for forty-eight hours.
2 After soaking, the Pottali is retrieved, and the seeds removed and washed in warm water. Once clean, the seeds are laid out on absorbent paper to dry.
3 Traditionally, the seed husk is removed; this is usually done by lightly grinding the seeds in an iron mortar, then placing them in a strong cloth and

abrading the remaining husk with course brick dust. In practice, this is not an easy job as the seeds are small and the husk highly resilient.

4 Another method for husk removal is to select a dozen or so seeds and lightly strike them individually using a mortar. Striking the seed splits the husk and allows the seed to be separated. Due to the potency of Dhattura seeds, only a small quantity is required. As husked seeds are likely to remain potent for some time, small batches of purified seeds can be kept and powdered as required.
5 The de-husked seeds are then boiled in milk using the Dolā-yantra method of purification. The seeds are continually boiled for three hours, occasionally topping up the reduced milk level. After boiling, the seeds are removed from the pottali, washed in cool water and dried.
6 After drying, the seeds are ground in a granite mortar.
7 The powdered Dhattura is stored in a glass jar. This material has a shelf life of twelve months without noticeable loss of potency.

Note: Seed oil is another way to harness the power of Dhattura, although this is used for external application only. For the extraction of oils from seeds, see Section 12.4/ Bhallātaka (bhilawan Nut).

Benefits of Dhattura

Dhattura offers a range of benefits for various health conditions. It can effectively alleviate abdominal swelling, asthma, bronchitis, intestinal parasites, eczema, lower back pain, earaches, rheumatism, gout, colic pain, fever, and hair loss (alopecia). Furthermore, it aids in improving liver function and has a calming effect on the nervous system.

When it comes to consuming Dhattura, the most commonly used Anupāna is raw milk sweetened with honey. This combination enhances the effectiveness and taste of Dhattura preparations.

12.3/ Aconite

> 'Like good and bad, day and night, light and darkness, gods and demons, 'Visha and Amṛīta' also have their origins from the same stem in creation and are like two diverse sides of the same leaf. After various Shodhana treatments, Visha is devoid of its evil, demerits and becomes restorer and stabiliser of health, the dispeller of disease.'
>
> *Visha in Āyurveda*

As expounded upon later in this book, snake venom dominates the hierarchy of poison; however, its nearest competitor resides in the plant kingdom. The potency of the claw-like tuber Vatsanābha (*Aconitum napellus*) was deemed comparable to that of cobra venom. As the cobra seen as the king of mobile poisons, so was aconite was the queen of immobile poisons.

This close association with cobra venom highlights the profound respect given to the use of aconite root in its medicinal form. Aconite stands out among poisons, and the comparison also serves as a cautionary reminder of its lethal effects if mishandled. Just like cobra venom, if used incorrectly, aconite was capable of bringing death, but it also possessed the potential to bestow life. In fact, certain texts even praise the potency of these two poisons, surpassing that of mercurial preparations when it came to restoring health to the afflicted. Like Dhattura, *Vatsanābha*, a name meaning 'like a calf's umbilicus' finds its way into numerous Rasa formulations, demonstrating its remarkable versatility as a healer.

Early observers from the Himalayan regions are believed to have been the first to understand how this plant could be utilised, documenting their observations of its medicinal properties. Today, aconite is classified within the family *Ranunculaceae* and can be found growing in temperate regions across the globe. Categorising aconite has revealed the existence of over 100 distinct varieties, varying from highly toxic to completely inert.

The traders in India who still supply this medicinal root offer a mixture known as *Bachhnāga*. Upon closer examination, this composition is revealed to be a combination of various tubers, including *Aconitum napellus, Ferox, Lycoctonum, Chasmanthum, Palmatum, and Heterophyllum*. Among these, three species particularly stand out and are associated with beneficial medicinal properties in the following order: *Aconitum ferox*, *Chasmanthum*, and *Napellus*.

Due to their inherent toxicity and requirement for Shodhana (purification) before use, many modern pharmacies no longer stock this ancient remedy.

Above, aconite tubers soaking for one season in mustard seed oil and rock salt

The roots of Vatsanābha are typically harvested in late winter or early spring. While new plants can propagate from seeds, they also renew themselves through 'daughter roots' that grow from the main root. These daughter roots remains dormant until spring and then re-establish the plant in the following year. Each new main root lasts only one season. As a general observation, other plants tend not to grow in close proximity to Vatsanābha. It thrives in slightly moist soil and seeks out the shade provided by larger overhanging trees.

All tubers appear whitish upon extraction but tend to darken during processing. The unprocessed tuber quickly deteriorates after harvesting, attracting fungus and insects. Wrapping the tuber in a fine cloth infused with red mustard seed oil is recommended to maintain its freshness.

Raw, unpurified aconite was recognised for its ability to depress the functioning of the heart. However, once purified, this effect was reversed. This reversal of roles is a characteristic often observed in Visha substances; they are poisonous in their raw state but in their purified state have the capacity to counteract the harmful effects of other poisons. The beneficial properties of aconite were also sought after for treating venomous bites from both scorpions and snakes. This honorary quality may have contributed to its elevated status as a Visha, second only to Sarpa-Visha, the venom of cobras.

*Above, Vatsanābha (*Aconitum napellus*) tubers being purified*

Most Visha medicines induce a certain level of heat in the body, and aconite is no exception. To undergo *Vishakalpa*, an extended therapeutic course, patients consume smaller doses of aconite over extended periods to purify and revitalise the bodily tissues. These types of highly

Above, aconite tubers being purified in cows urine, using the Dolā-yantra method

intense programs adhere to strict codes of conduct. Typically, it is advised that the treatment begin during the cooler seasons, accompanied by a diet of cooling foods. Patients are also instructed to avoid heating spices, and instead to consume ample cooling fluids. Following an initial purification phase, the depleted tissues are nourished with foods such as ghee, butter, milk, honey, rice, and wheat. The immune system is stimulated by gradually introducing toxins during this process.

Various purification methods are employed for Vishas, with cow urine being commonly favoured for its purifying properties. Raw milk is another popular medium for boiling these substances. Additionally, materials such as Mustā (*Cyperus rotundus*), Vaccha (*Acorus calamus*), Marica (*Piper nigrum*), Pippali (*Piper longum*), Adraka (*Zingiber officinale*), Yastimadhu (*Glycyrrhiza glabra*), Parsik yavani (*Hyoscyamus niger*), Gandhaka (sulphur) and Tankana (borax) are all well known for their neutralising effects on toxic materials.[173] Borax, in particular, possesses intriguing properties. The recommended approach involves mixing it in equal weight with any Visha and then subjecting it to the Laghu Puṭa method of incineration. This process yields a herbo-mineral Bhasma that can be administered in small doses. Moreover, it provides the added advantage of an extended shelf life.

Aconite in classical Chinese medicine

> 'There is no substance on earth that is more toxic than Chicken Poison (jidu; an ancient term for Chuanwu/Wutou aconite), yet a good physician collects and stores it away to be used for medicinal purposes.'
>
> *The Aconite Papers, Heiner Fruehauf*

173 Black pepper is the herb of choice to be combined in equal quantity with aconite. Black pepper, long pepper and ginger (equal quantities) are also frequently combined with aconite to aid in its effectiveness and help reduce any potential side effects.

Above, *freshly harvested tubers* (Aconitum nepellus)

Chinese medicine holds a deep reverence for the root of aconite, recognising its remarkable ability to expel internal coldness and pathogenic dampness while strengthening yang energy. In fact, aconite is often referred to as 'the flagship of Chinese medicine'.

Aconite has intense heat, bitterness, and extreme drying properties, but it is not without its risks. Proper preparation and meticulous dosage are considered of utmost importance, taking into account seasonal variations[174] and the differences in constitution of prospective patients. Authentic aconite is said to originate from the Sichuan Province, where the finest medicinal specimens of the plant (*Aconitum carmichaelii/kusnezoffii*) are cultivated. While the plant can be found in other provinces, this particular region has been dedicated to producing this highly specialised medicinal crop for over 2000 years.

The use of aconite in China has a long and storied history, steeped in ancient tales and legends that speak of its reputation as a potent and transformative herb. In the pursuit of physical immortality, Taoist aesthetics recognised aconite as a plant of immense potential, not only for its mystical qualities but also for its practical applications in daily pharmacopoeia. It was highly valued for its ability to maintain good health, particularly for a life in the wilderness. Before using it, it was necessary to subject aconite to various purification practices, with some accounts mentioning up to seventy different

Above, dried aconite tubers prior to processing

174 Harvesting at different times of the year captures a different energetic in the tuber. Late autumn, after a dry summer, increases alkaloid content. Harvesting tubers in the spring, after a wet winter, lessens its alkaloid content. The Chinese transplant seedlings on the winter solstice and harvest them just before the summer solstice to enhance the tuber's Yang energy. For more information see 'The Aconite Papers' by Heiner Fruehauf.

post-harvesting techniques. These methods aimed to reduce the concentration of its primary alkaloid, aconitum.

As a result of these intensive processing practices, aconite emerged as a premier herbal treatment of its time. It was renowned for its effectiveness in addressing a wide range of ailments, including rheumatic conditions, neuralgia, cholera, dysentery, paralysis, abdominal hernia, broken bones, headaches, eczema, contusions, bruises, and even burns when applied externally as a paste.

Aconite purification method

Ingredients

Vatsanābha tubers (*Aconitum nepellus*)
Rock salt
Mustard seed oil
Milk
Cow's urine

1. The harvested tubers are washed in warm water and dried.
2. The tubers are then submerged in mustard seed oil and ground rock salt. Usually, secondary processing can be undertaken after seven days. However, the tradition is to leave the tubers to soak for one whole season.
3. After soaking is completed, the tubers are taken out and any excess oil is removed. The tubers are then cut into discs and placed into a pottali.
4. Using the Dolā-yantra method, the pottali containing the sliced pieces of tuber is then immersed into either cow's urine or raw milk, and boiled. Using the urine method requires approximately three hours of heating; its milk equivalent is about six hours.
5. After cooking, the tuber pieces are removed from the pottali and thoroughly washed and dried.
6. When the pieces are completely dry, they can be ground in a granite mortar until fully powdered.
7. The final powder is sieved and stored in a glass jar. Powdered aconite has a shelf life of about a year without any loss of potency.

Benefits of aconite

Aconite is effective in treating conditions such as rheumatism, heart disease, gout, asthma, bronchitis, piles, fever, earache, sciatica, lumbago, and inflammation. It also aids digestion, eliminates toxins (āma), balances excessive Vāta and Kapha Dosha, and serves as a rejuvenating agent for the overall body (Rasāyana).

When formulated, aconite is mixed in equal quantity with Maricha (black pepper), or mixed with equal quantity of Shunthi (ginger), Yastimadhu (licorice) and Pippali (long pepper). Freshly prepared ginger juice, known as *Adraka* is a useful remedy to counteract the effects of improperly prepared Vatsanābha. Suitable Anupāna for aconite are milk, honey or Cyavana Prāśa.

12.4/ Bhallātaka (bhilawan nut)

> 'Fruits of Bhallātaka are Tīkṣṇa (sharp), Pākī (corrosive) and like fire (Agni Sama), but when prepared accordingly to prescribed methods they work like Amṛīta (ambrosia).'
>
> *Caraka Saṃhitā*
>
> 'Ripe fruits of Bhallātaka are sweet when digested, light, astringent, digestive, soothing, acrid, warm, a reducer of fat, purgative, increaser of retentive faculties and an increaser of hunger. They destroy phlegm, Vāyu, sores, Udara-roga, leprosy, piles, chronic diarrhoea, Gulma, dropsy, intestinal obstructions, fever and worms.'
>
> *Rasa Jala Nidhi*

Caraka Saṃhitā advises us that Bhallātaka is recognised as the most effective substance for igniting the digestive fire. It is frequently included in various formulations, particularly those aiming to provide rejuvenating benefits (Rasāyana). Bhallātaka is known to possess some Visha (toxic) properties; however, not everyone who encounters it will experience these effects. Individuals sensitive to its toxins may exhibit immediate signs of skin irritation, consequently, it is generally recommended to take precautions when handling Bhallātaka, such as wearing protective clothing and avoiding direct contact or accidental exposure.

When Bhallātaka is discussed in textbooks that focus on Rasa remedies, its beneficial qualities are often accompanied by a cautionary note about its potential for irritation. The following excerpt serves as a typical illustration of this:

> 'Swelling and irritation, due to Bhallātaka poison, is pacified by the external application of butter, sesame (Tila) milk and molasses (Khanda gurh).'
>
> *Rasa Jala Nidhi*

Bhallātaka belongs to the *Anacardiaceae* family and is closely related to the cashew. It is widely distributed throughout Asia, including parts of northern Australia. The Bhallātaka tree has historically served as a valuable source of medicinal substances and high-quality timber. Nutritional analysis of the Bhallātaka nut reveals its rich calorific content, along with significant concentrations of phosphorus, magnesium, and calcium.

To extract the black tarry oil from the ripe nut, the surface of the pericarp is broken, and pressure is applied. Freshly harvested nuts are then placed in water, and those that sink are retrieved while floating ones are discarded. Generally, sinking nuts are heavier and contain a higher quantity of the precious oil. This black exudate is known as Bhilawan oil or marking nut oil, since it was commonly used for inscribing identification marks on fabric. Although the oil is not soluble in water, it can be removed by using a mild solution of alcohol or detergent. To affix the oil onto fabric, it is immersed in a solution containing Tankana (borax), Kanksi (alum), or Sudhā (lime water).

Above, Bhallātaka or marking nut (Semecarpus anacardium*)*

Bhallātaka gained immense fame throughout the ancient world for its wide range of therapeutic applications, particularly its effectiveness in treating various skin disorders. Even in cases of leprosy, which was considered incurable in many ancient cultures, purified Bhallātaka remedies were known to yield successful outcomes. Additionally, Bhallātaka was highly regarded as a fertiliser, especially for the cultivation of Methi (fenugreek). Large quantities of the nuts would be harvested, crushed, and incorporated into the soil. The mashed nuts were also boiled in diluted cow's urine to create an irrigation agent and pesticide.

Due to its potent rejuvenating properties, Bhallātaka was frequently

Above
Bhallātaka black oil

included in various preparations; it was also consumed directly by individuals with robust constitutions. Like aconite, individuals living in close proximity to Bhallātaka often developed higher levels of resistance to its toxicity. Once purified, Bhallātaka oil was believed to enhance immunity, promote strength, and rejuvenate the body tissues. When combined with carrier oils like sesame, Bhallātaka oil was known to maintain glossy and grey-free hair. One common method of consuming the nut was by boiling it in milk. Other recipes involved removing the nut shells, lightly frying them in ghee or butter, mashing them, and serving them with rice Kanji (rice porridge). This mode of intake was recommended in cases of severe fatigue or long-term recovery periods.

Another method of extracting Bhilawan oil involved boiling pierced nutshells in milk. Once the oil was released and suspended in the milk the mixture was filtered, and the discarded nuts could be used as a fertiliser after maceration. The remaining milk was then combined with ghee and slowly heated. During this secondary cooking process, any remaining water and milk solids were separated, resulting in a mixture of ghee infused with Bhilawan oil. Finally, the oil was filtered and left to solidify, sometimes sweetened with jaggery.

Alternatively, the oil could be extracted by boiling pierced or sliced shells in hot water and carefully collecting the exudate from the water's surface. This collected oil could be further heated on a low flame or naturally dried under sunlight.

The sticky oil obtained through either method was added to melted butter and allowed to solidify. Jaggery could once again be added to sweeten the final mixture, counteracting the sharp nature of the oil. The ratio of Bhallātaka oil to butter was adjusted based on the desired therapeutic effect. However, it was essential to administer the resultant mixture cautiously, as higher dosages and increased frequency of use could lead to side effects in some patients.

Another technique for extracting Bhilawan oil is a Patala Yantra, as depicted in the illustration overpage. This method is commonly employed for various nuts and seeds, and although effective, it requires significant labor and time. *Patala* refers to below ground, while *yantra* denotes the apparatus used. The underlying principle involves applying concentrated external heat to nuts that

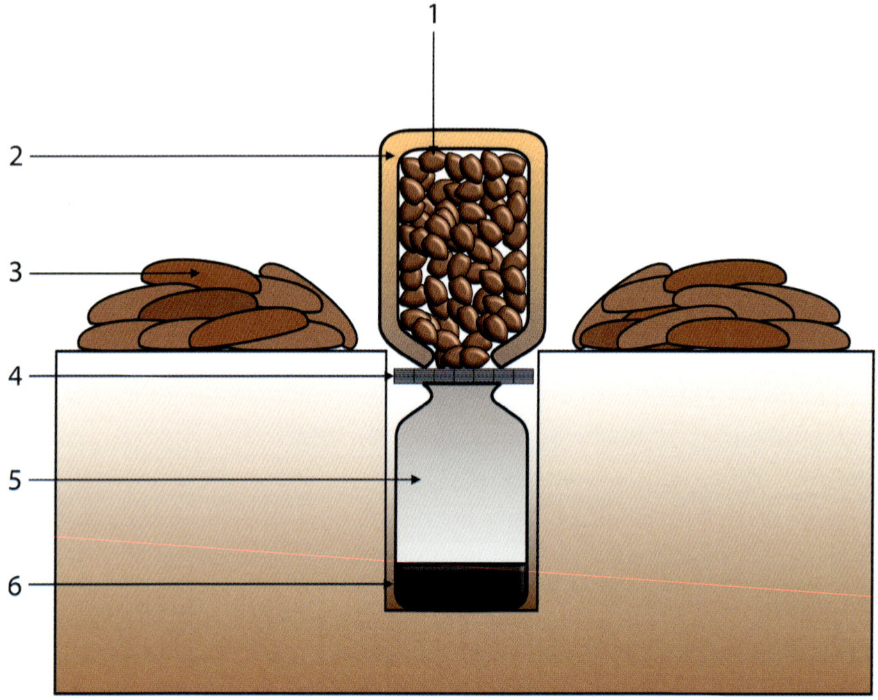

Right, Patala Yantra:

1) Bhallātaka nuts;
2) seven layers of clay and cloth surrounding the glass jar to protect from heat;
3) external heat source/cow dung cakes;
4) metal grill retaining nuts;
5) glass jar below ground level;
6) exudate oil

have been pierced. The heat causes them to soften and expand, releasing their oil which drips into a vessel situated below ground level. The oil, shielded from the heat source above, is harvested once the heating process is complete.

After extraction, the oil can be cooked with milk or combined with other oils like sesame, or with ghee. Both oil and ghee preparations can be applied topically or taken internally in the form of a Tailam or medicated ghee. Notable final bhilawan oil formulations include *Bhallātaka Taila* (medicated sesame oil), *Amṛītabhallātaka Grtha* (medicated ghee), and the Tibb formulation known as *Hab Diq-ul-atfal*.

The effects of Bhallātaka oil were known to be fast-acting. It was often said that its application to the soles of the feet could instantly be felt in the head, and vice versa. Purified forms of this oil were favoured for external application, usually in the form of a paste. When combined with sesame oil, Greek pitch (*Pix groeca*), and beeswax, it could be applied to the feet to alleviate dryness and cracked skin. Bhallātaka Taila, a medicated sesame oil infused with bhilawan oil, could be used both internally and externally. When taken orally, this oil was particularly effective in treating asthma and non-bleeding piles.

Left, this apparatus represents a crude form of Patala Yantra, in this case being used to extract birch tar, but the principle remains the same. Here firewood was used in place of cow dung cakes: far left: firewood was moved progressively closer to the clay pot containing shreds of pine bark; left: the clay pot is later removed to expose the subterranean chamber now holding the exudate oil

Modern scientific studies have been rediscovering the potential of this humble nut, including its anti-carcinogenic properties. Recent pharmacological and therapeutic studies conducted in India demonstrate that the use of purified Bhallātaka oil improved the life expectancy of cancer patients, including those with leukaemia, a type of blood cancer known to respond well to Bhallātaka. The following are two good Shodhana techniques for Bhallātaka:

Bhallātaka purification – method 1

Ingredients

Q.S. Bhallātaka (*Semecarpus anacardium*)

Cow's urine or milk

1 The Bhallātaka nuts are washed and sorted according to their ability to float or sink in water. Floating nuts are discarded, while nuts with the higher concentration of oil will sink.
2 Using heavy scissors or an abrasive flat stone, the stalk and skins are removed.

Note that Bhallātaka oil may react with some skin types, so the hands should be protected. In many cases, a coating of either ghee or coconut oil is usually sufficient. If not, vinyl gloves are advised.

3 The capped nuts are wrapped in a cotton cloth and a Pottali prepared. This is then suspended into cow's urine or milk for three hours, and boiled using the Dolā-yantra method of purification.
4 After cooking, the remaining liquid is poured off, while the remaining nuts are rinsed and air-dried.
5 After fully drying, the nuts are ground, sieved and stored in a glass jar.

Bhallātaka purification method 2

This second method is very quick, and yields excellent results

Ingredients

Q.S. Bhallātaka (*Semecarpus anacardium*)
Q.S. Kithul jaggery (date palm sugar)

1 Again, the Bhallātaka nuts are washed and sorted according to their ability to float or sink in water, with floating nuts being discarded, while nuts that have the higher concentration of oil are retained.
2 Using heavy scissors or an abrasive flat stone, the stalk and skins are removed. Again, because Bhallātaka oil may react with some types of skin the hands need to be protected with a coating of either ghee or coconut oil or, if necessary, vinyl gloves.
3 The remaining part of the nut is reduced to a soft pulp with an iron mortar, while an equal quantity of kithul jaggery is added to the pulp. To aid in the mix, the jaggery is usually grated prior to use. When pounded together, the two materials quickly form a black pliable mass.
4 Using a little ghee or coconut oil on the fingers, this mass can be separated and rolled into pills, which are then air-dried.
5 When complete, the pills are stored in a glass jar.

Benefits of Bhallātaka

Purified Bhallātaka offers a wide range of benefits. It is commonly used in the treatment of various conditions, including piles, splenic disorders, persistent skin diseases, abdominal bloating, and obesity, as well as being used as a Rasāyana (tonic) for the brain and nervous system. It is also effective for Majjā Dhātu (bone marrow health), asthma, poisonous bites, allergic reactions, leucoderma, osteoarthritis (swollen joints), parasites such as worms, and different types of cancer.

Above, Bhallātaka resin

Bhallātaka is particularly effective in addressing skin disorders, ranging from dry and scaly skin to discolourations, dermatitis, and warts. It can also be used internally to treat haemorrhoids; however, it is important to avoid excessive use here as it may worsen the condition.

Prolonged use of Bhallātaka can lead to itchiness and painful urination, and these are clear indications to immediately discontinue use. Interestingly, continual exposure to the oil of this nut can eventually result in cirrhosis, or skin conditions such as dermatitis.

The recommended Anupāna for Bhallātaka is sweetened milk; this is also the most commonly used remedy for any adverse effects caused by Bhallātaka. It is worth noting that Bhallātaka has a strong heating effect and tends to aggravate individuals of a Pitta constitution. Therefore, those individuals displaying signs of elevated Pitta Dosha are advised to avoid this medication.

12.5/ Langali (*Gloriosa superba Linn*)

The name *Gloriosa superba* translates to 'splendid amongst flowers', and true to its name, this climber/creeper is a visual marvel with its captivating beauty and intricate design. Under optimal conditions, the Langali plant can surpass fifteen feet in height, and its tuber can grow considerably. The shape of the tubers provides a distinguishing feature, with rounded tubers indicating the female plant and long flattened tubers the male plant.

Langali thrives in warmer climates and flourishes in abundant sunlight. It is commonly grown indoors in temperate regions, but with proper care and attention, it can also thrive outdoors. In India, it awakens just before the monsoon season, rapidly growing in the ensuing humidity and drying up as the rains subside. Its preferred habitats include Southern India, Sri Lanka, Bengal Bali, and Africa. Notably, Langali holds the honour of being the State Flower of Tamil Nadu and the national flower of Zimbabwe.

*Above, flame lily (*Gloriosa superba*)*

Like several others in the lily family, Langali is a toxic plant. Even a small amount of ingestion can lead to severe stomach cramping, sweating, vomiting, convulsions, and diarrhoea. Larger doses can be fatal, especially among elderly or young individuals. Incidents of poisoning have often been reported when the tuber of Langali, which resembles a sweet potato, has been mistakenly consumed.

Sometimes referred to as Swarnapushpa (gold plant) or raktapushpa (blood red plant), Langali earns these colour-related epithets due to its remarkable ability to change hues quickly. When its flowers bloom, they exhibit a golden-yellow hue in the central area. Soon, however a new pinkish-red colour emerges from the edges of the petals, gradually spreading towards the centre.

Properties of Langali

Like many Upavisha herbs used in Āyurveda, Langali possesses bitter, pungent, and astringent tastes. It also exhibits a quality of lightness while promoting strength. Overall, Langali has a heating effect, which aids in reducing Vāta and Kapha Dosha. Despite its energetically heating nature, Langali also assists in reducing certain aspects of Pitta Dosha. Additionally, Langali has a strong affinity for blood and blood-related disorders.

Far left, Langali soaked in cow's urine; left, after soaking Langali, it is rinsed in water and dried

From a chemical perspective, this plant possesses potent anticoagulant properties. Its tuber and seeds contain various resins and starches, among which the alkaloid colchicine is the most significant. Colchicine, as an active ingredient, has a long history of use, including its application as a liniment for painful joints. Today, it is primarily used as an anti-inflammatory agent and recommended for the treatment of gout. Colchicine was first isolated in the early 1900s but continues to be of interest to the pharmaceutical industry due to its anti-parasitic properties and its potential for tumour shrinkage. It achieves this by binding to specific types of proteins, thus inhibiting metastasis. Another important constituent of Langali is gloriosine, which exhibits powerful analgesic properties.

Langali purification method

Ingredients

- 5–6 Langali (*Gloriosa superba*) tubers
- Fresh cow's urine
- Fresh cow's milk

Langali undergoes a purification process before it is used internally or externally. There are two commonly practiced methods of purification. The first method involves soaking the tubers in fresh cow's urine for twenty-four hours. For the second, the tubers are skinned, sliced, and boiled in cow's milk using the Dolā-yantra method. Whether they have been soaked or boiled, the tubers or slices are afterwards washed and left to sun-dry. Once fully dried, they are either powdered or stored as whole or sliced pieces.

Benefits of Langali

Langali, when consumed in small doses, stimulates digestion while also acting as a laxative. It has rejuvenating effects on the skin and was historically used in the treatment of leprosy. It proves effective as a sweating agent in cases of typhoid and malaria, aiding in the recovery of strength following fevers. Langali reduces joint inflammation, alleviates sores and abdominal pain, and is even utilised to help expel intestinal parasites.

Above, Langali is dried and sliced into sections; these are used whole in decoctions, or powdered

When the tubers are crushed into a paste and applied externally, they effectively reduce haemorrhoidal masses and closing anal fissures. The paste also provides relief from itching. In cases of sciatica, the paste can be applied to the lower back area. Like many poisonous plants in Āyurveda, Langali's toxicity can neutralise other toxins such as scorpion and insect bites.

One notable property of Langali lies in its affinity with the uterus and uterine blood. However, higher doses can induce foetal abortion, making it known as an effective abortifacient in India. On the other hand, Langali brings relief during labor when the pasted tuber is applied to the palms, soles, and lumbar region. It is also believed that grasping a peeled tuber by the hand during delivery aids in the expulsion of the placenta and umbilicus.

Although Langali is not a commonly used ingredient in Āyurvedic formulas, it can be found in Kalakuta Rasa, Nirgundhi Tailam, Chitrakadi Tailam, and Brihat Marichadi Tailam. The preferred medium or 'vehicle' for Langali is cow's milk or sesame oil, the latter being used for external applications.

In the event of an adverse reaction after administering Langali, the prescribed antidote is warm, sweetened cow's (or goat's) milk infused with ghee.

12.6/ Guggulu (resin)

'Disease obstructs him not, a curse attains him not, whom the agreeable odour of the healing Guggulu attains. From him, diseases scatter away, like antelopes from a wild beast. If, O Guggulu, thou art from the river, or if also from the ocean, the name of both have I taken, that this man may be uninjured.'

Hymn 38: Atharvaveda

Guggulu, derived from the Sanskrit term meaning 'having excellent benefits', holds significance in various formulations despite not being categorised as an alchemical ingredient. Additionally, it plays a vital role in purifying other ingredients of Rasa Shāstra. Found in the arid regions of India, Pakistan, Bangladesh, Arabia, and Africa, this small, thorny tree stands at a height of 1.5–2 meters. Its reddish-brown flowers, serrated waxy leaves, and rounded, pulpy fruits characterise its appearance.

Guggulu is an exudate or niryasa, obtained through the tapping method, This method involves stripping or piercing the tree's bark to enable the flow of its gum. *Commiphora mukul* contains numerous ducts rich in oleo-gum-resin within its bark. Large leaves or coconut shells are affixed around or just below the excision point to collect this exudate effectively. Over time, these receptacles gradually fill up and the gum/resin is subsequently collected.

*Above, Guggulu (*Commiphora mukul*) raw resin*

Upon retrieval, the material undergoes hardening and darkening. It is not uncommon for each collection to contain contaminants such as leaves, insects, and bark, necessitating a cleaning process known as the Shodhana process of Guggulu. Harvested typically between May and June, each tree yields approximately 250g to 500g of resin during the harvest season. Noteworthy chemical constituents of Guggulu, in terms of quantity, include gums, essential oils, sucrose, fructose, amino acids, flavonoids, α-camphorene, ellagic acid, sterols (such as guggulsterone, β-sitosterol, Z- and E-guggulsterone), carbohydrates, and various inorganic ions.

*Right, Guggulu (*Commiphora mukul*) being purified in a Triphala decoction; far right, final stages of purification. Guggulu is cooked until a smooth paste is formed*

Guggulu, an important resin in Āyurveda, is categorised based on its colour, which determines its applications. For example, Mhashya (blackish/brown) is considered a medicinal-grade variety, while Mahaneed (bluish) is of indeterminate use. Kumud (pale or whitish) finds application in alchemy, while Padma (brown) is more suitable for animal medicines. Lastly, Kanak (yellow) is regarded as the highest medicinal-grade for human patients. In the marketplace, two commonly found varieties of Guggulu are *Kana-guggulu* (yellow/green) and *Mhasha-guggulu* (brown/black).

Guggulu purification – method 1

Ingredients

500g Guggulu (*Commiphora mukul*) resin
Q.S. Triphala decoction
Q.S. Ghee (clarified butter)

1. The resin is carefully sifted by hand to remove any impurities such as twigs, leaves, soil, and insects that may have been collected during the harvesting process.
2. The Guggulu resin is placed inside a cloth and tied to form a rounded bolas. The bolas is then secured with cotton cord around its neck.
3. A Triphala decoction (prepared at an 8:1 ratio) is made, reduced to one third of its original volume, and subsequently filtered.
4. The next step involves the use of a Dolā-yantra, or swing device. This apparatus allows the Guggulu resin to be suspended just above the Triphala decoction and steamed.
5. The decoction is heated and the Guggulu resin is steamed for a duration of two to three hours.
6. A heavy wooden board is prepared, cleaned, and lightly coated with ghee. The contents of the cloth parcel are emptied onto this board and quickly sifted

by hand. This process removes any unwanted material, while the resinous material is mixed with ghee.

7 After sifting, the Guggulu resin is placed in a mortar and pounded until a homogeneous mass is achieved. To prevent sticking, the mortar is lightly coated with ghee. It is essential to work in a warm environment, as cooler temperatures can make the Guggulu resin harder to work with.

Note: This first method yields the largest quantity of end product. Subsequent methods produce smaller quantities, but the end material is more medicinally potent.

Guggulu purification – method 2

This second method closely resembles the first, with the difference being that the cloth parcel is immersed in Triphala decoction. After cooking for approximately two hours, the cloth is squeezed to extract all the resin before removing it, leaving behind the discarded mass. The decoction that now contains Guggulu *Sattva*, or essence, is retained. To retrieve the essence, the liquid is gradually reduced over a low heat until all the water evaporates, leaving only the resin behind. Towards the end of this process, the resin will start to cling to the surface of the pan, indicating the removal of water content. The final resin is collected from the pan, it is then mixed with ghee and thoroughly beaten to form a homogeneous lump.

Left, Guggulu resin fried in ghee

Please note that while Triphala decoction is commonly used in this second method, cows' urine or milk can also serve the same purpose. The resin is only partially soluble in water, so the clean-up process after processing can be messy and time-consuming.

Guggulu purification – additional procedure

Following either of these methods, Guggulu resin can be further processed into powder through an additional procedure. This involves frying the resin in ghee at a higher temperature, which transforms it into a solid mass that can be powdered after cooling.

1. Using a fairly high heat, Guggulu resin is fried in ghee. An iron Kadai (Indian cooking pot) is ideal for this purpose, heated with a small amount of ghee inside.
2. As the resin heats up, it will liquefy and may darken or start to stick to the pan. At this stage the resin needs to be worked with quickly; a broad wooden spatula is recommended for this job.
3. At this point, it is crucial to discontinue the heat or the resin may carbonate. Care should be taken to prevent overheating.
4. Once cooled, the cooked resin mass will harden. It can then be easily powdered using suitable methods.

Popular formulas containing Guggulu:

1. **Kanchanara Guggulu** – reduction of glandular swellings and skin diseases;
2. **Goksuradi Guggulu** – reduction of osteoarthritis and worn joints;
3. **Kaisora Guggulu** – reduction of gout and raised uric acid levels;
4. **Yogaraj Guggulu** – strengthens the muscular-skeletal system, reduces pain;
5. **Simhanada Guggulu** – reduction of swellings and joint pain (rheumatoid arthritis);
6. **Maha-Yogaraj Guggulu** – reduction of urinary stones, respiratory and digestive dysfunctions;
7. **Lakshadi Guggulu** – strengthening and replenishment of bone tissue.

Note: Guggulu resin has excellent longevity, with a twenty-plus-year shelf-life.

Benefits of Guggulu

Guggulu is a valuable substance in various therapeutic applications. Below are some of the many benefits associated with it.

1. **Anti-inflammatory and anti-arthritic properties** Guggulu exhibits excellent anti-inflammatory effects, making it beneficial in treating arthritis and reducing joint inflammation.
2. **Thyroid regulation** Guggulu is known to help regulate thyroid function, which is crucial for maintaining overall hormonal balance.
3. **Antioxidant activity** Guggulu is a potent antioxidant that protects the body against oxidative stress and promotes overall well-being. It also has cardio-protective properties, contributing to heart health.
4. **Cytotoxic properties** Guggulu has cytotoxic effects, which means it can inhibit the growth of specific cells, making it potentially helpful in managing certain types of cancers.
5. **Antimicrobial action** Guggulu exhibits antimicrobial properties, helping to combat various infectious microorganisms.
6. **Treatment of skin diseases** Guggulu effectively manages a range of complicated and challenging skin conditions, offering relief from symptoms and promoting skin health.
7. **Internal benefits** When taken internally, Guggulu has antiseptic, bitter, and astringent effects. It stimulates appetite, aids digestion, and provides a calming effect on the mind and body. Āyurveda commonly utilises Guggulu to address obesity, arthritis (both osteo and rheumatoid), paralysis, diabetes, sciatica, gout, haemorrhoids and constipation. It is also used for combating inflammation, cysts, coronary thrombosis, anaemia, and urinary calculus.
8. **External applications** When used externally in the form of a lotion or paste (Lepa), Guggulu has shown efficacy in treating stubborn skin conditions.
9. **Tincture or herbal wine** Guggulu can be prepared as either and proves beneficial for conditions such as sore throats, tonsillitis, gum infections, and toothache.
10. **Inhalation** Inhaling the fumes of Guggulu resin (similar to incense) is recommended for hay fever, nasal catarrh, bronchitis, and laryngitis.

Contraindications of Guggulu

Āyurveda advises that freshly harvested gum/resin can potentially worsen skin conditions, disrupt menstruation and fertility, induce nausea, trigger headaches and vertigo, and cause dryness. In certain cases, it may even lead to diarrhoea and liver dysfunction. Due to these potential risks, it is recommended to subject fresh resin (less than twelve months old) to some level of purification. Patients who experience any of the aforementioned conditions should avoid its use.

Based on my experience, mature batches of Guggulu (over two years old) may not necessarily require Shodhana (purification). However, since this process generally enhances the potency and ensures the purity of the medicine, it is advisable to perform Shodhana regardless of the resin's age.

Special use: Mitra-Pañchaka

Guggulu plays a pivotal role in the formulation of Mitra-pañchaka, a significant blend utilised to extract the *Sattva*, or essence, from specific preparations. This intriguing combination serves as a means to extract impurities from minerals, metals, and animal products. Among the materials involved are copper and iron sulphate, mica and Shilajit, and peacock feathers, to name a few.

Along with Guggulu, Mitra-Pañchaka consists of borax, a natural fluxing agent, honey, purified powder of Gunja (*Abrus precatorius*), and jaggery (coconut sugar). To these, ghee is then added to form a pliable paste. Note: in some instances, ghee becomes the fifth ingredient with borax being removed from the list.

Equal proportions of these components are skilfully beaten together, forming a waxy pulp that is then blended with the aforementioned materials. Following this, the final mixture is subjected to a high-temperature, reaching approximately +1000°c, while enclosed within a sealed crucible. Under such

Right, ingredients for Mitra-panchaka; far right, resin of Mitra-panchaka when mixed.

intense heat, an otherwise unattainable essence, known as *Sattva*, emerges. For instance, in the case of copper, the resulting essence embodies the properties of copper, making it highly assailable by the body. Similarly, peacock feathers yield an essence resembling copper.

Section 13

A living tradition of herbo-mineral-metalic medicines

14.1/ Rasa formulae

> 'Visha destroys life, is Vyavāyī (spreads quickly all over the body), Vikāśī (causes loss of Ojas and looseness of joints), Āgneya (acts like fire), mitigates Vāta and Kapha, Yogavāhi (increases the qualities and potency of a substance with which it is combined), and Madāvaham (causes profound metal disorders).'
>
> *Bhāvaprakāśa Pūrvakhanda, Bhāvamiśra*

Disclaimer: The formulations discussed in this subsection are presented solely for academic interest. They are not intended to be replicated, either partially or entirely.

The following remedies are integral to a living tradition of herbo-mineral medicines that remain actively practiced in Asia and the Middle East, within both the Āyurvedic and Unani Tibb traditions. These medicinal practices have been passed down through generations and continue to play a significant role in healthcare within these regions.

Within this subsection, we delve into how various individual materials have been integrated into traditionally prepared formulas, some of which have been discussed in previous sections of Part II. Each material within a formula possesses unique properties, which often require balancing or counteracting based on their individual strengths or synergistic effects. When used together, certain materials like Sangeyasab (jade) and Akika (agate) mutually support each other, as do Shilajit (bitumen) and Guggulu (*Commiphora mukul*). Conversely, there are instances where certain materials, such as Gandhaka (sulphur) and Loha (iron), tend to counteract each other's effects. However, this counteraction can

Far left, herbal ingredients of medicated ghee. Ingredients include: Ashwagadha, Shatavari, Amalaki and Kapikachhu

be mitigated by first preparing Kajjali (black sulphide of mercury) as a carrier medium, allowing for a more harmonious interaction between these substances. It is essential to recognise that both selecting and combining materials in traditional formulas is based on their specific properties and desired therapeutic effects. The careful balance and synergy among these materials contributes to the overall efficacy of the formulations in various traditional medicinal practices.

In the realm of Rasa formulation, each material possesses its own general indications and properties. However, specific formulae were developed to target and treat specific conditions.[175] These formulations provide an interesting opportunity to examine and understand the individual energetics of different ingredients, thereby fostering familiarity with herbo-metallic preparations.

It's important to note that this subsection only scratches the surface of Rasa formulation, as there exist hundreds, if not thousands, of formulae with their own cultural variations, practitioner preferences, availability, and material grades.[176] From this vast diversity, I have chosen a selection of formulae that not only incorporate each sample material but also play a crucial role in enhancing their therapeutic value.

Please be aware that all Rasa materials indicated with an asterisk (*) would have undergone purification before being included in any formula, however, it is worth noting that not all purification methods for these have been detailed in this particular book.

For further information on the individual practices of these purification techniques, I refer readers to my previous works, particularly *Rasa Shāstra: The Art of Vedic Alchemy, volumes 1–4.*

175 Sometimes a formula designed for a specific complaint will be used to treat an entirely different ailment. This practice is specific to different lineages of Vaidya/ healers who have developed a special relationship with a number of medicines and are able to augment the remedy via Anupāna.

176 Rasa materials have a number of qualitative tests that are to be applied before use; these include reaction to heat, colour, taste, feel and general appearance, i.e. looking for shiny particles or contaminants.

Formula 1 – Hiṅgula (cinnabar)			
Formula: Hiṅguleśwara Rasa			
Qty	**Sanskrit**	**English**	**Latin**
1 part	Hiṅgula*	Cinnabar (mercuric sulphide)	
1 part	Vatsanābha*	Aconite (root)	*Aconitum napellus*
1 part	Pippali	Long pepper (fruits)	*Piper longum*
Method	All ingredients are ground together and bound with a little acacia gum pasted in water. Small pills are individually rolled, then left to dry in sunlight.		
Anupāna	Honey		
Dosage	125mg		
Uses	Vāta Jabra, a Vāta type of fever that accumulates between the seasons of spring and autumn.		
Signs	Erratic symptoms, general fatigue, dryness, stiffness of limbs, feeling cold, constipation and dry stools. Dry skin and darkness about the eyes.		
Hiṅgula	VPK, improves digestion, Rasāyana, reduces fever, and destroys āma Vāta.[177]		
Vatsanābha	Balances VPK, improves digestion, heating, reduces fever.		

Formula 2 – Gandhaka (sulphur)			
Formula: Gandhaka Rasāyana			
Qty	**Sanskrit**	**English**	**Latin**
1 part	Gandhaka	Sulphur (element)	
1 part	Jaggery	Palm sugar	

177 Āma Vāta indicates distributed toxins in the body, due to poor digestion and an excess of Vāta dosha.

Bhavana materials (×7 each)			
1 part	Godugdha	Milk	
1 part	Nāgakesara	Cobras saffron	*Mesua ferrea*
1 part	Ela	Cardamom	*Elettaria cardamomum*
1 part	Ardraka	Ginger (juice)	*Zingiber officinale*
1 part	Tamalpatra	Bay leaf (juice)	*Cinnamomum tejpata*
1 part	Āmalakī Harītakī Bibhītaki	Triphala	*Emblica officinalis* *Terminalia chebula* *Terminalia belerica*
1 part	Bhringaraj	False daisy	*Eclipta alba*
1 part	Tvak	Cinnamon bark	*Cinnamomum zeylanicum*
Method	Sulphur is subjected to Bhavana, first with milk infused with saffron and cardamom, then ginger juice and fresh bay leaves, followed by a decoction of Triphala, Bhringaraj, Sunthi and Tvak. The resulting Gandhaka is then mixed with an equal quantity of jaggery, rolled into pills and dried under sunlight.		
Anupāna	Milk		
Dosage	125mg		
Uses	Toxic blood, skin diseases, high āma, poor digestion and assimilation; it reduces arthritic conditions and Pitta.		
Sulphur	Improves digestion, burns up toxics and bonds with free radicals. Sulphur is an excellent Yoga vāhin, has aphrodisiac properties, reduces Vāta and Kapha, and is a general Rasāyana.		

Formula 3 – Shilajit (bitumen)			
Formula: Chandraprabha Vati (per 250mg tablet)			
Qty	**Sanskrit**	**English**	**Latin**
57.6mg	Shilajit*	Bitumen	
57.6mg	Guggulu*	Indian myrrh	*Commiphora mukul*
28.8mg	Kalkandu	Rock candy	

14.4mg	Loha Bhasma*	Iron (oxide)	
7.2mg	Tabasheer	Bamboo manna	*Bambusa arundinacea*
7.2mg	Nāgadandi	Wild muell (root)	*Baliospermum montanum*
7.2mg	Trivruta	Morning glory (root)	*Operculina turpethum*
7.2mg	Tamalpatra	Indian bay leaf	*Cinnamomum tejpata*
7.2mg	Ela	Cardamom (fruit)	*Elettaria cardamomum*
7.2mg	Tvak	Cinnamon (bark)	*Cinnamomum zeylanicum*
5.4mg	Āmalakī Harītakī Bibhītaki	Triphala (churna)	*Emblica officinalis* *Terminalia chebula* *Terminalia belerica*
1.8mg	Sarji Ksara	Sodium carbonate	
1.8mg	Dhanyaka	Coriander (seeds)	*Coriandrum sativum*
1.8mg	Daruharidra	Indian burberry (stem)	*Berberis aristata*
1.8mg	Kalanamak	Black salt	
1.8mg	Swarna Maksika*	Chalcopyrite (copper sulphide)	
1.8mg	Yavaksara	Potassium carbonate	
1.8mg	Saindhava Lavaṇa	Rock salt	
1.8mg	Haridra	Turmeric (rhizome)	*Curcuma longa*
1.8mg	Bhunimba	Kirata (whole plant)	*Swertia chirata*
1.8mg	Kachura	Zedoary (rhizome)	*Curcuma zedoaria*
1.8mg	Chitraka	Ceylon leadwort (root)	*Plumbago zeylanica*
1.8mg	Chavya	Thai long pepper (fruit)	*Piper chaba*
1.8mg	Mustaka	Nutgrass (rhizome)	*Cyperus rotundus*
1.8mg	Ativisha	Aconite (root)	*Aconitum heterphyllum*
1.8mg	Gajapippali	Large long pepper (fruit)	*Scindapsus officinalis*
1.8mg	Devadaru	Cedar wood (bark)	*Cedrus devdara*
1.8mg	Pippali	Long pepper (fruit)	*Piper longum*
1.8mg	Pippalimula	Long pepper (root)	*Piper longum*

1.8mg	Vaccha	Sweet flag (rhizome)	*Acorus calamus*
1.8mg	Maricha	Black pepper (fruit)	*Piper nigrum*
1.8mg	Shunthi	Dried ginger (rhizome)	*Zingiber officinale*
1.8mg	Vidanga	Embilia (fruits)	*Ebelia ribes*
Method	Plant materials are first powdered, and decoctions prepared as required. The final mass is ground and kneaded into pills until they can be rolled without sticking to the hand. The rolled tablets are then placed to dry in a shady but warm area.		
Anupāna	Warm water or milk		
Dosage	250mg		
Uses	Diabetes, obesity, anaemia, piles, skin diseases and eye disorders, poor digestion/ assimilation, and kidney stones. Chandraprabha is also an excellent Rasāyana of the urinary tract.		
Shilajit	A general Rasāyana, but working specifically on the kidneys and urinary system. It is beneficial in cases of diabetes, gallstones, kidney stones, anaemia, tuberculosis, oedema, gout and heart disease.		
Guggulu	A general Rasāyana, it reduces and lightens the body, removes arthritic pain, gout, scrapes āma from the joints, reduces swelling and excess mucus, and is useful in cases of diabetes.		
Loha Bhasma	Blood Rasāyana, reduces free radicals and improves vision. It treats anaemia, consumption, worms, obesity, oedema, liver disease, diabetes, lipoma, and piles.		
Swarna Maksika Bhasma	Cancer, anaemia, skin disease, piles, tuberculosis, loss of taste, digestive disorders, abdominal pain, Rasāyana, promotes bone marrow, Yoga vāhin.		

Formula 4 – Rajavarta (lapis lazuli)

Formula: Rajavarta Rasa			
Qty	**Sanskrit**	**English**	**Latin**
1 part	Rajavarta*	Lapis lazuli	
1 part	Rasa Sindoora*	Mercuric sulphide	
1 part	Tamra Bhasma*	Copper oxide	
1 part	Yastimadhu	Liquorice	*Glycyrrhiza glabra*

Method	Rajavarta, Rasa Sindoora, Tamra Bhasma and Yastimadhu are ground together and lightly roasted in ghee in an iron frying pan. The paste formed is rolled into a pill the size of a Raktika seed.
Anupāna	Honey, ghee or jaggery
Dosage	125mg
Uses	The removal of toxins, restoration of tissues, breaking of addictions (such as alcoholism) and cleansing of thoughts.
Rajavarta	A brain tonic, promotes digestion, heals diabetes, urinary disorders, and tuberculosis. It is useful in cases of asthma, and controls vomiting.
Rasa Sindoora	An aphrodisiac, it promotes intelligence, reduces the effects of ageing, and is a general tonic.
Tamra Bhasma	Ascites, anaemia, piles, skin disease, bronchitis, asthma, tuberculosis, chronic rhinitis, gastritis, colic pain, oedema, liver disease and goitre.

Formula 5 – Haritāla (arsenic trisulphide)			
Formula: Rasa Maanikya			
Qty	**Sanskrit**	**English**	**Latin**
1 part	Haritāla*	Arsenic trisulphide	
Method	Pieces of purified Haritāla are sandwiched between thin sheets of Abhraka (mica) and heated in a medium flame until their sulphur content is liquefied. Once cooled, the red crystallised content can be removed and ground. The word *maanikya* implies its similarity to ruby in appearance. In some cases, it is recommended that it be mixed in equal quantity with Harītakī (*Terminalia chebula*).		
Anupāna	Honey or ghee		
Dosage	125–250mg		
Uses	Chronic skin diseases; it also reduces high fever, gout and chronic bronchitis. Used in the treatment of syphilis, fistula and sinusitis. Rasa Maanikya has anti-cancer properties; it is also a brain tonic.		
Haritāla	Fevers, haemophilia, syphilis, gout, fistula, skin disease, fever and urticaria, diabetes, facial paralysis, rheumatism and arthritis.		

Formula 6 – Nāgapashana (serpentine)			
Formula: Jawahar Mohra Bhasma			
Qty	**Sanskrit**	**English**	**Latin**
1 part	Nāgapashana*	Magnesium silicate	
1 part	Kutki	Hellebore (rhizome)	*Picororhiza kurrao*
1 part	Arka*	Crown flower (latex)	*Calotropis gigantea*
1 part	Arjuna	Arjun tree (Kwatha)	*Terminalia arjuna*
Method	Once purified, Nāgapashana is ground with Kutki juice, Arka latex or Arjuna decoction. Cakrika are prepared and calcined. This process is repeated seven times until a fine reddish Bhasma is achieved.		
Anupāna	Apple juice		
Dosage	125–250mg		
Uses	Cardiac and liver tonic. It is also an excellent Rasāyana for the heart and has good anti-Visha properties.		
Nāgapashana	A Rasāyana for heart and brain, liver tonic, and antidote for Visha/snake bite/poisons. It reduces vomiting, and promotes happiness and the strength of bodily organs.		
Arka	Anti-inflammatory. It is also a spermicidal agent used in treating bronchial asthma, lesions, sores, inner ear infections, ulcers, abdominal diseases, and digestive parasites. A general rejuvenative.		

Formula 7 – Tamra (copper)			
Formula: Tamra Parpati			
Qty	**Sanskrit**	**English**	**Latin**
3 parts	Tamra Bhasma*	Copper (oxide)	
9 parts	Kajjali*	Mercuric sulphide	
Added equal quantity			
1 part	Vatsanābha*	Aconite (root)	*Aconitum napellus*

Method	Kajjali is prepared from Pārada and Gandhaka, and ground with Tamra Bhasma. The ingredients are heated in an iron ladle over a mild flame until melted. The resulting contents are then poured onto a flattened banana leaf and pressed flat. This method of Rasa manufacture is known as *Parpati* (see Part II). Having ground the Parpati, an equal quantity of Vatsanābha is added and mixed. To consume, add a small amount of honey to the final powder, it can then be formed into a paste with one's finger and licked.
Anupāna	Varies according to condition
Dosage	125–250mg
Uses	Chronic colitis (taken with Triphala and honey), urinary pain and anaemia (taken with castor oil), skin conditions including leucoderma and fungal infections (taken with powdered Vaccha (*Acorus calamus*)).
Individual ingredients	These have been previously covered. See Section 9.2/ Copper

Formula 8 – Vanga (tin)

Formula: Madukadi Curna

Qty	Sanskrit	English	Latin
125mg	Vanga Bhasma*	Tin oxide	
2g	Yastimadhu	Liquorice (Churna)	*Glycyrrhiza glabra*
2g	Haridra	Turmeric (rhizome)	*Curcuma longa*
Method	The ingredients are ground together and mixed with the juice of pulped Arka leaves (*Calotropis procera*) or Kumārī (*Aloe vera*) gel.		
Anupāna	Milk		
Dosage	As above		
Uses	Severe anaemia, urinary tract infections, vertigo, and tinnitus.		
Vanga	Paralysis, urinary tract infections, premature ejaculation, consumption, worms, diabetes, excessive sweating, dermatitis, inner ear infections, vertigo, and tinnitus.		

Formula 9 – Yasada (zinc)			
Formula: Kushta Musallus (Unani)			
Qty	**Sanskrit**	**English**	**Latin**
10g	Yasada*	Zinc oxide	
10g	Vanga*	Tin oxide	
10g	Nāga*	Lead oxide	
250g	Khasabeeja*	Opium poppy	*Papaver somniferum*
Method	Equal quantities of the three metals are heated at high temperature until liquid, which is then poured into an iron container filled with mustard seed oil. This process is repeated seven times. The resulting mixture is then heated again in an iron pan until liquid, with dried khasabeeja (opium poppy) heads added a little at a time until the metallic mass is mixed in amongst the ash from the dried plant material. Heating is discontinued and the ash stirred until all the metal is absorbed. The resulting ash is washed and dried. Yoghurt is then added to the ash and pasted enough to form cakrika. After drying, the cakrika are sealed into a Sharaava and calcined. Upon cooling, the cakrika are removed and powdered. Yoghurt is again added and new cakrika again formed and dried. These are then sealed into a Sharaava and again calcined. This process is repeated five to seven times or until a whitish-yellow Bhasma is formed.		
Anupāna	Butter or *Majun Arad Khurma*;[178] see the ingredients of this formula below:		
Formula: Majun Aarad Khurma (10g serving approx)			
Qty	**Unani**	**English**	**Latin**
0.5 kg	Khurma	Date palm (dried fruit)	*Pheonix dactylifera*
0.5 kg	Samagh Arabi	Acacia gum	*Acacia arabica*
0.5 kg	Singhara Khushk	Buffalo nut (dried fruit)	*Trapa bispinosa*
50g	Maghz Badam Sheereen	Almond (fruit kernel)	*Prunus amygdalus*
50g	Maghz Chilghoza	Chilghoza pine (seed kernel)	*Pinus gerardiana*

178 A medicine for the treatment of low sperm count, specific to Unani Tibb.

50g	Maghz Funduq	Hazelnut (seed kernel)	*Corylus avellana*
10g	Maghz Pamba Dana	Levant cotton (seed kernel)	*Gossypium herbaceum*
6g	Qaranfal	Cloves (dried bud)	*Syzygium aromaticum*
3g	Javitry	Nutmeg (fruit shell)	*Myristica fragrance*
3g	Jaifal	Nutmeg (fruit)	*Myristica fragrance*
250mg	Turanjabeen	Salt cedar gum	*Tamarix indica*
250mg	Asl	Honey	
Dosage	60–125mg		
Uses	Sexual debilitation, premature ejaculation, low sperm count, nocturnal emissions. Zinc increases Ojas.		
Yasada	Eye diseases, diabetes, anaemia, bronchial asthma, all types of skin disease, throat infections, chronic wounds, tonsillitis, fistula and stomatitis. Zinc also reduces Pitta and Kapha types of diseases.		
Nāga	Menorrhagia, piles, diabetes, urinary infections and seminal debility.		
Khasabeeja	Calms the nervous system, analgesic, reduces diarrhoea and restlessness, and improves sleep.		

Formula 10 – Sphatika (quartz)

Note: Clear quartz can be substituted for diamond; however, the potency is significantly reduced. Chemically, quartz is ($SiO4$) and diamond is (C).

Formula: Bhutankusa Rasa

Qty	Sanskrit	English	Latin
⅓ part	Hiraka Bhasma*	Diamond	
1 part	Kajjali*	Mercuric sulphide	
1 part	Loha Bhasma*	Iron oxide	
1 part	Abhraka Bhasma*	Mica	
1 part	Tamra Bhasma*	Copper oxide	
1 part	Mukta Bhasma*	Pearl	
1 part	Manah śhilā*	Arsenic disulphate	

1 part	Haritāla*	Arsenic trisulphide	
1 part	Sasyaka*	Copper sulphate	
1 part	Rasanjana*	Antimony sulphide	
1 part	Samudra Phena*	Cuttle fish bone	
1 part	Shilajit*	Bitumen	
1 part	Souviranjana*	Antimony sulphide	
1 part	Saindhava Lavaṇa	Rock salt	
Bhavana materials (x7 each)			
Qty	**Sanskrit**	**English**	**Latin**
Q.S.	Bhringaraj	False daisy (juice)	*Eclipta alba*
Q.S.	Chitraka	Ceylon leadwort (juice)	*Plumbago zeylanica*
Q.S.	Snuhi	Milk hedge (latex)	*Euphorbia ligularis*
Method	All materials are given Bhavana in the juice of Bhringaraj, Chitraka and Snuhi latex. From this mixture a balled mass is prepared and placed into a crucible and calcined. The resultant material is ground into a fine powder for medicinal usage.		
Anupāna	Ginger juice		
Dosage	125–250mg		
Uses	Mental seizures or madness (attributed to restless earthbound spirits).		
Hiraka	Increases longevity, destroys disease, improves digestion, removes diabetes, fatigue and wasting. Quartz cures anaemia, vertigo and tuberculosis, it reduces inflammation and improves overall immunity. Hiraka reduces all three Dosha.		
Kajjali	Increases longevity and has a powerful Rasāyana effect.		
Abhraka	Relieves cough, asthma, anaemia, tuberculosis, fever and all types of diabetes. It is useful in cases of gastritis, piles, heart disease, vertigo, colic pain and urticaria.		
Mukta	Controls fever, promotes healthy bones, heals tuberculosis, asthma, bronchitis and heart disease. It promotes vision, improves digestion, cures respiratory diseases and is a general Rasāyana.		
Manah śhilā	Cures chronic bronchitis and reduces chronic fever. It is also useful for skin diseases, itching, anaemia, tuberculosis and worms.		
Sasyaka	Is useful in all cases of eye diseases, poisoning and skin disease, especially leucoderma. It is also good for parasitic infestation (gastric worms) and ulcers.		

Samudra Phena	Heals skin, reduces mineral deficiency in silica, iron and phosphoric acid, and heals earache.
Souviranjana	Cures myopia, hypermetropia, cataracts, conjunctivitis, glaucoma, ulcers, bleeding disorders, menorrhagia, excessive hiccups and poisoning.

Formula 11 – Akika (agate)			
Formula: Akika Bhasma (agate, jade and pearl)			
Qty	**Sanskrit**	**English**	**Latin**
1 part	Akika	Agate Bhasma	
1 part	Sangeyasab Bhasma	Jade Bhasma	
1 part	Mukta Pisti	Pearl Pisti	
Method	Akika and Sangeyasab are ground together and mixed with a small amount of apple juice, made into Sharaava and calcined in a sealed crucible. The cakrika are then retrieved and re-ground with Mukta Pisti. The ground material is then mixed with a little apple juice and made into pills.		
Anupāna	Jaggery		
Dosage	60–125mg		
Uses	A heart and intestinal tonic with anti-consumptive properties. A powerful Rasāyana for the eyes. Gives strength to the jaw and teeth.		
Akika	Reduces Pitta and is an excellent heart tonic. Stops bleeding, removes urinary calculi, and prevents haemorrhages. Akika promotes good vision and strong teeth.		
Sangeyasab	Heart disease, colic pain, urinary disorders, urinary calculus, gastric pain and dysentery. Jade is a general Rasāyana for the reproductive organs.		

Formula 12 – Nilama (blue sapphire)

Notably, this formula incorporates all nine major gemstones, each being represented by one of the nine planets. Such formulations would have required a significant accumulation of ingredient stocks over time, not to mention large amounts of time to prepare.

Although this type of formulation can be costly to produce, there are still certain manufacturers who undertake the task of creating it using traditional methods.

Ingredients: all ingredients, except deer musk, camphor and saffron are initially mixed together.

Formula: Nava Ratna Raja Mrganka Rasa

Qty	Sanskrit	English	Graha (planet)
1 part	Vaiduryam Bhasma*	Cat's eye oxide	Ketu (southern node)
2 parts	Kajjali*	Mercuric sulphide	
1 part	Swarna Bhasma*	Gold oxide	
1 part	Rajata Bhasma*	Silver oxide	
1 part	Rasaka Bhasma*	Zinc oxide	
1 part	Tarksya Bhasma*	Emerald oxide	Budha (Mercury)
1 part	Hiraka Bhasma*	Diamond oxide	Shukra (Venus)
1 part	Pushparaga Bhasma*	Topaz oxide	Brihaspati (Jupiter)
1 part	Nilama Bhasma*	Sapphire oxide	Shani (Saturn)
1 part	Maanikya Bhasma*	Ruby oxide	Surya (the Sun)
1 part	Chandrakanta*	Moonstone oxide	Chandra (the Moon)
1 part	Pravala Pisti*	Red coral	Kuja (Mars)
1 part	Gomeda Bhasma*	Hessonite oxide	Rāhu (northern node)
1 part	Kanta Pashana*	Magnetic iron ore oxide	
1 part	Vanga Bhasma*	Tin oxide	
1 part	Nāga Bhasma*	Lead oxide	
1 part	Vimala Bhasma*	Iron pyrite oxide	
1 part	Shankha Bhasma*	Conch shell oxide	
1 part	Tamra Bhasma*	Copper oxide	
1 part	Sukti Bhasma*	Mother-of-pearl oxide	

1 part	Haritāla*	Arsenic trisulphide	
1 part	Abhraka Bhasma*	Mica	
1 part	Hiṅgula*	Mercuric sulphide	
1 part	Manah śhilā*	Arsenic disulphate	
1 part	Kapura	Camphora	
1 part	Kastūrī	Deer musk	
1 part	Nāgakesara	Cobra saffron (Mesua ferrea)	
Method – Part 1	This process involves giving the materials seven rounds of Bhavana, incorporating all the listed ingredients. Following the final Bhavana, the mixture is shaped into small pills and dried thoroughly.		
Bhavana materials (×7 each)			
Qty	**Sanskrit**	**English**	**Latin**
Q.S.	Goksura	Small caltrops juice	*Tribulus terrestris*
"	Nāgavalli	Betel leaf juice	*Piper betel Linn*
"	Vasā	Malabar nut juice	*Adhatoda beddomei*
"	Mundi	Globe thistle juice	*Sphaeranthus indicus*
"	Pippali	Long pepper decoction	*Piper longum*
"	Chitraka	Ceylon leadwort decoction	*Plumbago zeylanica*
"	Iksu	Sugar cane stem	*Saccharum officinarum Linn*
"	Guduchi	Tinospora juice	*Tinospora cordifolia*
"	Dhattura	Dhattura juice	*Datura stromonium/metel*
"	Agnimantha	Premna juice	*Premna mucronata*
"	Draksa	Grape decoction	*Vitis vinifera*
"	Shatāvari	Wild asparagus decoction	*Asparagus racemosus*
"	Punarnava	Mountain ebony decoction	*Boerhavia diffusa*
"	Shatapatrika	Rose water	*Rosa damascena*
"	Yastimadhu	Liquorice decoction	*Glycyrrhiza glabra*
"	Salmali	Cotton tree decoction	*Bombax malabaricum*
"	Dhataki	Fire-flame bush decoction	*Woodfordia fruticosa*
"	Jatisasya	Nutmeg decoction	*Myristica fragrans*

"	Bala	Country mallow decoction	*Sida cordifolia*
"	Nāgabalā	Arrowleaf decoction	*Sida spinosa*
"	Atibala	Indian mallow decoction	*Abutilon indicum*
"	Hribera	Iruvale decoction	*Coleus zeylanicus*
"	Tvak	Cinnamon bark decoction	*Cinnamomum zeylanicum*
"	Devapuspa	Clove decoction	*Syzygium aromaticum*
"	Kankola	Tailed pepper decoction	*Piper cubeb*

Method – Part 2	The dried pills are then placed inside a kūpī and subjected to heat in a sand bath for approximately twelve hours. Once the process is complete, the kūpī is allowed to cool down, and opened. The pills are retrieved from the bottom of the jar and subjected to another round of Bhavana, incorporating the same set of herbs, and again repeated seven times. Once this Bhavana is concluded, the mixture is combined with deer musk, camphor, and saffron, and everything is ground together. The resulting blend is subsequently dried and carefully stored in a glass jar for preservation.
Anupāna	Honey, ghee, butter, each taken with a little ground Pippali.
Dosage	125–250mg
Uses	As a powerful Rasāyana with aphrodisiac properties. It is also helpful in cases of consumption, cough, and anorexia, as well as Vātarakta (wind in the blood), anaemia, epilepsy and abdominal disorders (gas and distension). This medicine aids in restoring digestion and increasing assimilation.
Indications for individual ingredients	
Nilama Bhasma	Arthritis, brain dysfunction, improves Agni, Vāta and Pitta diseases, tuberculosis, infertility, skin diseases, consumption, poisons and general debility.
Vaiduryam Bhasma	Anaemia, tuberculosis, poor appetite, eye diseases and bleeding disorders, psychic and paranormal disturbances.
Swarna Bhasma	A Rasāyana for the immune system, it also treats cancer, anaemia, skin disease, piles, tuberculosis, loss of taste, digestive disorders, and abdominal pain. Promotes bone marrow.
Rajata Bhasma	Useful for consumption, diabetes, anaemia, blood disorders and piles. It helps promote digestion, and heals urinary disorders and other ailments such as tuberculosis, asthma, vomiting and high fever.

Tarksya Bhasma	Reduces high fever, vomiting (it has excellent anti-Visha properties), asthma, piles, anaemia, oedema, immune dysfunction and diseases of the central nervous system.
Kanta Pashana	Anaemia, consumption, obesity, parasitic infestation and oedema. Cures diseases of the liver and spleen while strengthening the heart and nervous system. Useful in conditions such as nephritis, leucorrhoea and diabetes.
Pravala Pisti	Reduces chronic fever, tuberculosis, bronchitis, bleeding disorders and excessive sweating. It is also useful in cases of abdominal bloating and gastritis, and hair and tooth loss.
Vimala Bhasma	Anaemia, skin diseases, piles, tuberculosis, loss of taste, digestive disorders, abdominal pain and a Rasāyana for bone marrow. It is also a good Yoga vāhin.
Maanikya Bhasma	A heart tonic and aphrodisiac, it promotes circulation but controls bleeding and haemorrhaging. Maanikya increases Agni (digestive power) and removes consumption.
Chandrakanta	It heals duodenal ulcer and heart palpitations; it also reduces high fever, nervous disorders, difficult menstruation and burning sensations.
Pushparaga Bhasma	Destroys skin diseases, and helps with impaired digestion and poor appetite. It neutralises poisoning and general weakness in the limbs.
Shankha Bhasma	Removes indigestion, diarrhoea, gastritis and duodenal ulcer. It heals eye problems, lung diseases, peptic ulcers, ear discharges and IBS (irritable bowel syndrome).
Sukti Bhasma	Useful in cases of colic, urinary stones, asthma and heart disease. Sukti is a digestive stimulant and useful in the treatment of diseases affecting the spleen.

Formula 13 – Mukti (pearl)

Formula: Kanchanadi Rasa

Qty	Sanskrit	English	Latin
1 part	Mukti*	Pearl oxide	
1 part	Swarna*	Gold	
1 part	Rajata*	Silver	
1 part	Rasa Sindoora*	Mercuric sulphide	
1 part	Loha*	Iron oxide	

1 part	Abhraka*	Mica	
1 part	Pravala*	Red coral	
1 part	Manah śhilā*	Realgar (arsenic disulphide)	
1 part	Mrga Made	Musk	
1 part	Harītakī	Yellow myrobalans	*Terminalia chebula*
Method	All ingredients are mixed together and made into pills, using acacia gum and water as binding agents. Once formed, the pills are dried under sunlight.		
Anupāna	Honey		
Dosage	250mg		
Uses	Cures consumption, increases Ojas and strengthens immune functioning.		

Formula 14 – Mrga Śrnga (deer horn)

This particular method of purifying Mrga Śrnga (deer horn) results in the production of a white Bhasma, rather than the typical dark brown colour. The white Bhasma derived from deer horn shares similarities with its dark brown counterpart, but it is widely acknowledged to possess heightened therapeutic effectiveness, particularly in the treatment of heart conditions.

Formula: Mrga Śrnga white Bhasma

Qty	Sanskrit	English	Latin
1 part	Mrga Śrnga*	Deer horn	
1 part	Agastya	Hummingbird tree juice	*Sesbania grandiflora*
1 part	Godugdha	Milk	
Method to prepare the whitish-grey Bhasma from Mrga Śrnga (deer horn)	1 Short lengths of Mrga Śrnga are cut and split into thin slivers. 2 These slivers are soaked in the juice extracted from Agastya leaves (*Sesbania grandiflora*) for three days. 3 After soaking, the slivers are removed, dried, and then placed into a Sharaava. 4 The slivers are subjected to calcination, where they are exposed to high temperatures until they become ash. 5 Once the ash has cooled down, the slivers are removed and ground into a fine powder.		

	6 The powder is then triturated with milk and cakrika is prepared. 7 The mixture is calcined again; the procedure is repeated a total of three times. 8 After the third repetition, the deer horn will have been transformed into a whitish Bhasma, ready for use.
Anupāna	Butter, milk or ghee
Dosage	250mg
Uses	An excellent heart tonic. It alleviates *Angina pectoris*, pleurisy, and pain in the sides of the chest. Helps with sinus problems, migraine, cough and chronic hiccups. This Bhasma helps restore hair loss.

Formula 15 – Mayūr piccha (peacock feather)

Formula: Maheshvara Dhupa

Qty	Sanskrit	English	Latin
1 part	Mayūr Piccha*	Peacock feather	
1 part	Devadaru	Cedar resin	*Cedrus deodara*
1 part	Musta	Nutgrass juice	*Cyperus rotundus*
1 part	Kutki	Hellebore juice	*Picrorhiza kurroa*
1 part	Nāgakesara	Cobra saffron stamen	*Mesua ferrea*
1 part	Sarpkankoli	Snake skin	
1 part	Goshrnga	Cow's horn	
1 part	Kapas	Cotton seeds	*Gossypium indicum*
1 part	Joye Kasattu	Barley seeds	*Hordeum vulgare*
1 part	Dhani/Dhany	Paddy husk	
1 part	Rajeeka	Indian mustard seeds	*Brassica juncea*
1 part	Neem	Margosa juice	*Azadirachta indica*
1 part	Gardabha Mutra	Donkey urine	
1 part	Aja mala	Goat stool	
1 part	Ghee	Clarified butter	

Method	All of the above ingredients are mixed and given Bhavana in the cow's urine. The mixture is allowed to dry and burnt, like incense, on an open flame and the fumes inhaled.
Uses	The treatment of insanity, madness and possession.
Mayūr Piccha	An excellent anti-Visha material; it reduces Vāta and Kapha. It is helpful in cases of chronic bronchitis and hiccups, chronic colitis, asthma and other breathing difficulties. Generally this material aids in the strengthening of the lungs.

Formula 16 – Vatsanābha (aconite)

Formula: Jvara Mrityunjaya Rasa

Qty	Sanskrit	English	Latin
1 part	Vatsanābha*	Aconite tuber	*Aconitum napellus*
2 parts	Kajjali*	Mercuric sulphide	
1 part	Pippali	Long pepper	*Piper longum*
1 part	Maricha	Black pepper	*Piper nigrum*
1 part	Tankana*	Borax	
1 part	Ardraka	Ginger juice	*Zingiber officinale*

Method	All materials are ground well and mixed with the juice of fresh ginger rhizome. The paste formed from this mixture is rolled into pills and dried under sunlight.
Anupāna	Ginger juice, buttermilk, coconut water or honey.
Dosage	250mg
Uses	For the treatment of chronic fever, related to imbalanced Dosha.
Vatsanābha	Useful in cases of rheumatism and heart disease, gout, asthma, bronchitis and piles. It reduces high fever, improves digestion and rejuvenates the body.
Tankana	Beneficial for nervous disorders, high fever, asthma, itching and ulcers.

Formula 17– Dhattura (*Datura stramonium/metel*)

Formula: Brihat Jvarankusha Rasa

Qty	Sanskrit	English	Latin
3 parts	Dhattura*	Dhattura seeds	*Datura stramonium/metel*
2 parts	Kajjali*	Mercuric sulphide	
1 part	Vatsanābha*	Aconite tuber	*Aconitum napellus*
2 parts	Maricha	Black pepper	*Piper nigrum*
2 parts	Sunthi	Ginger	*Zingiber officinale*
2 parts	Pippali	Long pepper	*Piper longum*
Method	All the ingredients are mixed well and made into pills, using water and acacia gum to paste and bind. Once formed, these pills are dried under sunlight.		
Anupāna	Honey, lime juice or ginger juice.		
Dosage	250mg		
Uses	Removal of chronic fever by two or more Dosha.		
Dhattura	Reduces abdominal swelling, asthma, bronchitis, intestinal parasites, eczema, lower back pain, earache, rheumatism, gout and colic pain. It also helps reduce high fever and hair loss. Dhattura improves liver function, as well as calming the nervous system.		

Formula 18 – Bhallātaka (bhilawan nut/marking nut)

Formula: Bhallātaka Rasāyana

Qty	Sanskrit	English	Latin
1 part	Bhallātaka*	Marking nut	*Semicarpus anacardium*
1 part	Loha Bhasma*	Iron oxide	
1 part	Sunthi	Ginger (powdered)	*Zingiber officinale*
1 part	Vidanga	False black pepper seeds	*Embelia ribes*

Method	The ingredients are mixed well and made into pills, again using water and acacia gum for binding and pasting. These pills are also dried under sunlight.
Anupāna	Honey and ghee
Dosage	500mg–1g
Uses	Reduces abdominal swelling and improves digestive ability. Bhallātaka Rasāyana is a powerful Rasāyana for bone marrow. It also treats anaemia, Alzheimer's disease and senility.
Bhallātaka	Useful in the treatment of piles, splenic disorders, persistent skin diseases, abdominal bloating and obesity. It is especially useful for nervous system disorders and is a general brain tonic.

14.2/ Rasa Shāstra in therapeutic application

During my studies in Sri Lanka, one of the local doctors, who I worked with in Colombo, invited me along on a visit he made to the Memorial Herbal Immunotherapy Research and Cancer Information Centre in Niripola.

This clinic had gained considerable recognition on the island for its specialisation in cancer treatments based on the pioneering work of the late Dr. Bernard Randeniya. Although the centre was not far from the facility in Dompe, the journey time on the island tended to be longer than expected. After arriving and taking a brief tour, I had the opportunity to speak with the director, several staff members, and a few resident patients.

During my visit, I learned that the clinic was thriving and had garnered significant political support, particularly from Chandrika Bandaranaike Kumaratunga, the Sri Lankan president at that time. The director mentioned that despite being a small facility, interest in their services was growing, given the rising incidence of cancer on the island.

Residential patients at the centre shared their success stories with the treatments, and it appeared that more patients were seeking their services. When I inquired about the types of cancer that were more prevalent, I was informed that brain tumours and oral cancers were particularly common.

The primary treatment at the centre involved herbal remedies, typically consumed as teas multiple times a day, accompanied by a simple yet nourishing diet. Additionally, a Rasa medicine formula was used in conjunction with the teas. While the tea recipes varied, a quick examination of the favoured ingredients revealed the inclusion of six Upavisha herbs (all toxic plants) and Guggulu resin. Notably, Bhallātaka (*Semicarpus anacardium*) appeared in both the teas and the Rasa medicine. This particular plant had garnered interest in recent years due to its observed effects on certain types of cancers, such as B16 melanoma and leukaemia L-1210 cells.

Continuing my conversation with the director, he presented numerous scans, most of which depicted brain tumours that, over time, had stopped growing and instead developed what he referred to as a 'containment membrane' surrounding the tumour. According to him, consistent consumption of the tea led to a more defined membrane. However, if the treatment was halted, the membrane would diminish, and the tumour would start growing again.

As a non-expert in this field, I cannot comment on the long-term effectiveness

of this treatment plan. However, I did observe that the clinic also functioned as a Pañcakarma clinic. This practice, it seemed, was encouraged for patients as they regained strength or reached a stage where they could undergo some level of purification before commencing their cancer therapy.

14.3/ Anti-cancer herbs used by the Niripola clinic

Upon entering the main reception area of the building, a prominent feature was a large display case presenting all 41 herbs utilised at the clinic. While I encourage the reader to explore these herbs in greater detail, it is worth mentioning that the majority of them were readily available on the island or were commonly used Āyurvedic herbs. Many of these herbs have a longstanding tradition of being blood cleansers, tonics, and Rasāyana drugs, known for their rejuvenating properties. The inclusion of six Upavisha plants is particularly intriguing, as these plants are generally less commonly featured in formulas.

Above, herbal display in the cancer centre in Niripola

14.3/ Anti-cancer herbs used by the Niripola clinic

41 Herbs			
Note (*) are all Upavisha plants			
No.	**Sanskrit**	**English**	**Latin**
1	Bhallātaka*	Marking nut	*Semicarpus anacardium*
2	Guggulu	Indian bdellium	*Commiphora mukul*
3	Gotu Kola	Asiatic pennywort	*Centella asiatica*
4	Bṛāhmi	Water hyssop	*Bacopa monnicri*
5	Kanchanara	Mountain ebony	*Bauhinia variegatya*
6	Devadaru	Himalayan cedar	*Cedrous deodara*
7	Gorakshganja	Mountain knotgrass	*Aerva lanta*
8	Arka*	Crown flower	*Calotropis gigantea*
9	Dhattura*	Devils trumpet	*Datura metel*
10	Manjistha	Indian madder	*Rubia cordifolia*
11	Parijata	Indian coral tree	*Erythrina variegata*
12	Guduchi	Heart-leaved moonseed	*Tinospora cordifolia*
13	Gunja*	Crab-eyed peas	*Abrus precatorius*
14	Shatāvari	Wild asparagus	*Asparagus racemosus*
15	Ashwagandha	Indian ginseng	*Withania somnifera*
16	Punarnava	Spreading hogweed	*Boerhavia diffusa*
17	Bhringaraj	False daisy	*Eclipta prostrata*
18	Bhuiamla	Niruri	*Phyllanthus debilis*
19	Āmalakī	Indian gooseberry	*Phyllanthus emblica*
20	Kutki	Hellebore	*Picrorhiza kurroa*
21	Draksa	Common grape	*Vitis vinifera*
22	Nirgundi	Five-leafed chaste tree	*Vitex negundo*
23	Dadima	Pomegranate	*Punica granatum*
24	Gandhamoolaka	Aromatic ginger	*Kaempferia galanga*
25	Kshudrabeeja	Indian almond	*Terminalia catappa*
26	Haridra	Turmeric	*Cucuma longa*

27	Vaccha	Sweet flag	*Acorus calamus*
28	Rukmal	Ceylon champaca	*Horsfieldia iryaghedhi*
29	Parpataka	Threadstem carpetweed	*Mollugo cervina*
30	Rakta chitrak	Red Indian leadwort	*Poumbago indica*
31	Harītakī	Chebulic myrobalan	*Terminalia chebula*
32	Kapikacchu	Cowhage	*Mucuna pruriens*
33	Chopchini	China root	*Smilax china*
34	Tilaparni	Pseudo China root	*Gybura pseudo china*
35	Musta	Purple nutsedge	*Cyperus rotundus*
36	Nyantara	Madagascar periwinkle	*Catharanthus roseus*
37	Langali*	Flame lily	*Gloriosa superba*
38	Ankenda	Indian aspen	*Acronychia pedunculata*
39	Snuhi Ksheera*	Antique spurge	*Euphorbia antiquorum*
40	Sugandi	Indian sarsaparilla	*Hemidesmus indicus*
41	Taalisa	East Himalayan fir	*Abies spectabilis*

The original test results for this formula were given as follows:

The patients were divided into five groups to evaluate the effects of the formulation. It was observed that those who received the Āyurvedic drug as an adjuvant therapy demonstrated the most significant response. Notably, there was a significant improvement in haemoglobin percentage, body weight, and overall lifespan. Patients who received only the Āyurvedic drug also exhibited results comparable to those who underwent chemotherapy. Moreover, the drug was found to be particularly effective in cases of splenomegaly (enlarged spleen), lymphoma (cancer of the lymph nodes), squamous cell carcinoma (a type of skin cancer), and adenocarcinoma (a form of tissue cancer).

Ingredients for the popular Rasa drug used at the clinic:

- Bhallātaka (*Semicarpus anacardium)*
- Rohitaka (*Tecomella undulata*)
- Yastimadu (liquorice)
- Tamra Bhasma (copper oxide)

Appendix

Calalogue of materials and their use

Note: The true identities of some materials have been lost; that is to say, there is no complete agreement on what the ancients were referring to. Where this is the case, I have highlighted it with an asterisk.* The proposed identity of the material selected should, therefore, be treated with caution. It should also be noted that not all fields in the following tables are relevant to every material and have been deleted where appropriate.

Terminology: The concept of VPK (Vāta/Pitta/Kapha) is an integral part of Rasa Shāstra, particularly in understanding the actions of certain ingredients. Wherever possible the English equivalents have been used, but in places the original Sanskrit terms have been left, as they better convey the energetics of the material concerned.

Table template	
English	English name(s)
Sanskrit	Sanskrit name(s)
CCM	Classical Chinese medicine name(s)
Tibb	Unani Tibb name(s)
Chemical formula	Common chemical formula
Mohs	Mohs hardness scale
Quality	Qualitative description
Shodhana	Materials associated with the purification of that material
Saṁpuṭa	Size and number of puṭa required to convert to Bhasma (medicated ash)
Amṛītakarana	Amṛīta = nectar. Final purification procedure designed to give extra potency, also referred to as 'nectarisation'
Anupāna	Vehicle of delivery
Attributes	Energetics and effects on Vāta/Pitta/Kapha, hereafter (VPK)
Taste	Six tastes
Post-digestive	Post-digestive effect
Dynamics	Heating or cooling
Dosage	Suggested guide to dosage
Treatment of	Commonly prescribed conditions
Poisoning	Effects of poisoning due to improper processing
Antidote	Treatment(s) to counteract improper purification

1/ Mercury	
English	Mercury, hydrargyrum, quicksilver
Sanskrit	Pārada, Rasendra, Rasa, Chapala, Rasa Raja Any of the 108 names of Lord Shiva are acceptable
CCM	Dān
Tibb	Simab, Para, Rasa
Chemical formula	Hg
Quality	Deep blue colouration at its centre, bright white toward its periphery, without imperfections in its sheen. Discoloured, yellow, pale, or smoky examples should be avoided for medicinal applications
Shodhana	Himalayan garlic pearls (juice), betel leaf juice, Triphala decoction, Bhringaraj decoction, brick dust and turmeric, lemon juice, *Aloe vera*, slaked lime powder
Taste	All five tastes, except salty
Dynamics	Yoga vāhin, Rasāyana, tonic, extends life, heals wounds, cures all disease
Treatment of	Pārada (mercury) mixed with Gandhaka (sulphur) forms Kajjali, the base ingredient for most Rasa medicines
Poisoning	Skin disease, burning sensations, impotency, lethargy, vomiting, bloody stools, digestive pains, abdominal bloating, vertigo, collapse, coma, death
Antidote	Sulphur, betel leaf juice, lime juice, goat's milk, egg white, milk, ghee

2/ Mercuric sulphide	
English	Mercuric sulphide
Sanskrit	Kajjali
Chemical formula	HgS
Quality	Yoga vāhin (vehicle for other medicines)
Anupāna	Betel leaf juice, milk, honey, jaggery. Generally, sweet and oily foods are increased when taking this medicine; sour and salty foods are minimised
Information	Kajjali forms the base of many Rasa Shāstra formulae. Although not commonly prescribed alone, there are cases when Kajjali is solely recommended as a Rasāyana to counteract the effects of ageing

Maha Rasa

3/ Mica	
English	Mica, biotite (black), muscovite (white)
Sanskrit	Abhraka, Girijabija, Gagana, Vajra, Koku, Akashagarbha
CCM	Bái yŭn mü
Tibb	Abhraq, Kabubulars
Chemical formula	$K(MgFe)_3(AlSi_3)O_{10}(OHF)_2$ (biotite) $KAl_2\ (AlSi_3O_{10})\ (F,OH)_2$ (muscovite)
Mohs	Muscovite (2.0–2.5) Biotite (2.5/3.0)
Quality	Mica should be black, heavy, clear and hard. The sheet should be wide and easily separated; it should be thermally stable in fire. Mica is a powerful Rasāyana, it gives strength, promotes good vision, improves metabolic function, and increases longevity
Purification	Triphala decoction, milk, castor oil, turmeric (fresh) juice
Saṁpuṭa	10–100 (Gaja Puṭa)
Amṛītakarana	Triphala decoction and ghee
Anupāna	Honey, milk or ghee
Attributes	Unctuous/oily
Taste	Sweet
Dynamics	Cold
Dosage	125–250mg
Treatment of	Tuberculosis, asthma, persistent cough, heart disease, diabetes, anaemia, gastritis, reduced vision, piles, vertigo, colic pain, urticaria, jaundice, chronic diarrhoea, dysentery, abdominal diseases, leprosy, high fever, loss of appetite
Poisoning	Improperly purified mica is said to aggravate the stomach like the fur of a tiger. It aggravates Vāta/Kapha, gives piercing pain, disturbs or lowers Agni and promotes intestinal parasites
Antidote	Warm milk with a little turmeric (powder or fresh juice)

4/ Fluorite*	
English	Fluorite, fluorspar, blue john or Derbyshire spar (violet)
Sanskrit	Vaikrānta, Kuvajra, Jīrna vajraka
CCM	Yíng shí
Chemical formula	CaF_2
Mohs	4.0
Quality	Fluorite, feldspar, tourmaline, preferably clear without blemish, and octahedral
Shodhana	Lemon juice, sour orange, sulphur
Saṁpuṭa	8 (Gaja Puṭa)
Anupāna	Milk or honey
Attributes	VPK, promotes longevity, complexion and strength, increases Agni and intellect. Called the monarch of Rasa Shāstra, Vaikrānta makes the body like Vajra (diamond). It has excellent antioxidant properties
Taste	All six tastes
Post-digestive	Sweet
Dynamics	Cold
Dosage	125–250mg
Treatment of	Poisoning, anaemia, high fever, skin diseases, asthma, bronchitis, diabetes, urinary infections, abdominal disorders

5/ Copper pyrite	
English	Chalcopyrite, copper pyrite
Sanskrit	Swarna Maksika, Tapija, Tapya, Garuda, Paksi, Brhadvarna
CCM	Huâng tông kuâng
Tibb	Suvarnamakshika
Chemical formula	$CuFeS_2$
Mohs	3.5–4.0
Quality	Swarna (having golden flecks), its surface should be shiny and radiating with a blackish tinge. Kansya (bronze-looking variety) is also acceptable

Shodhana	Lemon juice, sour orange, rock salt, castor oil, sulphur, Vasā decoction (*Adhatoda vasica*), plantain leaf juice (*Musa paradisiaca*)
Saṁpuṭa	10–14 (Gaja Puṭa)
Amṛītakarana	Kidaram/Suran root (elephant yam)
Anupāna	Milk, honey or ghee
Attributes	–P –K, Rasāyana, Yoga vāhin (a vehicle for other medicines)
Taste	Bitter/sweet
Post-digestive	Pungent
Dynamics	Cold
Dosage	125mg
Treatment of	Eye diseases, tuberculosis, anaemia, skin diseases, abdominal diseases, immune stimulant, parasites, digestive disorders, haemorrhoids, fistula, loss of appetite, loss of taste
Poisoning	Constipation, anaemia, impairment of vision, eye diseases, swelling of abdomen, skin diseases, carbuncles, loss of appetite, nausea
Antidote	Pomegranate juice, Kulatha decoction (horse gram) sweetened

6/ Iron pyrite	
English	Iron pyrite, fool's gold
Sanskrit	Vimala, Tara vimala, Raupya maksika
CCM	Huâng tië kuâng
Tibb	Raupya maksika
Chemical formula	FeS_2
Mohs	6.0–6.5
Quality	Tara (containing silver) Vimala has similar properties to Swarna Maksika, but of a lower potency. This material is often substituted in place of Swarna Maksika
Shodhana	Lemon juice, sour orange, rock salt, castor oil, sulphur, Vasā decoction (*Adhatoda vasica*), plantain leaf juice (*Musa paradisiaca*)
Saṁpuṭa	10–12 (Gaja)
Anupāna	Honey

Attributes	–V –P
Taste	Bitter/sweet
Post-digestive	Pungent
Dynamics	Cold
Dosage	125–250mg
Treatment of	Eye diseases, tuberculosis, anaemia, skin diseases, abdominal diseases, immune stimulant, parasites, digestive disorders, piles, fistula, loss of appetite, loss of taste
Poisoning	Constipation, anaemia, impairment of vision, eye diseases, swelling of abdomen, skin diseases, carbuncles, loss of appetite, nausea
Antidote	Pomegranate juice, Kulatha decoction (horse gram) sweetened

7/ Bitumen	
English	Asphaltum, bitumen
Sanskrit	Shilajit, Adrija, Sūryathapi, Giriniryasa, Girija and Sila sveda
Tibb	Shilajit, Hajrul Musa, Momiai, Faqrul yahud
Chemical formula	$C_{14}H_{10}N_4O_5$ Trace minerals include: fulvic acid, humic acid, iron, zinc, magnesium, copper, nickel, potassium, manganese, silicon, silver, sodium, sulphur, iodine etc
Quality	Different grades of Shilajit are associated with different metals: Swarna (gold), Roupya (silver), Tamra (copper), Loha (iron), Nāga (lead) and Vanga (tin). Loha grade is said to be the most therapeutic in value. Shilajit should be black, soft and heavy, having a smell reminiscent of cow's urine
Shodhana	Triphala decoction, milk
Saṁpuṭa	No puṭa required. However, it has been mentioned that a preparation of Shilajit combined with Haritāla and Manah Śhilā is subjected to Kapota Puṭa
Anupāna	Warm sweetened milk
Attributes	VPK, makes the body like a stone, strong Rasāyana for the kidneys and urinary system, excellent Yoga vāhin (vehicle for other medicines). Avoid consuming fish or Kulatha (horse gram) while taking Shilajit
Taste	Pungent, bitter
Post-digestive	Pungent
Dynamics	Hot

Dosage	1–5g *Caraka Saṃhitā* recommends a daily dosage of 12g
Treatment of	Kidney stones, urinary calculi, irregular blood sugar, high Vāta conditions, irregular blood, leprosy, cysts/growths, oedema, persistent asthma/cough, haemorrhoids, persistent nausea
Poisoning	Joint inflammation, vertigo, constipation, loss of appetite, nausea, internal haemorrhaging
Antidote	Milk with ghee and black pepper, administered for seven consecutive days

8/ Copper sulphate	
English	Copper sulphate, blue vitriol
Sanskrit	Sasyaka, Hemasara, Mayuraka, Tamaragarbha, Amṛītasanga, Sikhigriva, Tuttha
CCM	Liú suān tóng
Tibb	Nila Thotha, Zajul Akhsar, Tobal Mis
Chemical formula	$CuSO_4$ (copper sulphate) Cu_5FeS_4 (bornite)
Mohs	3.0
Quality	Two types of Sasyaka are available: artificially produced (copper sulphate) and naturally occurring (bornite)
Shodhana	Sulphur, borax, Triphala decoction, lemon juice, cow's urine
Saṁpuṭa	4–8 Laghu (copper sulphate), 8–12 Gaja (bornite)
Amṛītakarana	Fried in ghee and honey
Anupāna	Milk, honey, or butter
Attributes	Rasāyana
Taste	Pungent, astringent
Dynamics	Hot and light (–P –K)
Dosage	75–125mg
Treatment of	Poisoning, diseases of the eye, heart disease, diabetes, consumption, skin diseases/leprosy, vertigo, gastritis, vomiting, ulcers, itching, haemorrhoids
Poisoning	Loss of Ojas, vomiting, diarrhoea, vertigo, weight loss
Antidote	Sulphur, lemon juice, lime water, fried paddy rice

9/ Bismuth sulphide*	
English	Bismuth sulphide, bismuthinite, bismuth glance
Sanskrit	Chapala, Shaila, Rasaraj sahay
CCM	Huî bì kuâng
Chemical formula	Bi_2S_3
Mohs	2.0–2.5
Quality	Goura (yellow), Swetha (white), Krishna (black), Rakta (red) Yellow and white varieties are used to restrain mercury (solidification) in Lohasiddha practice
Shodhana	Ginger juice, Triphala decoction, lemon juice, Kanji
Anupāna	Triphala decoction
Attributes	Aphrodisiac, strengthens bodily tissues
Taste	Bitter, sweet
Dynamics	VPK, unctuous, heavy, purgative Chapala is used in mercurial operations as a fixative or heat-stabilising agent Chapala means 'to move quickly'
Treatment of	Tuberculosis, gonorrhoea, menorrhagia, high fever, excess mucus, obesity Purgative and general Rasāyana

10/ Calamine	
English	Calamine, smithsonite
Sanskrit	Rasaka, Kharpara, Reetikrita, Tamraranjaka, Netraogari, Yasada karana
CCM	Líng xîn kuâng
Tibb	Kharpara, Sange Basari
Chemical formula	$(Fe, Mn)ZnCO_3$
Mohs	4.5
Quality	Dardura (fibrous and layered), Karavellaka (crystalline, compact and without layers). Karavellaka is the better medicinal grade
Shodhana	Cow's urine, butter milk, lemon juice, Kanji, horse urine
Anupāna	Honey, ghee, aloe gel

Attributes	Rasāyana for the lungs
Taste	Pungent, astringent
Dynamics	Cold and light, –P –K
Dosage	125–250mg
Treatment of	Tuberculosis, diabetes, trachoma, eye disease, anaemia, consumption, leprosy, chronic fevers, asthma, menstrual disorders, urinary disorders, diarrhoea, excess bleeding of the uterus, skin diseases, abscesses, persistent hiccups
Poisoning	See 35/ Zinc
Antidote	Bala (*Sida cordifolia*) decoction, Harītakī (*Chebulic myroblan*) decoction, jaggery, Kanji, cow's urine

Uparasa

11/ Sulphur	
English	Sulphur
Sanskrit	Gandhaka, Gandupashana, Keetaghna, Pamari, Shulvari, Bali, Keetanashana, Sulvari, Puti gandha, Sugandhika
CCM	Liú huáng
Tibb	Gandhak, Kibreet
Chemical formula	S
Mohs	1.5–2.5
Quality	Yellow (highest medicinal value), red (used in Lohasiddha), white (external use only), black (unused – toxic)
Shodhana	Warmed milk and ghee, Bhringaraj decoction, aloe gel, red onion juice
Anupāna	Milk, ghee
Attributes	Bonds with mercury, purifies all Loha (metal) Sulphur is naturally found in eggs, carrots, garlic, onions, milk, blood and bile
Taste	Pungent
Post-digestive	Sweet
Dynamics	Hot, –V –K, Yoga vāhin (a vehicle for other medicines)
Dosage	125–250mg
Treatment of	Leprosy/skin diseases, high fever, mercury poisoning, removal of āma (toxins), itching, loss of appetite, excess mucus formation, intestinal parasites, and digestive disorders
Poisoning	Skin disease, vertigo, burning pain, reduced strength
Antidote	Milk and ghee

12/ Red iron oxide	
English	Iron oxide, hematite, red ochre, kidney ore, limonite
Sanskrit	Gaireeka, Raktadhātu, Girija, Dhātu, Raktapashana, Giri mrt, Loha dathu, Giri mrdbhava
CCM	Chî tiĕ kuâng
Tibb	Gerumitti, Gil-e-surkh, Teen-i-rumi
Chemical formula	Fe_2O_3
Mohs	5.5–6.5
Quality	Swarna Gaireeka (gold/soft), Pashana Gaireeka (hard stone) Swarna Gaireeka has the better medicinal quality
Shodhana	Milk
Saṁpuṭa	Calcined with Kumārī (*Aloe vera*) gel in a Musa (open crucible)
Anupāna	Milk, honey or Kanji
Attributes	Builds Rakta (blood)
Taste	Sweet and astringent
Post-digestive	Astringent
Dynamics	Cold/astringent
Dosage	750mg–1.5g
Treatment of	Heart disease, anaemia, poisoning, snakebite, vomiting, impaired vision, abdominal bloating, itching, boils, urticaria, pustule eruptions, stomatitis, sores, ulcers, burns, and bleeding haemorrhoids

13/ Ferrous sulphate	
English	Ferrous sulphate, green vitriol, copperas
Sanskrit	Kasisa, Khechara, Khanga, Pamsuka, Puspa Kasisa
Tibb	Hira Kasis, Tutiya-e-sabz
Chemical formula	$FeSO_4 \cdot 7H_2O$
Mohs	2.0
Quality	Vālukā (sand coloured), Pushpa (artificial) is green or green-yellow in colour

Shodhana	Lemon juice, Bhringaraj or Triphala decoction
Saṁpuṭa	4 (Varaha)
Anupāna	Triphala Churna and honey
Attributes	Builds blood
Taste	Astringent, sour
Dynamics	Hot, –V –K
Dosage	125–250mg
Treatment of	Tuberculosis, poisoning, skin diseases, kidney stones, painful urination, hair that is greying or falling-out, loose teeth. Kasisa builds blood, tonifies the liver and spleen. It is useful in cases of anaemia, leucoderma, uterine or rectal prolapse
Poisoning	See 32/ Iron
Antidote	Triphala decoction, ghee and honey

14/Alum	
English	Potassium aluminium sulphate, alum, potash alum, kalanite
Sanskrit	Kanksi, Saurastri (originally mined in Saurastri/North Gujarat), Sphatika (rock crystal)
CCM	Bái Fán
Tibb	Phitkari, Shibb-e-yamani, Gulabi Phitkari (pink alum)
Chemical formula	$KAl(SO_4)_2{\cdot}12H_2O$
Mohs	2.0–2.5
Quality	Kanksi (white), Fatkadi (red ferric alum)
Shodhana	Heating and hydration. Alum can also be triturated with bile from a cow and then hydrated (this method is favoured in mercurial solidification practices)
Anupāna	Rosewater, honey, or sugar water
Attributes	Hot
Taste	Sweet, sour, and astringent
Post-digestive	Sweet
Dynamics	VPK

Dosage	250–500mg
Treatment of	Leucoderma, bleeding gums, malaria and stomatitis (inflammation of the mouth and gums). Poisoning through animal or insect bites, prolapsed uterus and rectum, scabies, ulcers, eye diseases, epilepsy and cracked, dry skin

15/ Arsenic trisulphide	
English	Arsenic trisulphide, orpiment
Sanskrit	Haritāla, Tala, Ala, Talaka, Natabhooshana, Pinjaka, Pita, Lomahrita, Malla gandhaja
CCM	Cî huâng
Tibb	Hartal Warql, Zirnikh-e-zard
Chemical formula	As_2S_3
Mohs	1.5–2.0
Quality	Golden colouration, heavy, oily and leafed
Shodhana	Juice of Kushmanda (ash pumpkin), lime water, Kanji, sesame oil, calcined sesame plant (Ksara), Triphala decoction, lemon juice, borax
Saṁpuṭa	Haritāla Bhasma can be prepared; however, it is usually heated between sheets of mica. The resulting crystals are then powdered
Anupāna	Ghee or milk with honey
Attributes	Patra (layered leaves), Pinda (without leaves), Tabaki (artificial) Patra is considered the best medicinal grade
Dynamics	Hot, –V –K (+P). Pitta-increasing foods and lifestyles should be avoided whilst taking this remedy
Dosage	10–20mg
Treatment of	Cancers, leprosy, poisoning, high fevers, Pitta type skin diseases, gout, piles, fistula, asthma, persistent cough
Poisoning	Vomiting, burning sensations, diarrhoea, skin eruptions, pain, darkening of the fingernails over longer periods of exposure
Antidote	Kushmanda juice, garlic, fennel seeds, purified sulphur, ground cumin mixed with raw cane sugar, and goat's milk sweetened with jaggery

16/ Arsenic disulphide	
English	Arsenic disulphide, realgar,
Sanskrit	Manah śhilā, Nāga jihva, Nāga mata, Kunnati, Manogupta, Manohvā, Śhilā, Gola, Nāga jihvika
CCM	Xióng huâng
Tibb	Hartal Tabqi, Zirnikh-e-surkh, Rahj-al-ghar
Chemical formula	As_2S_2
Mohs	1.5–2.0
Shodhana	Ginger juice, lime water, juice of Agastya leaf (*Sesbania grandiflora*), Harītakī (*Terminalia chebula*) or Bhringaraj decoction
Anupāna	Ghee mixed with honey
Attributes	Syamangi (yellow/red and heavy), Kanaviraka (coppery), Khandakhya (powdery and vermilion in colour, resembling Hiṅgula). Of these three grades Khandakhya is considered to have the highest medicinal value
Taste	Bitter and pungent
Dynamics	Hot
Dosage	10–20mg
Treatment of	Tuberculosis, skin diseases, high fever, leprosy, intestinal parasites, gout, chronic bronchitis, bronchial asthma, itching, anaemia. Improves vision
Poisoning	Vomiting, burning sensations, diarrhoea, skin eruptions and pain. Darkening of fingernails may occur over long periods of use
Antidote	Ginger juice, liquorice decoction, goats milk, milk and ghee sweetened with honey

17/ Antimony sulphide	
English	Antimony sulphide, stibnite, antimony glance, grey antimony, antimonite
Sanskrit	Anjana, Sauvira, Krsnanjana
CCM	Huî tî kuâng
Tibb	Surma, Sang-e-surma, Sang-e-basri
Chemical formula	Sb_2S_3

Mohs	2.0–2.5
Quality	Srotonjana, Souviranjana, Rasanjana, Nilanjana pushpanjana
Shodhana	Triphala or Bhringaraj decoction, lemon or lime juice
Saṁpuṭa	8 (Laghu)
Anupāna	Honey or rice water
Attributes	Rasāyana
Taste	Sweet and astringent
Dynamics	Cold, heavy, and unctuous
Dosage	125mg
Treatment of	Improves vision, reduces blood toxicity, reduces obesity. A general Rasāyana, it has excellent anti-Visha properties. Helps with chronic bronchitis, bleeding disorders, ulcers, and excessive vomiting

18/ Galena	
English	Lead sulphide, galena, blue lead
Sanskrit	Nilanjana
CM	Fâng oiân kuâng
Tibb	Surma, Kajal, Krishna surma
Chemical formula	PbS
Mohs	2.0–2.5
Shodhana	Bhringaraj decoction, lemon juice
Anupāna	Tandulodaka (water from washed rice)
Attributes	Heavy and oily, –VPK
Taste	Sweet and astringent
Dynamics	Cold, heavy, and unctuous
External treatment of	Eye diseases, ulcers, haemorrhaging and bone fractures. Galena is a Rasāyana for the eyes and used as a collyrium in the form of makeup (kohl). Nilanjana (pasted) is applied to bone fractures to aid recovery

19/ Malabar tamarind *	
English	Himalayan rhubarb (*Rheum emodi*) or malabar tamarind (*Garcinia Cambogia*)
Sanskrit	Kankushta, Recaka, Varanga, Kolavaluka
Quality	Nilika (yellow, heavy and unctuous), Renukā (yellow/black and light), Swetha (white and liquid). Nilika has the better medicinal effect
Shodhana	Fresh ginger juice
Attributes	Strong purgative
Taste	Bitter, pungent
Dynamics	Hot
Dosage	75–125mg
Treatment of	Skin diseases, abdominal bloating, enlargement of the spleen, excess mucus, colic, fistula and haemorrhoids, water retention, weight loss
Poisoning	Avoid betel leaf juice when using this remedy
Antidote	Babul root (*Acacia arabica*) mixed with Tankana (borax) and Jeeracam (*Cuminum cyminum*)

Sadharana Rasa

20/ Monkey face fruit	
English	Monkey face tree fruits (*Mallotus philippensis*)
Sanskrit	Kampilla, Kampillaka, Rajanaka
Shodhana	Water, juice of Matulunga (*Citrus medica*)
Anupāna	Curd, milk, honey
Dynamics	Hot, purgative
Dosage	250mg
Treatment of	Skin diseases, boils, parasites (has powerful anti-fungal properties), fistula, haemorrhoids and constipation. Reduces excess phlegm

21/ Arsenic trioxide	
English	Arsenic trioxide, white arsenic, asenolite
Sanskrit	Gauri pashana, Somala, Malla, Mooshak, Sambala, Dara musa, Mallaka
Tibb	Sankh Visha, Sammulfar, Sankhiya
Chemical formula	As_2O_3
Mohs	1.5–2.0
Quality	White, heavy, semi-transparent, and oily
Shodhana	Karawella (*Momordica charantia*), milk
Anupāna	Honey or ghee
Attributes	Hot, –V –K (+P) Pitta-increasing foods and lifestyle should be avoided whilst taking this remedy
Taste	Pungent
Dynamics	Hot, –V –K
Dosage	1–2mg

Treatment of	Cancer, heart disease, consumption, leprosy, scorpion bite, filariasis, bronchitis, asthma, high fever, enlargement of the spleen, ring worm, itching, carbuncles, anaemia, rheumatism, skin disease, general swellings, and general debility. Arsenic has aphrodisiac properties
Poisoning	Vomiting, burning sensations, diarrhoea, skin eruptions, pain. Longer periods of exposure lead to darkening of the fingernails
Antidote	Karawella (*Momordica charantia*), ginger juice, liquorice decoction, goat's milk, milk and ghee sweetened with honey

22/ Ammonium chloride	
English	Ammonium chloride, sal ammoniac
Sanskrit	Navasadar, Narasara
Tibb	Armina, Noshadar
Chemical formula	NH_4Cl
Quality	Naturally occurring mineralogical form or artificially prepared from plant ash or animal dung
Shodhana	Boiling water
Anupāna	Honey or ghee
Treatment of	Mouth ulcers and bleeding gums, heart disease, eye diseases, poisoning, headaches, acid reflux, liver and splenic diseases, excess Pitta, bone fractures, animal bites, and insect bites and stings

23/ Cowrie shells	
English	Cowrie/cowry shells
Sanskrit	Kapardika, Charachara, Kapardaka, Varatika, Varata
Chemical formula	$CaCO_3$
Quality	Yellow, white and grey shells; orange, white and brown shells Yellow variety is the best medicinal grade
Shodhana	Vinegar, milk, Bhringaraj decoction
Saṁpuṭa	4 (Laghu)

Anupāna	Hot water, lemon juice, buttermilk When topically applied, powder is mixed with lemon juice and pasted
Taste	Pungent
Dynamics	Hot, –V –K
Dosage	250–500mg
Treatment of	Diminished eyesight, cataracts, hearing loss, high fever, heart disease, urinary stones, tuberculosis, hyperacidity, peptic ulcer, improves Agni (balances digestive power), reduces headaches and cough

24/ Ambergris	
English	Ambergris, (*Ambra grasea*)
Sanskrit	Agnijara, Vahnijara, Ambar, Tundamaya
Tibb	Kahrabaa, Misbah al-room
Quality	Opaque and waxy, strong odour, white, grey, grey-brown, or light yellow in colour with black spots upon its surface
Anupāna	Milk, ghee, honey
Attributes	Highly aromatic, an aphrodisiac
Taste	Pungent
Dynamics	VPK
Treatment of	Tetanus, reduction of Vāta, irregular digestion, loss of appetite, colitis, convulsions, paralysis, stabbing pains, diseases of the head and brain Ambergris is a powerful Rasāyana

25/ Mercuric oxide	
English	Red mercuric oxide
Sanskrit	Giri Sindoora, Girisiaduram
Chemical formula	HgO
Purification	Ginger juice
Anupāna	Honey

Attributes	Heavy fine red powder Nāga Sindoora (red lead) is sometimes substituted but it is chemically very different (PbO)
Dynamics	VPK, purgative
Dosage	25–50mg
Treatment of	External wounds, skin infections, burns, poisoning, leprosy, eye diseases, digestive dysfunction, ulcers, herpes, chronic itching, infirmity of limbs, general weakness. Giri Sindoora is a powerful Rasāyana
Poisoning	Skin disease, burning sensations, impotency, lethargy, vomiting, bloody stools, digestive pains, abdominal bloating, vertigo, collapse, coma, death
Antidote	Sulphur, betel leaf juice, lime juice, goat's milk, egg white, milk, ghee

26/ Cinnabar	
English	Cinnabar, dragon's blood, Chinese red, vermilion
Sanskrit	Hiṅgula, Darada, Mlechcha, Chitranga, Rakta, Hingola, Curna Pārada, Ranjana
CCM	Zhū shâ
Tibb	Hingul, Sinjiraph
Chemical formula	HgS
Mohs	2.0–2.5
Quality	Red like the feathers on a parrot's head. It can be artificially prepared by mixing 1 part mercury to 4 parts sulphur and subjecting to high heat, resulting in HgS
Shodhana	Ginger juice, goat's milk, lemon juice, Kanji
Saṁpuṭa	None required, although the Unani Tibb method for obtaining cinnabar Bhasma has been given. See Preparation of Kushta Sangraf
Anupāna	Honey, betel leaf juice
Attributes	Powerful Rasāyana, the best treatment for old age
Taste	Pungent
Dynamics	Hot, VPK
Dosage	25–50mg

Treatment of	Leprosy, old age and debility, hepatitis and splenomegaly, rheumatism, chronic fevers, diabetes, impotency, chronic rhinitis, skin diseases. Hiṅgula is a powerful Rasāyana and balances all three Dosha
Poisoning	Skin disease, burning sensations, impotency, lethargy, vomiting, bloody stools, digestive pains, abdominal bloating, vertigo, collapse, coma, death
Antidote	Sulphur, betel leaf juice, lime juice, goat's milk, egg white, milk, ghee

27/ Lead monoxide	
English	Litharge, lead monoxide, massicot
Sanskrit	Mrddara Śrnga
Tibb	Murdar sang, Usrab, Seesha, Sufaida
Chemical formula	PbO
Mohs	3.0–4.0
Quality	Pita (yellow), Pitapandura (yellow/white), Kritrima (artificial) Pita is considered the most medicinal
Shodhana	Lemon juice, ginger juice
Anupāna	Honey
Attributes	Cold
Dynamics	–V –K, Rasāyana
Dosage	None recommended (external use only)
Treatment of	Bone fractures, greying or falling of hair, improves fertility, eczema, scabies, external ulcers, syphilitic sores, herpes

Dhātu

28/ Gold	
English	Gold
Sanskrit	Swarna, Suvarama, Kanaka, Hiranya, Hema, Hataha, Tapaniya, Satakumba, Kanchana, Chamikara, Hataka
Tibb	Swarna, Tila
Chemical formula	Au
Purification	Sesame oil, curd, cow's urine, Kanji and Kulatha (5 liquid method), Kanchanara decoction (*Bauhinia variegata*) and saffron
Saṁpuṭa	12–15 (Gaja)
Anupāna	Honey, ghee or buttermilk
Taste	Sweet, astringent and bitter
Dynamics	Increases Ojas, strong Rasāyana, boosts the immune system, VPK
Dosage	10–25mg
Treatment of	Heart disease, lowered immunity, tuberculosis, diminished eyesight, skin diseases, diabetes, general debility, senility, learning difficulties, fertility issues, improves taste, digestion and assimilation
Poisoning	Reduces strength, loss of fertility, loss of intellect
Antidote	Warm milk with saffron

29/ Silver	
English	Silver
Sanskrit	Rajata, Roupya, Tara, Chandra, Kaladhātu, Chandi, Rupa
Tibb	Rajat, Nuqra, Fizza, Chandi
Chemical formula	Ag
Purification	Sesame oil, curd, cow's urine, Kanji and Kulatha (5 liquid method) Agastya (*Sesbania grandiflora*), Snuhi (*Euphorbia ligularis*)

Saṁpuṭa	12–15 (Gaja)
Anupāna	Honey or milk
Taste	Sour and astringent
Post-digestive	Sweet
Dynamics	Cold, –P –V +K, Rasāyana
Dosage	50–75mg
Treatment of	Heart disease, stomach disorders, diabetes, vertigo, anaemia and senility Also used in the treatment of madness and general debility. Silver improves Agni and appetite, aids in weight loss and is a general Rasāyana
Poisoning	Diabetes, stomach disorders, loss of appetite, general weakness
Antidote	Goat's milk sweetened with honey

30/ Copper	
English	Copper
Sanskrit	Tamra, Shulva, Bhaskar, Mleccha, Vaktra, Nāgamatha, Ravibriya, Rataka, Sūrya Loha
Tibb	Nuhas, Tamba
Chemical formula	Cu
Shodhana	Rock salt, lemon/lime juice, Kanji, Hiṅgula (cinnabar), Gandhaka (sulphur), sesame oil, curd, cow's urine, Kanji and Kulatha (5 liquid method), buttermilk
Quality	Nepalaka (from Nepal, reddish and resistant to heating), Mleccha (white or blackish). Nepalaka is considered the best medicinal grade
Saṁpuṭa	12–15 (Varaha)
Amṛītakarana	Kidaram/Suran root (elephant yam)
Anupāna	Honey
Taste	Astringent and bitter
Post-digestive	Sweet
Dynamics	Hot, –P –K, antibacterial
Dosage	50–100mg
Treatment of	Consumption, eye diseases, liver disorders, anaemia, abdominal diseases (gastritis and colic pain), chronic fever, leprosy, asthma, cough, piles, oedema, hyperacidity, catarrh, parasitic infections (worms), colic, obesity, poisoning

Poisoning	Painful urination, pain in abdomen, general weakness, poor complexion, vertigo, nausea, vomiting, death
Antidote	Goat's milk, buttermilk with honey, Kulatha (horse gram) soup

31/ Iron	
English	Iron
Sanskrit	Loha, Lauha
Tibb	Hadida, Faulad, Ahna
Chemical formula	Fe
Quality	Kanta Loha (best) – magnetic iron ore Tikshna Loha (medium) – bluish/black, higher temperature, resilient Munda Loha (low) – high carbon content, lower melting point, brittle and unstable See also 32/ Rust of iron (Mandura)
Shodhana	Heated and quenched in Triphala decoction, lemon or lime juice. Cow's urine and five salts are also used in the purification/conversion of iron
Saṁpuṭa	30–100 (Gaja)
Anupāna	Honey, Triphala churna or ghee
Taste	Astringent, bitter
Post-digestive	Sweet
Dynamics	Cold, heavy, –P –K, blood Rasāyana
Dosage	75–175mg
Treatment of	Blood disorders, anaemia, leprosy, eye diseases, abdominal disease, enlarged spleen or liver, dysentery, colic pain, consumption. Iron destroys intestinal parasites, restores impaired vision, and reduces obesity, diabetes, and infertility. Helps with loss of vitality
Poisoning	Diseases of the skin, heart disorders, impotency, constipation, burning sensations in the digestive tract
Antidote	Triphala decoction, ghee, goat's milk, cow's urine

32/ Rust of iron	
English	Iron
Sanskrit	Mandura, Kitta, Loha mala, Loha bhava
Tibb	Manduram, Zang-e-ahana, Khabsul hadeed, Lohe ka zang
Chemical formula	Fe_2O_3
Quality	Mandura (rusted iron) should be 100 years in age or obtained from iron sheets heated, beaten, cooled and allowed to oxidise. Rust is graded by its source: Kanta (best), Tikshna (medium) and Munda (lowest). See 31/Iron for more information
Shodhana	Heated and quenched in Triphala decoction, lemon or lime juice. Cow's urine and five salts are also used in the purification/conversion of rust
Saṁpuṭa	10–12 (gaja)
Anupāna	Honey, Triphala churna
Taste	Astringent, bitter
Post-digestive	Sweet
Dynamics	Cold, –P –K
Dosage	50–250mg
Treatment of	Childhood anaemia, intestinal parasites, jaundice, abdominal disease, enlarged liver and spleen, dysentery, colic pain, diabetes, bleeding piles, menorrhagia
Poisoning	Diseases of the skin, heart disorders, impotency, constipation, and burning sensations in the digestive tract
Antidote	Triphala decoction, ghee, goat's milk, cow's urine

33/ Tin	
English	Tin
Sanskrit	Vanga, Ranga, Pichchata, Trapusa, Kutila, Ranya, Sukraloha, Trapu
Tibb	Vanga, Qalai
Chemical formula	Sn
Quality	Khuraka – jersey tin; Mishraka – commercial tin; Krishna – black tin, not suitable for medicinal use

Shodhana	Primary – Milk, turmeric, Nirgundi (*Vitex negundo*), Apamarga (*Achyranthes aspera*) Secondary – Sesame oil, curd, cow's urine, vinegar and horse gram decoction (5 liquid method)
Saṁpuṭa	12–15 (gaja)
Anupāna	Honey, cream, milk, betel leaf juice
Taste	Bitter
Dynamics	Cold
Dosage	75–125mg
Treatment of	Diabetes, urinary disorders, premature ejaculation, infertility, anorexia/ emaciation, brain dysfunction, senility, learning difficulties, skin diseases, chronic fevers, burning sensations, piles
Poisoning	Diabetes, anaemia, swellings, urinary disorders, kidney stones, abdominal bloating, piles, skin disease
Antidote	Sweetened goat's milk with turmeric

34/ Lead	
English	Lead
Sanskrit	Nāga, Seesa, Kurana, Kuuvanga, Sisaka, Sindoorakara, Kuranga
Tibb	Nāga, Sisa
Chemical formula	Pb
Quality	Kumara – quick melting, heavy, black and shiny, foul-smelling upon heating (best quality) Samala – contaminated by other metals, melts slowly, light coloured (not suitable for medicinal use)
Shodhana	Primary – Milk, turmeric, Nirgundi (*Vitex negundo*), Apamarga (*Achyranthes aspera*) Secondary – Sesame oil, curd, cow's urine, vinegar and horse gram decoction (5 liquid method)
Saṁpuṭa	12–15 (Gaja)
Anupāna	Honey
Taste	Sweet, bitter

Dynamics	Hot, heavy, –V –K
Dosage	50–125mg, dose not to exceed 1mg or be administered for more than 2–3 days (high risk factor – lead poisoning)
Treatment of	Diabetes, rheumatism, digestive disorders, piles, poisoning, skin diseases, abdominal bloating, IBS. Nāga gives strength to the tissues
Poisoning	Diabetes, skin disease, arthritis, jaundice, oedema and tuberculosis
Antidote	Harītakī with goat's milk, Swarna Bhasma (gold ash)

35/ Zinc	
English	Zinc, spelter
Sanskrit	Yasada, Jashada, Ritihetu, Kharparaja
Tibb	Putty, Jist, Tutiya
Chemical formula	Zn
Quality	Yasada (extracted from calamine ore) Puspanjana (zinc oxide)
Shodhana	Milk, turmeric, Nirgundi (*Vitex negundo*), Apamarga (*Achyranthes aspera*), sesame oil, curd, cow's urine, vinegar and horse gram decoction (5 liquid method), dried lime and cannabis leaf, poppy heads/leaf
Saṁpuṭa	5–7 (Gaja)
Anupāna	Milk, honey
Taste	Bitter
Dynamics	Cold, –P –K
Dosage	75–125mg
Treatment of	Diabetes, eye diseases, high fever, skin diseases, reproductive and urinary disorders. Promotes fertility and sharp vision
Poisoning	Abdominal distension, diabetes, urinary disorders, skin diseases, consumption, nausea, and loss of appetite
Antidote	Bala (*Sida cordifolia*) decoction, Harītakī (*Chebulic myroblan*) decoction, jaggery, Kanji, cow's urine

36/ Brass	
English	Brass (Cu: 3 pts, Zn: 2 pts)
Sanskrit	Pittala, Pittalasa, Ritika, Aarakuta, Kapiloha
Quality	Ritika (turns red upon heating) Kakatundi (turns black upon heating) Ritika is considered to have the better medicinal value
Shodhana	Sesame oil, curd, cow's urine, vinegar and horse gram decoction (5 liquid method) Nirgundi (*Vitex negundo*), turmeric, Manah śhilā, Gandhaka, Pancha Lavaṇa (five salts) method
Amṛītakarana	Kidaram/Suran root (elephant yam)
Saṁpuṭa	6–8 (Gaja)
Anupāna	Honey
Taste	Bitter
Dynamics	Cold and drying, –P –K, Yoga vāhin
Dosage	60–125mg
Treatment of	Hepatitis, anaemia, intestinal parasites, blood disorders skin diseases, enlargement of liver and spleen. Pittala has a strong Lekana action, meaning it scrapes and cleans āma from the tissues

37/ Bronze	
English	Bronze (Cu: 8 parts, Sn: 2 parts)
Sanskrit	Kansya, Kansyaka, Ghoshaka and Vahniloha
Quality	Pushpaka (bright, high tin content) Tailika (dull, low tin content)
Shodhana	Sesame oil, curd, cow's urine, vinegar and horse gram decoction (5 liquid method) Gandhaka, Haritāla, lemon juice Pancha Lavaṇa (five salts) method
Saṁpuṭa	6–8 (gaja)
Amṛītakarana	Kidaram/Suran root (elephant yam)
Anupāna	Honey

Taste	Bitter and pungent
Dynamics	Hot, –V –P All foods are recommended to be stored in bronze containers, except ghee This metal imparts health to stored goods
Dosage	60–125mg
Treatment of	Diseases of the skin and eyes, blood disorders, obesity, poor digestion, and intestinal parasites
Poisoning	Painful urination, pain in abdomen, general weakness, poor complexion, vertigo, nausea, vomiting, death Bronze is considered toxic when taken in conjunction with Pārada medicines, or using utensils and plates made of bronze, while consuming Pārada medicines
Antidote	Goat's milk, buttermilk with honey, Kulatha (horse gram) soup

38/ Tri-Loha (three metals)	
English	Three metal alloy (25pts Au, 16 pts Ag, 10 pts Cu)
Sanskrit	Tri-Loha
Quality	As gold
Shodhana	Nirgundi (*Vitex negundo*), Apamarga (*Achyranthes aspera*), turmeric, milk, sesame oil, curd, cow's urine, vinegar and horse gram decoction, lime water and *Aloe vera* juice
Saṁpuṭa	12–15 (gaja)
Anupāna	Honey, ghee, Trikatu and Triphala
Taste	Sweet and sour
Dynamics	Hot, Rasāyana, increases longevity
Dosage	125mg
Treatment of	Heart disease, low immunity, senility, fertility issues, loss of vitality, poor digestion and assimilation, loss of taste. Helps with learning difficulties

39/ Varta Loha (five metals)	
English	Five metal alloy, equal quantities of bronze, copper, lead, brass and iron, also known as Pasloha or bell metal
Sanskrit	Varta Loha, Pancha Loha, Pasloha
Saṁpuṭa	6–8 (Gaja)
Anupāna	Honey
Taste	Pungent and sour
Dynamics	Cold, –P –K
Dosage	125mg
Treatment of	Skin diseases, reduced vision, insufficient digestive power, intestinal parasites, constipation and urinary disorders. Varta Loha promotes Bala (physical strength)

Ratna

40/ Ruby	
English	Ruby, corundum
Sanskrit	Maanikya, Padmaraga, Ravi Ratna, Lohita
CCM	Hóng bäo shí
Tibb	Yaqoot surkh, Sona Ratna
Chemical formula	Al_2O_3
Mohs	9.0
Jyotish	Primary gemstone for Sūrya (the Sun)
Quality	Padmaraga (pink, lustrous and resembling the colouration of the lotus), it is considered to have the highest medicinal quality Kuruvindaja (similar type of corundum, deep red in colour) Saugandhaka (found with spinel and having a yellow/red colouration) Nilagandhi (turbid red/blue, obtained from riverbeds) Rubies should be free of fissures and blemishes
Shodhana	Lemon/lime juice, rosewater, orpiment, realgar, sulphur, rock salt and lotus seed decoction
Saṁpuṭa	8–10 (Gaja)
Anupāna	Honey, milk or musk
Taste	Sweet
Dynamics	–V –P, oily
Dosage	125–175mg
Treatment of	Heart disease, brain disorders, eye diseases, tuberculosis, reproductive issues, digestive weakness, removal of toxins, psychic disturbances relating to one's paternal ancestors

41/ Pearl	
English	Pearl
Sanskrit	Mukta, Chandraratna, Jeevaratna, Muktaphala, Sasi Ratna, Muktaphala
CCM	Zhēn zhū
Tibb	Mauktika, Moti
Chemical formula	$CaCO_3$
Mohs	3.5–4
Jyotish	Primary gemstone for Chandra (the Moon)
Shodhana	Rosewater, milk, Kanji, Bhringaraj decoction (*Eclipta alba*), buttermilk, yoghurt
Saṁpuṭa	3–4 (Varaha)
Anupāna	Milk, butter, cream, jaggery
Dynamics	Cooling, –V –P, Rasāyana
Dosage	50–75mg
Treatment of	Heart disease, poisoning, stomach disorders, Asthi (bone) disorders, impaired digestion, inflammation, diminished eyesight, excessive cough, asthma and bronchitis, general debility, high fever, psychic disturbances relating to one's maternal ancestors

42/ Red coral	
English	Red coral
Sanskrit	Pravala, Vidruma, Lathamani, Ambhodhivallabha, Sagargarbhakeeta, Bhauma Ratna, Kuja Ratna
Chemical formula	$CaCO_3$
Mohs	3.5–4
Jyotish	Primary gemstone for Kuja (Mars)
Quality	Red is preferable to white, grey or black varieties
Shodhana	Kanji, milk, *Aloe vera* juice, rosewater (Pisti)
Anupāna	Milk, butter, cream, jaggery
Taste	Sweet and astringent

Dynamics	Cold
Dosage	125–250mg
Treatment of	Bleeding/blood disorders, tuberculosis, atrophy of Māmśa (muscle) and sinew, diminished eyesight, cough, poisoning, impaired digestion, nervous disorders, high fever, excess sweating

43/ Emerald	
English	Emerald
Sanskrit	Tarksya, Marakata, Panna, Budha Ratna, Harita Ratna
Tibb	Zamrud, Zammarrud, Harid Ratna
Chemical formula	Al_2O_3 (oriental emerald) – beryl ($Be_3Al_2(SiO_3)_6$)
Mohs	7.5–8.5
Jyotish	Primary gemstone for Budha (Mercury)
Quality	Deep green, free of fissures and blemishes
Shodhana	Lemon/lime juice, rosewater, orpiment, realgar, sulphur, rock salt, Triphala decoction, milk and lotus seed decoction
Saṁpuṭa	8–10 (Gaja)
Anupāna	Honey
Dynamics	VPK, increases Ojas
Dosage	100–175mg
Treatment of	Lung diseases, asthma, bronchitis, immune dysfunction, CNS diseases, senility, poisoning, vomiting, anaemia, piles, oedema, digestive problems, general debility

44/ Topaz	
English	Topaz, corundum
Sanskrit	Pushparaga, Guru Ratna
CCM	Huâng Yù
Tibb	Yaqoot, Afsar, Yaqoot zard

Chemical formula	$Al_2SiO_4(F,OH)_2$
Mohs	8.0
Jyotish	Primary gemstone for Brihaspati (Jupiter)
Quality	Pushparaga (dense pale yellow), Kaurantaka (deep yellow/red), Kaṣāya (transparent yellow/red); Pushparaga has the highest medicinal value
Shodhana	Lemon/lime juice, rosewater, orpiment, realgar, sulphur, rock salt and lotus seed decoction
Saṁpuṭa	8–10 (gaja)
Anupāna	Honey and ghee
Dynamics	–V –K
Dosage	100–175mg
Treatment of	Poisoning, inflammation, skin diseases, vomiting, digestive disorders, improves appetite, removes heaviness, reduces burning sensations; Pushparaga is a Rasāyana and aphrodisiac

45/ Diamond	
English	Diamond
Sanskrit	Hiraka, Vajra, Kulisa, Bhidura, Hira, Shukra Ratna
CCM	Jîn Gân Shí
Tibb	Vajra, Heera
Chemical formula	C
Mohs	10
Jyotish	Primary gemstone for Shukra (Venus)
Shodhana	Lemon/lime juice, orpiment, realgar, sulphur, rock salt, Triphala decoction and Kulatha decoction
Saṁpuṭa	10–12 (gaja)
Anupāna	Ghee and milk, cream or jaggery
Dynamics	Yoga vāhin
Dosage	5–10mg
Treatment of	Cancer, tumours, immunodeficiency diseases (HIV), consumption, diabetes, anaemia, impotency, diminished eyesight, urinary tract infections, skin diseases, oedema and learning disabilities

Poisoning	Skin diseases, heart pains, burning sensations, vertigo, chest pains
Antidote	Milk sweetened with sugar

46/ Blue sapphire	
English	Blue sapphire, corundum
Sanskrit	Nilama, Nila Ratna, Maha nila, Shani Ratna
CCM	lân bäo shí
Chemical formula	Al_2O_3
Mohs	9.0
Jyotish	Primary gemstone for Shani (Saturn)
Quality	Indranila (deep blue/purple), Jalanila (whitish/blue)
Shodhana	Lemon/lime juice, rosewater, orpiment (asenic trisulphide), realgar, sulphur, rock salt, and lotus seed decoction
Saṁpuṭa	8–10 (Gaja)
Anupāna	Honey or milk
Dynamics	VPK
Dosage	100–175mg
Treatment of	Tuberculosis, arthritis, poisoning, brain dysfunction, diseases of the CNS, infertility, low Agni, skin diseases, general debility and weakness Nilama reduces –V –P types of disease
Poisoning	Skin diseases, heart pains, burning sensations, vertigo, chest pains
Antidote	Milk sweetened with sugar

47/ Garnet	
English	Garnet, hessonite, cinnamon stone
Sanskrit	Gomeda, Pinga spatika, Rāhu Ratna, Tamo mani
Chemical formula	$Ca_3Al_2(SiO_4)_3$
Mohs	7.5
Jyotish	Primary gemstone for Rāhu (northern lunar node)

Quality	Should resemble Go-meda, the colour of cow's fat
Shodhana	Lemon/lime juice, rosewater, orpiment, realgar, sulphur, rock salt, Triphala decoction or lotus seed decoction
Saṁpuṭa	8–10 (gaja)
Anupāna	Honey or sweetened milk
Dynamics	–P –K
Dosage	100–175mg
Treatment of	Consumption, skin diseases, anaemia, digestive disorders, high fever, learning difficulties, poisoning, mental disorders, loss of taste

48/ Cat's eye	
English	Cat's eye, chrysoberyl, alexandrite
Sanskrit	Vaiduryam, Marjakanetra, Lahasuniya, Ketu Ratna, Viduraja, Vayaja
Chemical formula	$BeAl_2O_4$
Mohs	8.5
Jyotish	Primary gemstone for Ketu (southern lunar node)
Quality	Best examples are a light golden brown in colouration, resembling milk and honey. Alexandrite displaying red to green colourations are also highly prized
Shodhana	Lemon/lime juice, rosewater, orpiment, realgar, sulphur, rock salt, Triphala decoction and lotus seed decoction
Saṁpuṭa	8–10 (gaja)
Anupāna	Honey or sweetened milk
Taste	Sweet
Dynamics	Cold, –P
Dosage	100–175mg
Treatment of	Blood diseases, heart disease, senility, loss of brain functionality, general debility, constipation, poisoning. Vaiduryam promotes longevity

Uparatna

49/ Red agate	
English	Red agate, chalcedony
Sanskrit	Akika, Rudhirapalanka
CCM	Mänäo
Tibb	Aqeeq surkh
Chemical formula	SiO_2
Mohs	6.5–7.0
Quality	Deep unblemished blood red or vermilion examples
Shodhana	Lemon/lime juice, rosewater, sulphur and rock salt, Bhringaraj decoction, milk, lotus seed decoction, or Reetha (soap nut)
Saṁpuṭa	7–8 (Gaja)
Anupāna	Honey, milk, or butter
Dynamics	Cold, dry
Dosage	125–250mg
Treatment of	Heart disease, internal bleeding, infertility, insanity, menorrhea, general debility, urinary calculi, weak gums, teeth, or jaw

50/ Sunstone[179]	
English	Sunstone, orthoclase feldspar
Sanskrit	Sūryakanta, Agnigarbha, Suryopala, Sūryamani
CCM	Rì cháng shí
Tibb	Sūryamani, Vahni Garbha, Jvalanopala

179 Sunstone is part of the Feldspar group. In Sri Lanka red spinel (balas ruby) may be substituted for either sunstone or red coral in astrological jewellery. The Spinel group (MgAl2O4) is largely a mix of aluminium and magnesium.

Chemical formula	$KAlSi_3O_8$ traces FeO_3
Mohs	6.0–7.0
Jyotish	Secondary gemstone for Sūrya (the Sun)
Quality	Red or yellow/orange colouration, shimmering like burning coals
Shodhana	Rock salt, rosewater, sulphur, realgar, lemon/lime juice, lotus seed decoction
Saṁpuṭa	7–8 (Gaja)
Anupāna	Honey
Dynamics	Hot, –V –K
Dosage	125–250mg
Treatment of	Heart disease, blood disorders, brain dysfunction, poor digestion, senility and V-K imbalances

51/ Moonstone	
English	Moonstone, orthoclase feldspar
Sanskrit	Chandrakanta, Chandramani, Chandropala, Sasi kanta, Indu kanta
CCM	Yuè cháng shí
Chemical formula	$KAlSi_3O_8$ opalescent
Mohs	6.0–7.0
Jyotish	Secondary gemstone for Chandra (the Moon)
Quality	Lustrous, oily and bluish white, resembles moonlight on water
Shodhana	Rock salt, rosewater, sulphur, realgar, *Aloe vera* juice, lotus seed decoction
Saṁpuṭa	7–8 (gaja)
Anupāna	Honey
Dynamics	Cold, –P
Dosage	125–250mg
Treatment of	Stomach disorders, blood diseases, heart disease, digestive disorders, high fever, Pitta imbalances

52/ Lapis lazuli	
English	Lapis lazuli, lazurite
Sanskrit	Rajavarta, Nilasma, Avarta mani, Lajavarda
Tibb	Lajward
Chemical formula	$3(NaAlSiO_4)\cdot Na_2$
Mohs	5.5
Quality	Without calcite or pyrite contamination, heavy and deep peacock blue
Shodhana	Lemon/lime juice, cow's urine, sulphur, Bhringaraj decoction
Saṁpuṭa	7–8 (Gaja)
Anupāna	Honey, milk, or butter
Taste	Pungent, bitter
Dynamics	Cold, –V –K, Rasāyana
Dosage	50–175mg
Treatment of	Diabetes, consumption, urinary disorders, blood disorders, excess Pitta, tuberculosis, anaemia, brain dysfunction, insanity, piles, irregular digestion, asthma, vomiting. Rajavarta promotes feelings of wellbeing

53/ Jade	
English	Jade, jadeite, loin stone
Sanskrit	Sangeyasab, Vyomāshma, Bhimapashana
CCM	Yìng yù
Tibb	Sang-e-yasara
Chemical formula	$NaAl(Si_2O_6)$
Mohs	6.5–7.0
Quality	Heavy, lustrous and oily, deep green or green-blue in colour
Shodhana	Lemon/lime juice, rosewater, orpiment, realgar, sulphur and rock salt, Triphala decoction, milk and lotus seed decoction
Saṁpuṭa	8–10 (Gaja)

Anupāna	Honey, milk or butter
Taste	Sweet
Dynamics	Heart and blood Rasāyana
Dosage	125–250mg
Treatment of	Heart disease, blood diseases, infertility, stomach disorders, gastric pain, urinary calculi, internal bleeding or internal wounds, bruises, colic pain, gastric pain and dysentery

54/ Turquoise	
English	Turquoise
Sanskrit	Pirojaka, Peroja
Chemical formula	$CuAl_6(PO_4)_4(OH)_8 \cdot 4(H_2O)$
Mohs	6.0
Quality	Without calcite or pyrite contamination, heavy, deep blue/green in colour
Shodhana	Lemon/lime juice, cow's urine, sulphur, Bhringaraj decoction
Saṁpuṭa	7–8 (gaja)
Amṛītakarana	Kidaram root, Suran (elephant yam)
Anupāna	Honey, milk or butter
Taste	Sweet and astringent
Dynamics	Cold
Dosage	50–175mg
Treatment of	Heart diseases, palpitations, eye diseases, poisoning, renal calculi, duodenal ulcer, infertility, constipation, learning difficulties

55/ Black tourmaline	
English	Tourmaline, schorl
Sanskrit	Vaikrānta, Kuvajra, Curna vajra, Jīrna vajraka
Chemical formula	$(Na,Ca)(Mg,Fe_2+,Al,Li)_3B_3(Al,Fe_3+)_6O_{27}(OH,F)_4$

Mohs	7.0–7.5
Quality	Krishna (black) and Marakata (emerald-coloured) are best for longevity Black is considered the best medicinal grade Swetha (white) is best for silver-making and Pita (yellow) for gold-making Rakta (red) and Nila (blue) have good healing properties Paravātaprabha (pigeon grey) and Syamala (brown) are inferior Tourmaline readily agglutinates with impurities in molten iron to form slag
Shodhana	Salt water, Kulatha decoction, lemon/lime juice, orpiment, realgar and sulphur
Saṁpuṭa	7–8 (gaja)
Anupāna	Milk and honey
Dynamics	Tourmaline has all six tastes. Can be used in place of diamond Vaikrānta is a powerful Rasāyana, it balances VPK
Dosage	125–250mg
Treatment of	Skin diseases, tuberculosis and weakness of the lungs, asthma, bronchitis, diabetes and urinary disorders, anaemia, high fever, ascites (abdominal disease)
Poisoning	Skin diseases and anaemia
Antidote	Ghee and Kulatha decoction

56/ Amber	
English	Amber, copal
Sanskrit	Kaharuba, Trinakanta, Trnakanta mani
Chemical formula	$C_{10}H_{16}O$
CCM	Hǔ Pò
Mohs	2.0–2.5
Quality	Lemon yellow; when rubbed it should smell lemony
Shodhana	Rosewater
Anupāna	Honey, milk, or cream
Taste	Sweet
Dynamics	Heart and brain tonic, warming and drying

Dosage	75–150mg
Treatment of	Heart disease, blood circulation, internal bleeding, bleeding piles, dysentery, menorrhea, gastro-intestinal weakness, sensitive stomach, nervous disorders

57/ Quartz	
English	Clear quartz
Sanskrit	Shiva Ratna, Spatika, Amala mani, Dhātu śāli, Sphatikopala
CCM	Shí yîng
Chemical formula	SiO_2
Mohs	7.0
Jyotish	Secondary gemstone for Shukra (Venus)
Quality	Clear, bright, free from fissures or blemishes
Shodhana	Lemon/lime juice, rosewater, sulphur and rock salt, Kulatha decoction
Saṁpuṭa	7–8 (gaja)
Anupāna	Honey and ghee
Taste	Sweet
Dynamics	Cold, –P
Dosage	125–250mg
Treatment of	Low immunity, blood disorders, high fever, burning sensations, internal bleeding, infertility, memory loss. Shiva Ratna is an excellent Rasāyana and is used as a substitute for diamond

Miscellaneous Rasa

58/ Lead tetroxide	
English	Red lead, minium
Sanskrit	Nāga Sindoora, Nāgaja, Nāgagarbha, Mangalya
CCM	Qiān dān
Tibb	Murdar sang
Chemical formula	Pb_3O_4
Mohs	2.0–3.0
Shodhana	Lemon or lime juice; sometimes no purification is undertaken
Anupāna	Ghee, butter
Dynamics	Hot
Dosage	30–60mg, dose not to exceed 1mg or be administered for more than 2–3 days (high risk factor – lead poisoning)
External Treatment of	Eczema, bone fractures, ulcers, herpes, scabies, leprosy, swellings, inflammation, and ringworm
Poisoning	Skin disease, stiff joints, arthritis, jaundice, oedema, depression, erratic sleep patterns, lethargy, brain inflammation, seizures, and coma
Antidote	Harītakī and goat's milk, Swarna Bhasma (gold ash)

59/ Serpentine	
English	Serpentine, magnesium silicate
Sanskrit	Nāgapashana, Nāgasma
Tibb	Zeher mohra
Chemical formula	$Mg_3Si_2O_5(OH)_4$
Mohs	3.0–4.0
Shodhana	Salt water, rosewater

Saṁpuṭa	5–6 (Varaha)
Anupāna	Milk or cream
Dynamics	Hot, dry, anti-Visha, Rasāyana
Dosage	125–500mg
Treatment of	Poisoning, excessive vomiting, snake bite, heart disease, infertility, learning difficulties. Nāgapashana is an excellent tonic for general debility and weakness

60/ Sodium borate	
English	Sodium borate, borax, tinkal
Sanskrit	Tankana, Tanga, Ranga, Saubhagya, Loha Shodhana
CCM	Péng shā
Tibb	Suhaga, Tinkar
Chemical formula	$Na_2B_4O7 \cdot 10H_2O$
Mohs	2.0–2.5
Quality	Pure white granular crystal mass
Shodhana	Heating and dehydration
Anupāna	Honey/water
Attributes	Anti-Visha, cleansing, cleaning. An excellent fluxing agent for metal
Taste	Pungent
Dynamics	Hot, –V –K
Dosage	75–125mg
Treatment of	Bleeding disorders, severe itching, ulcers and insect/animal bites, CNS disorders, high fevers, asthma, bronchitis, eye diseases

61/ Jew's stone	
English	Jew's stone, lime silicate, fossil encrinites
Sanskrit	Badarasma
Tibb	Sang-e-yahood, Hajrul yahood

Chemical formula	$CaSiO_3$
Shodhana	Salt water, rosewater, Kumārī (*Aloe vera*) gel, Arka latex (*Calotropis gigantea*) and sandalwood water
Saṁpuṭa	3–4 (Varaha); Unani recommends its use as Pisti
Anupāna	Butter, ghee, warm water, traditional wines, rosewater
Dynamics	Cold
Dosage	500mg–1g
Treatment of	Kidney stones, polycystic kidney disease, renal calculi, poisoning, snake bite, skin diseases, itching, high Pitta conditions

62/ Lodestone	
English	Lodestone, magnetite
Sanskrit	Kanta Pashana, Chamak
CCM	Ci shi, cí tiě kuàng
Chemical formula	Fe_3O_4
Mohs	5.5–6.5
Quality	Pita (yellow – Brahma), Krsna (black – Vishnu) and Rakta (red – Shiva) Yellow variety is used in Lohasiddhi Red variety is used in mercurial operations Black variety is the best Rasāyana
Shodhana	Rock salt, Triphala decoction, *Aloe vera*, sulphur, ghee and lemon juice
Saṁpuṭa	7–8 (gaja)
Anupāna	Milk sweetened with honey
Taste	Bitter
Dynamics	Cold, VPK, general Rasāyana
Dosage	125–250mg
Treatment of	Heart disease, consumption, diabetes, anaemia, obesity, parasitic infestation, nephritis, oedema, leucorrhoea, colic, dysentery, piles, fistula and jaundice Strengthens the liver, spleen and nervous system

63/ Asbestos	
English	Asbestos (magnesium silicates), actinolite, crysotile
Sanskrit	Kauseyasma
Chemical formula	$Mg_3Si_2O_5(OH)_4$
Mohs	2.5–3
Shodhana	Kumārī (*Aloe vera*) gel, rosewater or saltwater
Saṁpuṭa	7–8 (Gaja); other texts advocate Kauseyasma as Pisti
Anupāna	Milk, honey
Dynamics	Cooling, improves circulation
Dosage	125–250mg
Treatment	Diabetes, epilepsy, anaemia, urinary disorders, burning sensations, blood disorders, pyorrhoea, and gum disease

Miscellaneous animal products

64/ Peacock feather	
English	Peacock feather, pavo, Indian peafowl
Sanskrit	Mayūr piccha, Mayura piccha, Mayura pankha, Sikhi, Sikhandi, Kalapi
Chemical formula	$CaCO_3$
Quality	Brightly coloured tail feathers with a prominent 'eye' and strong spine
Shodhana	Ghee, honey, Kumārī (*Aloe vera*) gel
Saṁpuṭa	1 (20–30 cow dung cakes)
Anupāna	Ghee, Pippali and honey
Attributes	Anti-Visha
Taste	Bitter
Dynamics	Hot, –V –K
Dosage	125–250mg
Treatment of	Chronic bronchitis, asthma, chronic hiccups, breathing difficulties, poisoning, general weakness of the lungs, and chronic colitis An excellent lung Rasāyana

65/ Laksha	
English	Lac (insect)
Sanskrit	Lakshana
Latin	*Laccifer lacca* (belonging to the *Coccoidea* family)
Tibb	Luk
Quality	Reddish dye, pungent wax, and resin (non-water soluble)
Shodhana	Boiled in water to remove lac-dye or melted over a naked flame to extract shellac and button-lac
Attributes	Promotes strength and vitality, builds Asthi (bone tissue)

Taste	Cold and astringent
Dynamics	Cooling, but pungent in post-digestion, balances Pitta and Kapha
Dosage	1–2mg
Treatment of	Sciatica, osteoporosis, jaundice, asthma, epilepsy and heart palpitations. Laksha has excellent anti-inflammatory properties, is anti-bilious, stomatic, haemostasis and an aphrodisiac. It has anti-obesity properties, acts as an expectorant, is a tonic for the liver and kidneys, and destroys parasites

66/ Bezoar stone	
English	Gallstone (of cow), naturally occurring calcareous concretions massing in an animal's gastrointestinal system, gallbladder, or stomach
Sanskrit	Gorochana
Tibb	Hazrul bahr, Pedaru bazara, Fad zeher hewani
Chemical formula	Fatty acid calcium, and bilirubin calcium
Shodhana	Rosewater
Saṁpuṭa	4 (Varaha)
Anupāna	Milk
Dynamics	Cooling
Dosage	125mg
Treatment of	Heart palpitations, poisoning, pain, and stress Prevents miscarriage and preserves pregnancy

67/ Deer horn	
English	Deer horn, antelope horn
Sanskrit	Mrga Śrnga, Harina Śrnga, Ena Śrnga, Mrga vishanaka, Krishna Mrigam
Tibb	Qarn-ul-aiyal, Sankh gozan sokhta
Chemical formula	$Ca_{10}(PO_4)_6(OH)_2$ with ($CaCO_3$)
Shodhana	Kanji, milk, Katura murunga (*Sesbania grandiflora*), Arka (*Calotropis gigantea*) or Kumārī (*Aloe vera*) gel

Saṁpuṭa	4 (Gaja)
Anupāna	Butter, milk or ghee
Dynamics	Hot, drying –K Black variety of Bhasma is high in phosphate, the white variety higher in calcium
Dosage	250–500mg
Treatment of	Heart disease, pleurisy, pain in the sides of the chest, eye diseases, sinus and migraine conditions, cough and chronic hiccups

68/ Deer musk	
English	Musk gland (*Moschus moschiferus*) of Siberian deer
Sanskrit	Kastūrī, Kasturīmrig, Mriganābhi, Mringnaphā
CCM	She xiang
Tibb	Mushk
Purification	Drying in sunlight
Anupāna	Honey or Makara Dwaja
Taste	Acrid, aromatic
Dynamics	Highly pungent, heating, sharp, drying, –VK Kastūrī becomes less heating with age but more drying in nature
Dosage	75–150mg
Treatment of	Inflammation, nerve debility (paralysis), reduced vision, Rakta Pitta (bleeding), skin diseases, excess vomiting, poisoning, oral problems (diseases of the mouth etc.) Kastūrī is a general Rasāyana and aphrodisiac; it also increases the sperm count
Antidote	Kastūrī poisoning is reduced by Kapura (camphor oil), rosewater and Vaṃśalocana (bamboo manna)
Poisoning	Highly drying, headache, places strain upon the heart, causes memory loss and yellows the skin after excessive use

69/ Conch shell	
English	Conch shell
Sanskrit	Shankha, Kambu, Sunada, Dirgha nada, Kamboja
Chemical formula	$CaCO_3$
Quality	Shankha shell is a composite of calcium carbonate, iron, magnesium, phosphate and sulphate
Shodhana	Kanji, milk, Bhringaraj decoction
Saṁpuṭa	4 (Gaja)
Anupāna	Warm water or lime juice, Triphala decoction or the juice of Nimbu (lemons)
Dynamics	Cold
Dosage	125–250mg
Treatment of	Gastritis, indigestion, peptic ulcer, IBS, diarrhoea, duodenal ulcer, eye diseases (cataract), lung diseases and discharges from the ear canal

70/ Eggshell	
English	Eggshell
Sanskrit	Kukkutanda tvak
Tibb	Bezae murgh
Chemical formula	$CaCO_3$
Shodhana	Saline water (rock salt and Nara Sara), lemon juice or Bhringaraj decoction Milk (to prepare greyish Bhasma), Kumārī (*Aloe vera*) gel to prepare whitish Bhasma. Other methods include the combined use of Kukkutanda and Hiṅgula to produce Syama Varna Bhasma
Saṁpuṭa	4 (Varaha)
Anupāna	Honey, ghee or butter
Dynamics	VPK, builds Asthi Dhātu
Dosage	125–250mg
Treatment of	Osteoporosis, sciatica, asthma, leucorrhoea, bronchitis, diarrhoea, rickets, arthritis, and bone fractures

71/ Cuttlefish bone	
English	Cuttlefish bone
Sanskrit	Samudra Phena
CCM	Hăi piāo xiāo
Tibb	Kafe darya
Chemical formula	$CaCO_3$
Shodhana	Lemon juice, salty water
Anupāna	Milk
Dynamics	Builds Asthi Dhātu, VPK
Dosage	500mg–1g
Treatment of	Osteoporosis, mineral deficiency (silica), low phosphoric acid, bone fractures, arthritic joints, and discharges from the ear canal

72/ Mother-of-pearl	
English	Mother-of-pearl
Sanskrit	Sukti, Shuktika, Toutika, Durnama, Dirghakoshika, Muktamata, Maha sukti
CCM	Zhēn Zhū Mŭ
Tibb	Sadaf sadiq
Chemical formula	$CaCO_3$
Shodhana	Kanji, milk, Bhringaraj decoction and lemon juice
Saṁpuṭa	3–4 (Gaja)
Anupāna	Warm water, lemon juice, ghee or honey
Dynamics	Cooling Mukta sukti (pearl-bearing oyster), Jala sukti (edible oyster); of these, Mukta sukti is considered the more medicinal
Dosage	500mg–1g
Treatment of	Heart disease, poisoning, splenic disorders, colic, urinary stones, asthma, high fever, impaired digestion

73/ Goat bone	
English	Goat bone (usually femur)
Sanskrit	Ajasthi
Tibb	Bakri
Chemical formula	$CaCO_3$
Shodhana	Kumārī (*Aloe vera*) gel, milk, vinegar
Saṁpuṭa	3–4
Anupāna	Honey or milk
Dynamics	Improves Agni, –V
Dosage	250–500mg
Treatment of	Osteoporosis, rickets, bone fracture, calcium deficiency during pregnancy and early childhood, loss of hair and teeth, inflammation of gums

74/ Gypsum	
English	Gypsum, alabaster, calcium sulphate
Sanskrit	Godanti Haritāla, Godantika, Godanta
CCM	Shí gâo
Tibb	Godanti Haritāla
Chemical formula	$CaSO_4{\cdot}2H_2O$
Mohs	1.5–2.0
Shodhana	Hot water, Katumba (*Leucas cephalotes*) or Nimbu (medicinal lemon juice), Kumārī (*Aloe vera*) gel
Saṁpuṭa	3–4 (Gaja)
Anupāna	Milk or honey
Dynamics	Cooling, –P
Dosage	500mg–1g
Treatment of	Tuberculosis, asthma, anaemia, rickets, osteoporosis, high fever, burning sensations, diarrhoea, and dysentery

75/ Calcite	
English	Calcite
Sanskrit	Surama sapheda
CM	Fâng jiě shí
Chemical formula	Ca CO_3
Mohs	2.5–3.0
Shodhana	*Aloe vera* juice, hot water
Saṁpuṭa	4 (Gaja)
Anupāna	Milk or honey
Dynamics	Cold, –P
Dosage	250–500mg
Treatment of	High fevers, anaemia, lung diseases, diarrhoea, excessive thirst, and calcium deficiencies such as osteoporosis and rickets

76/ Chalk	
English	Chalk
Sanskrit	Khatika, Khantini, Lekhanamrittika, Khati, Shukladhātu
Tibb	Gil-e-safed, Vilayati churna
Chemical formula	C_aCO_3
Mohs	1.0
Purification	Water
Anupāna	Milk or water Chalk is usually consumed as Katikadi churna, i.e. chalk, herbs and sugar
Taste	Sweet and bitter
Post-digestive	Sweet
Dynamics	Cold, –P. Green-coloured chalk is useful in cases of diarrhoea
Dosage	125–250mg
Treatment of	Excessive bleeding (external and internal) wounds, eye diseases, hyperacidity, burning sensations, reduces Kapha Dosha

77/ Limestone	
English	Quicklime, slaked lime
Sanskrit	Sudhā
Tibb	Aahak maghsool, Chuna maghsool, Kilsa, Nura
Chemical formula	CaO quick lime, $Ca(OH)_2$ hydrated lime
Mohs	1.5–2.0
Purification	Water
Anupāna	Lime water or milk
Dynamics	Strong anti-acid and anti-poison effects
Dosage	To suit
Treatment of	Hyperacidity, duodenal ulcers, intestinal parasites, painful joints, grazes and cuts, insect stings and bites, inflammation, and warts (removes); reduces high Pitta

78/ Soapstone	
English	Soapstone, talc, steatite
Sanskrit	Dugdha pashana, Kshiri, Madhavi, Vajrabhra, Medasannibha
CCM	Huá shí
Tibb	Sang-e-jerahat
Chemical formula	$Mg_3Si_4O_{10}(OH)_2$
Mohs	1.0
Purification	Salt water, *Aloe vera* juice, rosewater
Saṁpuṭa	2–3 (Varaha); some texts advise its use without heating
Anupāna	Milk or honey
Taste	Sweet
Dynamics	Cold
Dosage	125–250mg
Treatment of	Skin diseases, high fever, leucorrhoea, gonorrhoea, chest pains, bleeding disorders, diarrhoea, dysentery, menorrhagia, persistent cough

Visha

79/ Cobra venom	
English	Cobra venom
Sanskrit	Sarpa-Visha
Quality	Venom of black cobra, milked during the cooler seasons
Shodhana	Cow's urine, fresh ginger juice, mustard seed oil
Anupāna	Milk sweetened with honey
Taste	Pungent
Dynamics	Hot, VPK, a Rasāyana
Dosage	5–10mg
Treatment of	High fever, high blood pressure, virility, senility, and poor digestion Sarpa-visha has excellent Rasāyana properties
Poisoning	Vomiting, drowsiness, diarrhoea, coma, and death
Antidote	Jayapāla, goat's milk, ghee, milk with saffron and honey, saline water

80/ Aconite	
English	Aconite (*Aconitum napellus*, *Aconitum ferox*, *Aconitum lycoctonum*, *Aconitum palmatum*, *Aconitum carmichaelii* and *Aconitum kusnezoffii*)
Sanskrit	Vatsanābha, Vatsanāga, Visha, Hālāhala, Brahmaputra, Sourastraka, Predeepana, Darada, Kakola, Haimavāta, Kustha, Pippala, Patala, Sthavar, Gara, Amṛita, Nilapushpak
CCM	Zhì Fù Zǐ, Zhi Cao Wu
Tibb	Beesh, Atees and Mitha zahar
Quality	White root variety is highly laxative and fast acting; darker root colouration has the opposite properties
Purification	Mustard seed oil and rock salt, goat's milk, cow's milk, or cow's urine
Anupāna	Milk or honey
Taste	Pungent, bitter, and astringent

Post-digestive	Pungent
Dynamics	Hot, light, dry and sharp, –V –K, Rasāyana
Dosage	5–20mg
Treatment of	Heart disease, consumption, rheumatism, skin diseases (leprosy), mercury poisoning, high fever, inflammation, asthma and persistent cough, colitis, paralysis, facial palsy, and stabbing pains. Vatsanābha is an excellent Rasāyana
Visha	Paralysis, spasm and tremor, dilation of the pupils
Anti-Visha	Adraka (fresh ginger juice), Takana (borax in water), betel leaf juice (*Piper betel*) or Kuchala (*Nux vomica*)

Upavisha

81/ Crown flower	
English	*Calotropis gigantea/Calatropis procera*, crown flower, swallow wort or giant milkweed
Sanskrit	Arka, Alarka, Ksheera, Kheeradala, Nityapushpaka, Mandara
Tibb	Ark, Gule mudar
Quality	Red and white flowers (for internal use), latex or milk of Arka (for external use and purification of other Rasa base materials)
Purification	Ghee, honey, castor oil, sesame seed oil
Anupāna	Milk, butter
Taste	Pungent and bitter
Post-digestive	Pungent
Dynamics	Hot, –V –K
Dosage	5–10mg
Treatment of	Growths (tumours), syphilis, whooping cough, asthma, oedema, enlarged spleen or liver, rheumatism, poisoning, stiff limbs, leprosy, itching, haemorrhoids, and parasites. Arka has excellent purgative and laxative properties
Visha	Stomach pains, burning sensations, vomiting, convulsions, and vertigo
Anti-Visha	Water and Gaireeka (red ochre), goat's milk with ghee

82/ Milk hedge	
English	Milk hedge, *Euphorbia ligularis*, *Euphorbia neriifolia*, *Euphorbia antiquorum*
Sanskrit	Sehunda, Snuhi, Daluk, Snuka, Vajrī, Sudhā, Vajradruma, Bahukantaka, Netrari, Ksheera, Vātari, Kandaka
Purification	Tamarind leaf juice (*Tamarindus indica*)
Anupāna	Milk
Taste	Pungent
Post-digestive	Pungent

Dynamics	Hot, bitter, heavy and oily, –V –K
Dosage	5–10mg
Treatment of	Rheumatic disorders, gout, enlargement of the liver, leprosy, poor digestion, abdominal disorders, haemorrhoids, loss of appetite and toothache Milk hedge removes growths and is a laxative
Visha	Nausea and vomiting, lethargy, bloating and abdominal pain, vertigo, and diarrhoea
Anti-Visha	Tamarind leaf juice with water and jaggery

83/ Dhattura

English	Blackcurrant swirls, thorn apple, *Datura stramonium*, *Datura metel*, *Datura alba*, *Datura fastuosa*, *Datura tatula*
Sanskrit	Dhattura, Kitava, Unmatta, Shiva sekhara, Kharjughna, Mahamoni, Ghantapushpa, Kantakaphala and Kanaka
CCM	Yang jin hua, Nao yang hua
Tibb	Dhattura, Turkhm Dhattura Safed
Quality	Black Dhattura (*Datura stramonium*), also known as blackcurrant swirls, are considered king of Visha; the white flowering variety, known as thorn apple, are also medicinal
Purification	Cow's urine or milk
Anupāna	Milk and ghee
Taste	Sweet and astringent
Post-digestive	Bitter
Dynamics	Heavy, –P –K, warming, narcotic, +V
Dosage	10–20mg
Treatment of	High fevers, leprosy and other skin disorders, rheumatic pain, neuralgia, sciatica, intestinal parasites, colic, cough, and head lice. Dhattura is antispasmodic, with hepatic properties. It improves the complexion and digestion, and promotes hair growth
Visha	Paralysis, blindness, heaviness, coma, and death
Anti-Visha	Cow's milk sweetened with sugar, juice of eggplant (*Solanum melongena*), or saline water

84/ Flame lily	
English	Flame lily, night flower, glory lily, *Gloriosa superba*
Sanskrit	Langali, Vishalya, Swarnapushpa, Raktapushpa, Garbhapatini, Sarini, Kalihari
Purification	Cow's urine
Anupāna	Milk
Taste	Bitter, pungent, astringent
Post-digestive	Pungent
Dynamics	–V –K, light and sharp, Rasāyana, +P, causes abortion of foetus
Dosage	5–10mg
Treatment of	Blood impurity, skin diseases, ulcers and boils, inflammation, haemorrhoids, colic and abdominal parasites. This plant has a strong laxative effect, the tuber is held during childbirth to dull delivery pain
Visha	Burning sensations, vomiting, vertigo, drowsiness, coma, and death
Anti-Visha	Goat's milk with ghee

85/ Indian oleander	
English	Indian oleander, *Nerium indicum*
Sanskrit	Karaveera, Laguda, Shatakumba, Ashwamarak, Ashvamtaka, Hayari, Chandataka
Quality	Varieties of flowering plant include red (best), yellow, white, black, and violet
Purification	Milk
Anupāna	Milk and ghee
Taste	Pungent and bitter
Post-digestive	Pungent
Dynamics	–V –K +P, light, drying, narcotic, highly toxic
Dosage	5–15mg
Treatment of	Heart disease, toxic blood, high fevers, skin diseases, ulcers and boils, kidney and urinary stones, intestinal parasites

Visha	Contains strychnine; fatal for all livestock except goats Seeds contain oleandrin
Anti-Visha	Buffalo milk with sugar candy; Arka bark powder with milk

86/ Indian liquorice	
English	*Abrus precatorius*, Indian liquorice, rosary pea
Sanskrit	Gunja, Raktika, Choonamani, Swethagunja, Uchchata, Raktagunja, Krishna choornaka, Kakapilu, Kamboji
CCM	Hung tou
Tibb	Gowanchi
Quality	Red, white and black legumes are considered the most medically effective
Purification	Milk, Kanji and ghee
Anupāna	Milk or honey
Taste	Bitter, pungent and astringent
Post-digestive	Pungent
Dynamics	Hot, light and drying, –V –K
Dosage	30–50mg
Treatment of	Leprosy, boils, itching skin, infertility, hair loss (alopecia), vertigo, asthma, excessive thirst, and intestinal parasites. Gunja is known to have some anti-carcinogenic properties
Visha	Causes vomiting and diarrhoea; contains abrin (intravenous assimilation is more toxic than oral assimilation)
Anti-Visha	Tamarind, dates or grapes with honey, pomegranate. Āmalakī ground with prickly amaranth (*Amaranthus spinosus*), mixed with sugar and taken with milk is also an anti-Visha

87/ Opium poppy	
English	Opium poppy, *Papaver somniferum*
Sanskrit	Khasabeeja, Khas khas, Kasakasa, Ahiphena
CM	O fu jung (Yin tzū shu)

Tibb	Afyun
Purification	Ginger juice, milk, water
Anupāna	Honey
Taste	Bitter and astringent
Post-digestive	Pungent
Dynamics	Hot, –K –V (+P), aggravates Vāta in high or prolonged doses
Dosage	60–125mg
Treatment of	Pain, nerve spasms/convulsions, inflammation, restlessness, asthma, arthritis, and low appetite. Has a powerful analgesic property
Visha	Contains morphine, codeine, papaverine, narcotine and thebaine. Causes inflammation, spermatorrhea (involutary semen discharge) and constipation if used for extended periods. Reduces Ojas and strongly increases Vāta over longer periods
Anti-Visha	Rock salt, Pippali (*Piper longum*) and emetic nut (*Xeromphis spinosa*), ground and drunk with hot water. Sasyaka (copper sulphate) taken with Tankana (borax) and ghee

88/ Cannabis	
English	Cannabis, *Cannabis indica*, *Cannabis sattiva*, *Cannabis ruderalis*
Sanskrit	Vijaya, Madika, Bhanga, Jaya, Madini, Matula, Bhangi, Chapala, Ananda
CCM	Hou Ma Ren, Ta Ma
Tibb	Bhang
Quality	Indica, Sattiva and Ruderalis. Indica is considered to be the most medicinally effective
Purification	Decoction of gum arabic bark (*Acacia arabica*), milk and ghee
Anupāna	Milk
Taste	Bitter
Post-digestive	Pungent
Dynamics	Hot, –K +P; aggravates Vāta if used in high dosage or for prolonged periods
Dosage	250–500mg

Treatment of	Loss of appetite, poor digestion, poor blood circulation, intestinal spasm, heart disease, poor urine flow, impotency, tetanus, kidney disease, menorrhea, tinnitus
Visha	Contains cannabinoids
Anti-Visha	Juice of jackfruit leaves (*Artocarpus heterophyllus*), yoghurt with ginger juice

89/ Croton	
English	Croton, *Croton tiglium*
Sanskrit	Jayapāla, Maladravi, Rachaka, Vibhedana, Saraka, Kumbhini
CCM	Pa tou
Tibb	Aarand
Purification	Milk, lemon/lime juice or Kanji
Anupāna	Cold water
Taste	Bitter
Post-digestive	Pungent
Dynamics	Hot, –V –K
Dosage	15–25mg
Treatment of	Constipation, jaundice, skin diseases, abdominal pains, intestinal parasites, haemorrhoids, water retention, gallstones, kidney stones
Visha	Cramps, dehydration, excessive thirst, abdominal pain, burning sensations, and spasms
Anti-Visha	Yoghurt, coriander seeds and jaggery

90/ Marking nut	
English	Marking nut
Latin	*Semicarpus anacardium*
Sanskrit	Bhallātaka
Tibb	Beladur
Shodhana	Jaggery, milk, and cow's urine

Anupāna	Milk
Taste	Pungent, bitter and astringent
Post-digestive	Sweet
Dynamics	Hot, –V –K (+P), Rasāyana for (Majjā) marrow
Dosage	125–500mg
Treatment of	Haemorrhoids, splenic disorders, leprosy, skin diseases, abdominal bloating, obesity, intestinal parasites, and learning difficulties. Bhallātaka is known to have some anti-carcinogenic properties in certain types of blood cancer
Poisoning	Skin irritation, itching (anus and urethra), nausea and internal bleeding
Antidote	Pulped coconut flesh, taken with milk, ghee, or water

91/ Poison nut	
English	Poison nut, quaker buttons
Latin	*Nux vomica*
Sanskrit	Kuchala, Kuchila, Kupaka, Vishatinduka, Vishadruma, Ramyaphala, Kalakoota, Vishamushti, Jalapa, Kakapeeluka
CCM	Lū sung kuo (Fan mu pieh)
Tibb	Azaraqi, Kuchila
Shodhana	Boiled in Kanji (rice vinegar), milk, fried in ghee or roasted in a sand bath
Anupāna	Milk, ghee or butter
Taste	Bitter and pungent
Post-digestive	Pungent
Dynamics	VPK, dry, sharp and light
Dosage	25–50mg
Treatment of	Poor digestion, infertility, low urine flow, constipation, poor menstruation, hyperacidity, asthma, paralysis, abdominal bloating, heart weakness, oedema and high cholesterol. Has excellent antiseptic properties
Poisoning	Dilated pupils, protruding eyeballs, asphyxia, titanic spasms, high blood pressure, and mental illness
Antidote	Aconite (*Aconitum napellus*), warm goat's milk with honey, ghee camphor (*Cinnamomum camphora*), coffee

92/ Guggulu	
English	Indian bdellium, mukul myrrh tree
Latin	*Commiphora mukul, Commiphora wightii, Commiphora myrrha*
Sanskrit	Guggulu
CCM	Mo Yao
Tibb	Mogla, Muquil
Shodhana	Matured resin requires no purification. Newly acquired resin (less than twelve months old) is purified by steaming or boiling in Triphala decoction. The cleaned resin is then filtered and pounded with ghee
Anupāna	Milk, ghee, and sugar
Taste	Bitter, pungent, hot, light and dry
Post-digestive	Pungent
Dynamics	-VK + P, cleaning, scraping, restores the senses
Dosage	1000mg–3g
Treatment of	Heart conditions, blood toxicity, liver diseases, diabetes, obesity, thyroid conditions, joint problems (arthritis, rheumatism), infertility, respiratory ailments, cough, asthma, haemorrhoids, and intestinal parasites
Antidote	Saffron in curd, Triphala decoction, or sweetened milk

Bibliography

Bhishagratna, K.K.L. (2006) *The Susrutha Saṃhitā* (English Translation). Varanasi: Author.

Bhāvamiśra (1998) *Bhāvaprakāśa* (trans. Shri Kantha Murthy). Varanasi: Krishnadas Academy.

Birla, G.S. (2000) *Destiny in the Palm of your Hand*. Rochester, Vermont: Inner Traditions/Bear & Co.

Birla, G.S. (2007) *Introduction to Hast Jyotish: Ancient Eastern System of Palmistry*.

Choy, L.K. (1995) *Japan: Between Myth and Reality*. Singapore: World Scientific.

Cobb, C. and Goldwhite, H. (1995) *Creations of Fire: Chemistry's Lively History from Alchemy to the Atomic Age*. New York: Basic Books.

Dasgupta, A. and Hammett-Stabler, C. (2011) *Herbal Supplements: Efficacy, Toxicity, Interaction with Western Drugs and Effects on Clinical Laboratory Tests*. Hoboken, NJ: John Wiley.

Dash, Vaidya B. (1980) *Basic Principles of Āyurveda*. Delhi: Concept Publishing.

Dash, Vaidya B. (1980) *Materia Medica of Āyurveda*. Delhi: Concept Publishing.

Dash, Vaidya B. (1994) *Iatro-Chemistry of Āyurveda (Rasa Shāstra)*. Delhi: Concept Publishing.

Dash, Vaidya B. (1996) *Alchemy and Metallic Medicines in Āyurveda*. Delhi: Concept Publishing.

DeFouw, H. and Svoboda, R. (1996) *Light on Life: An Introduction to the Astrology of India*. London and New York: Arkana.

Dole, V.A. (2006) *Sri Vagbhatacharyas (Rasaratna Samuccaya)*. Delhi: Sri Satguru Publications.

Dole, V. and Paranjpe, P. (2004) *A Text Book of Rasa Shāstra*. Delhi: Chaukhamba Sanskrit Pratishthan.

Dutt, M.N. (1908) *The Garuda Purana*. Calcutta: Mahmatha Nath Dutti.

Ede, A. (2006) *The Chemical Element: A Historical Perspective*. Westport, CT: Greenwood Publishing.

Frawley, D. (1994) *Tantric Yoga and the Wisdom Goddesses*. Salt Lake City, UT: Passage Press.

Frawley, D. (2005) *Āyurveda and the Mind: The Healing of Consciousness.* Delhi: Motilal Banarsidass.

Gerke, B. (2013) 'Asian Medicine Tradition and Modernity: Mercury in Āyurveda and Tibetan Medicine.' *Asian Medicine*, Vol. 8.1, Leiden: Brill Academic Publishers.

Gerke, B. (2021) *Taming the poisonous: mercury, toxicity, and safety in Tibetan medical practice*. Heidelberg University.

Gogte, Vaidya V.M. (1982) *Āyurvedic Pharmacology & Therapeutic Uses of Medicinal Plants (Dravyagunavignyan).* Varanasi: Chaukhambha Bharti Academy.

Habu, J. (2004) *Ancient Jomon of Japan.* Cambridge: Cambridge University Press.

Hempen, C.H., and Fischer, T. (2009) *A Materia Medica for Chinese Medicine.* Philadelphia, PA: Elsevier Health Sciences.

Holmyard, E.J. (1990) *Alchemy*. New York: Dover Publications.

Hsu, H.-Y. and Peacher, W.G. (1977) *Chen's History of Chinese Medical Science*. Taipei: Modern Drug Publishers.

Joshi, D. (1991) 'Concept of Āyurvedic Shodhana method and its effects with reference to sulphur.' *Ancient Science of Life X*, 4, 214–222.

Joshi, D. and Rao, P. (2003) *Rasamritam* (English Translation). Varanasi: Chaukhamba Sanskrita Sansthan.

Kabir, H. (2003) *Shamsher's Morakkabat (Unani Formulations).* Delhi: Shamsher.

Kemp, C. (2012) *Floating Gold: A Natural and Unnatural History of Ambergris.* The University of Chicago Press; Illustrated edition (11 May 2012).

Khan, Hakim M.S. (2009) *An Introduction to Islamic Medicine*. Bahr Press.

Khare, C.P. (2004) *Indian Herbal Remedies, Rational Western Therapy, Āyurvedic and Other Traditional Usage, Botany*. New York: Springer.

Kohn, L. (2000) *Daoism Handbook*. Leiden: Brill.

Madhihassan, S. (1977) *Indian Alchemy or Rasāyana (In the Light of Asceticism and Geriatrics)*. Delhi: Motilal Banarsidass.

Mahajan, K.K. (2002) *The Secrets of Vāstu: A Guide to Harmonious Living.* Parts 1 and 2. Delhi: Alpha Publications.

Mason, A. (2009) *The Art of Vedic Alchemy Volume 1.* Dorchester: Neterapublishing.

Mason, A. (2010) *The Art of Vedic Alchemy Volume 2*. Dorchester: Neterapublishing.

Mason, A. (2011) *The Art of Vedic Alchemy Volume 3*. Dorchester: Neterapublishing.

Mason, A. (2012) *The Art of Vedic Alchemy Volume 4*. Dorchester: Neterapublishing.

Mason, A. (2013) *The Art of Japanese Alchemy*. Dorchester: Neterapublishing.

Masood, E. (2009) *Science and Islam: A History*. London: Icon Books.

Meulenbeld, G.J. (1999) *A History of Indian Medical Literature, Volumes 1 and 2*. Groningen: Egbert Forsten.

Miao, W.-W. (2008) *Chinese Pearls, Traditional Chinese Folk Wisdom* (Translated by Yue Chong-Xi).

Morio, M. (2006) *A Kaleidoscope of China*. Beijing: China Intercontinental Press.

Morris, R. (2003) *The Last Sorcerers: The Path from Alchemy to the Periodic Table*. Washington, DC: Joseph Henry Press.

Mukherji, B. (1998) *The Wealth of Indian Alchemy and its Medicinal Uses*. Volumes 1 and 2. Delhi: Sri Satguru Publications.

Nicholson, P.T. and Shaw, I. (2000) *Ancient Egyptian Materials and Technology*. Cambridge: Cambridge University Press.

Orzech, C.D., Sorensen, H.H., and Payne, R.K. (2011) *Esoteric Buddhism and the Tantras in East Asia*. Leiden: Brill.

Paranjpe, P. (2003) *Āyurvedic Medicine: The Living Tradition*. Delhi: Chaukhamba Sanskrit Pratishthan.

Pole, S. (2013) *Āyurvedic Medicine: The Principles of Traditional Practice*. London: Singing Dragon.

Pormann, P.E. and Savage-Smith, E. (2007) *Medieval Islamic Medicine*. Edinburgh: Edinburgh University Press.

Puri, H.S. (2002) *Rasāyana, Āyurvedic Herbs for Longevity and Rejuvenation*. CRC Press; 1st edition (17 Oct. 2002).

Ramanan, K.V. (1966) *Nāgarjuna's Philosophy*. Delhi: Motilal Banarsidass.

Ray, A.P.C. (2004) *History of Chemistry in Ancient and Medieval India*. Varanasi: Chowkhamba Krishnadas Academy.

Roy, M., with Subbarayappa, B.V. (1976) *Rasanavakalpa (Manifold Powers of the Ocean of Rasa)*.

Samarasekara, C. (2003) *Āyurveda Rasa Shāstraya.*

Sarma, S.S. (2015) *Rasa Tarangini.* Chaukhamba Surbharati Prakashan; 2015th edition (31 Mar. 2015).

Sato, T. (1996) *Kukai and Renkinjutsu/Alchemy* (Tankobon) Japan.

Sato, T. (1998) *The Mystery of Kukai* (Shuppan) Japan.

Sato, T. (2000) *The Buddha Mystery* (Shuppan) Japan.

Sato, T. (2006) *The World of Siddha Medicine* (Shuppan) Japan.

Satpute, A.D. (2003) *Rasendra Sāra Saṅgraha of Sri Gopal Krishna.* Varanasi: Chowkhamba Krishnadas Academy.

Schmieke, M. (2002) *Vāstu: The Origin of Feng Shui.* Hatchend, UK: Goloka Books.

Seth, K.N. and Chaturvedi, B.K. (2001) *Gods and Goddesses of India.* Delhi: Diamond Pocket Books.

Sharma, G.C. (1994) Maharishi Parasaras, *Brihat Parasara Hora Shāstra, Volumes 1 and 2.*

Shiba, R. (2003) *Kukai the Universal: Scenes from His Life.* New York: ICG Muse.

Singh, L.B. (1996) *Poisonous (Visha) Plants in Āyurveda.* Delhi: Chaukhamba Sanskrit Bhawan.

Singh, R.H. (2001) *Pañca Karma Therapy.* Varanasi: Chaukambha Orientalia.

Smith, V.A. (1999) *Practical Āyurveda: Secrets for Physical, Sexual & Spiritual Health.* European Institute of Vedic Studies. Available at www.atreya.com/Āyurveda/-Vaidya-Atreya-Smith-.html.

Smith, V.A. (2000) *Dravyaguna for Westerners.* European Institute of Vedic Studies. Available at www.atreya.com/Āyurveda/-Vaidya-Atreya-Smith-.html.

Smith, V.A. (2003) *Pancha Karma.* European Institute of Vedic Studies. Available at www.atreya.com/Āyurveda/-Vaidya-Atreya-Smith-.html.

Snodgrass, A. (1985) *Symbolism of the Stupa.* New York: Cornell University Press/SEAP Publications.

Southgate, P.C. and Lucas, J.S. (2008) *The Pearl Oyster.* Oxford: Elsevier.

Sudarshan, S.R. (1999) *Rasa-Dhātu-Kosha: Metallic and Mineral Drugs in Āyurveda.* Kalpatharu Research Academy; 1999a edición (1 Enero 1999).

Tirtha, Swami S.S. (1998) *The Āyurveda Encyclopaedia: Natural Secrets to Healing, Prevention and Longevity.* Āyurveda Holistic Center Press.

Unno, M. (2004) Shingon Refractions, Myoe and the Mantra of Light. Somerville, MA: Wisdom Publications.

Vidyalankar, Pandit S. (2009) *The Holy Vedas*. Delhi: Clarion Books.

Vohora, S.B. and Athar, M. (2008) *Mineral Drugs*. Delhi: Narosa Publishing House.

Wayman, A. (1992) *Ryujun Tajima – The Enlightenment of Vairocana. Book I: Study of the Vairocanabhisambodhi Tantra*. Delhi: Motilal Banarsidass.

Weeks, P. (2012) *Make Yourself Better*. London: Singing Dragon.

White, D.G. (1996) *The Alchemical Body: Siddha Traditions in Medieval India*. Chicago and London: University of Chicago Press.

Wu, J.-N. (2005) *An Illustrated Chinese Materia Medica*. Oxford: Oxford University Press.

Wujastyk, D. (2003) *Roots of Āyurveda*. London: Penguin Classics.

Wujastyk, D. (2018) *Histories Of Mercury In Medicine Across Asia And Beyond*. Motilal Banarsidass Publishers; First Edition (1 Jan. 2018).

Yadav, C.S. (2003) *Animal Drugs*. Varanasi: Chaukhambha Orientalia.

Yamasaki, T. (1998) *Shingon: Japanese Esoteric Buddhism*. Fresno, CA: Shingon Buddhist International Institute.

Yang, S.-Z. (1998) *A Translation of the Shen Nong Bencao Jing*. Boulder, CO: Blue Poppy Press.

Yoke, H.P. (2007) *Explorations in Daoism – Medicines and Alchemy in Literature*. Oxford: Routledge Curzon.

Resources

Āyurveda Resources

Vaidya Ātreya Smith (Āyurvedic training)

Offering a three-level training programme to anyone interested in learning Āyurveda through advanced learning methods on an e-learning platform. Ātreya has been teaching since 1989, and his programmes are available for students all over the world.

Website: www.atreya.com

College of Āyurveda (UK)

Accredited by the British Complementary Medicines Association (BCMA), the college provides Āyurvedic practitioner-level programmes in Milton Keynes in addition to pioneering Āyurvedic education in France, Germany, Belgium, The Netherlands, Croatia, and Japan. The College of Āyurveda was first established as an independent non-profit organisation in 1997 under the aegis of the Āyurvedic Medical Association (UK) and the leadership of Dr. Mauroof M. Athique.

Website: www.ayurvedacollege.org

Dr Venkata Narayana Joshi (Āyurvedic practitioner)

74 Warren Road, Croydon, Greater London, CRO 6PF

Website: www.croydonayurvedacentre.co.uk

Philip Weeks Clinic

58B Crawford Street, London, W1H 4JW

Website: www.philipweeks.co.uk

Āyurveda (UK)

A bispoke detox & stress management retreat, specialising in Āyurvedic massage, Yoga & meditation. Three to seven-day residential packages available.

Website: āyurveda.uk.com

Essential Āyurveda

Essential Āyurveda provide a range of organic Āyurvedic supplements, oils and teas, lovingly crafted from the highest quality herbs. Committed to supporting independent growers, small batch production enables care in every step of processing – blending the very best from traditional techniques with advances in modern lab testing and production. Practitioners themselves, Essential Āyurveda also train therapists in Amnanda therapies and are passionate in sharing Āyurveda to help people heal and live well.

Essential Āyurveda Ltd,
Unit 7D,
Vale Industrial Estate,
Spilsby, Lincolnshire,
PE23 5HE

Website: www.essentialayurveda.co.uk

Unani Tibb Resources

Mohsin Clinic of Natural Medicine

446 East Park Road, Leicester, LE5 5HH

Website: www.mohsinhealthproducts.co.uk

CUTAM (College of Unani Tibb and Alternative Medicine)

Based in Manchester and London and offering courses in Unani Tibb, and Western and Eastern Nutrition.

Website: cutam.org.uk

Glossary

Abhraka – mica, of two varieties, black and white sheets of silicate, highly resistant to high temperature.

Abhraka Bhasma – mica reduced to its oxide form.

Abhyaṅga – whole body Āyurvedic massage with the use of Dosha-specific herbally infused oils.

Adhahpatana Yantra – a device used in the purification of mercury, typically, heated earthenware vessels, sealed to prevent the loss of mercurial vapor.

Adraka – ginger root.

Adrija – alternative name for Shilajit, a naturally occurring, water soluble bitumen found largely in the mountainous regions of Indian, Nepal and Pakistan.

Agada Tantra – Āyurvedic toxicology.

Agastya – within the context of Rasa Shāstra, *Agastya* generally refers to the plant *Sesbania grandiflora*. The name Agastya is also synonymous with the South Indian sage, who is most associated with Siddha medicine.

Agni – refers to the body's internal fire that cooks (digests food). Agni is also a Sanskrit word meaning 'fire', or a popular fire deity in Hinduism.

Agnijara – ambergris, a rare substance recovered from the intestines of *Physeter catodon* (sperm whales), however, whether it is regurgitated or excreted remains unknown.

Agnitapta Bhasma – Pisti or ultra-fine powder achieved through prolonged grinding and sieving.

Ahiphena – Sanskrit name for opium poppy (*Papaver somniferum*).

Ajasthi – goat bone (usually femur).

Akash – one of the five states of matter, in this case, space.

Akash Moti – literally, 'sky pearls'. A disputed and unknown mineral recovered after thunderstorms with a pearlescent quality. It is a highly prized gemstone in India.

Akasha Gaṅgā – the heavenly river or Milky Way.

Akika – agate, a coloured type of quartz.

Amla – Indian gooseberry or *Phyllanthus emblica*.

Amla (taste) – one of the six actions of taste in Āyurveda, in this case, the sour taste.

Āma – undigested food matter, seen as the primary internal cause of disease in Āyurveda.

Amṛīta – divine nectar, similar to ambrosia as mentioned in Greek mythology.

Amṛītakarana – to potentise, final purification process that nectarises a material.

Añjali – an ancient weight measurement, equivalent to a cupped handful.

Anjana – stibnite, an antimony ore, sometimes substituted with galena, (lead sulphide).

Anupāna – a unique 'delivery system' specific to Āyurveda that helps ensure a medicine is directed toward the targeted location for healing in the body.

Apamarga – prickly chaff flower, *Achyranthes aspera*.

Apara-Ojas – that portion of Ojas that freely circulates within the body, and is constantly being used and replenished by the conversion of foods.

Apsaras – female celestial spirits of cloud and water, later to be associated with legends of nymphs, fairies etc.

Ārchārya Nāgārjuna – scholar, monk and philosopher, associated with alchemical and magical feats. This colorful, historical figure is surrounded by a number of myths and legends from India to Japan.

Ardra – one of twenty-seven Nakshatra (lunar mansions) that the Moon frequents during its monthly course. Ardra or Alpha Orionis is associated with Lord Shiva, as it is in the Kalāpurusha-mandala, i.e. the constellation of Orion.

Ariloha – metals antagonistic to other metals, i.e. likely to undermine their cohesion.

Arka – crown flower, *Calotropis gigantea*, a plant heavily associated with the Sun.

Arq gulab – rosewater, distilled from the Damascus rose.

Ashwagandha – winter cherry or *Withania somnifera*, a popular Āyurvedic herb that appears in many Rasāyana formulas.

Ashwin Twins – mythological horse-headed twins, famed for their medical and surgical arts on the battlefield.

Aṣṭāñga Hṛdayam – one of the three great ancient classics of Āyurvedic wisdom. This text is perhaps more popular with modern students as it is written in a voice that speaks to modern ears. In many ways this text presents the best elements of the previous classics *Caraka* and *Susrutha Saṃhitā* in one authoritative text.

Asthi Dhātu – one of seven Āyurvedic bodily tissues; *Asthi* means bone.

Asuras – powerful demons in the ancient world, whom you might liken to the Titans of Greek mythology.

Atharva Veda – the fourth Veda, and a later addition to the Vedic scriptures of Hinduism.

Atman – symbolically that part of the awareness that survives death, i.e. eternal essence, that which is unchanging.

Aushdhi – pharmacy, medicine, place of healing.

Avaleha – an Āyurvedic preparation that includes herbal materials combined with sugar and fruits to form a paste, not unlike a jam.

Āyurvedic Pañcakarma – the Āyurvedic flagship protocol for detoxification and rejuvenation of bodily tissues.

Badarasma – fossilised crinoid (ancient marine life) prepared as a mineral remedy. Also known as Hajrul Yahood.

Bala Roga – Āyurvedic delineation of childhood diseases.

Basti – Āyurvedic enemas.

Bhallātaka – important Upavisha (semi-toxic) plant, whose nuts produce a heavy and pungent black oil, also known as bhilawan nut.

Bhasma – alchemical ash produced using traditional alchemical techniques.

Bhavana – impregnation of liquid mediums to aid in the grinding and purification of materials.

Bhringaraj – *Eclipta alba* (false daisy) is an important Āyurvedic herb and appears in a large number of alchemical preparations. It has a higher than average density of plant salt.

Bhumi – Hindu goddess, also known as Bhudevi and Vasundhara, who is the personification of the earth.

Bhūtavidyā – Āyurvedic treatment of diseases, both psychic and supernatural in origin.

Bibhītaki – *Terminalia bellirica*, also known as Baheda, is one of three fruits used in the well-known Āyurvedic formula, Triphala.

Bṛāhmi – *Bacopa monnieri* (Indian pennywort) is an important and much used herb in Āyurvedic formula.

Brihaspati – Vedic deity closely aligned to the planet Jupiter.

Budha – Vedic deity closely aligned to the planet Mercury.

Budha Ratna – emerald, the primary gemstone related to the planet Mercury.

Cakrika – term used for the rounded, flat pillet prepared during the alchemical preparation of minerals, metals and gemstones.

Canopus – Prominent southern star in the constellation Carina, related to the sage Agastya.

Caraka Saṃhitā – one of three great ancient classics of Āyurvedic wisdom. Caraka is especially popular with students of internal medicine.

Chandra – Vedic deity closely aligned to the Moon.

Chandra moti – genuine marine pearl.

Chandrakanta – moonstone, a type of orthoclase feldspar.

Chapala – unsteady, in motion, a term applied to unfixed mercury. An unknown material also known as Chapala appears in Maha Rasa, it is sometimes associated with bismuth sulphide.

Cikitsāsathānam – Section Six in *Caraka Saṃhitā* deals with the study of preservation of health, prevention and management of various diseases.

Collyrium – medicines for the eye, usually a type of salve.

Cyavana Prāśa – well known Āyurvedic formula for rejuvenation of the respiratory system or convalescence.

Damaru Yantra – two-chambered device used in the sublimation of mercurial ores.

Dashamula – well known Āyurvedic formula combining the potent attributes of ten herbal roots; *dasa* = 10 and *mula*=root.

Dehasiddhi – acquisition or techniques that make the human body impervious to disease and ageing.

Devanagari – pre-Classical Sanskrit, an Indo-Aryan language dating back to around 1500 BCE.

Devas – exalted, shining ones, heavenly, divine beings.

Dhanvantari – Āyurvedic deity associated with the healing arts.

Dhattura – *Datura metel*, an Upavisha (toxic) plant, a favourite placatory plant of Lord Shiva.

Dhātu – meaning 'that which supports' can refer to any one of seven bodily tissues or one of the known seven Vedic metals.

Dinacharya – recommended daily routines to stave off illness and age.

Dolā-yantra – swing device, a steaming apparatus used in the purification of various minerals, metals and gemstones.

Dosha – means 'that which is prone to imbalance' or inherent flaws.

Dugdha Pashana – soapstone/talc.

Eranda – *Ricinus communis* or castor oil plant, is also known as the 'king of oils'.

Fajar – one of five important daily prayers in the Islamic tradition. Fajar is offered at dawn.

Fatkadi – alum or aluminium sulphate.

Gaireka – red iron oxide, soft enough to easily powder.

Gaja Puṭa – a large earthen pit whose dimensions are roughly equal to the impression left by the belly of a resting elephant.

Galena – lead sulphide.

Gandhaka – the element sulphur.

Garuda – an important Hindu deity (half man half bird), who primarily acts as the vehicle of Lord Vishnu.

Garuda Purana – one of eighteen important Mahapurana or great Hindu texts that expounds on a number of varied subjects, including palmistry, gemstones and healing.

Gauri Pashana – one popular name for white arsenic,.

Ghee – clarified butter, also written as *Grtha*.

Giri Sindoora – mercuric oxide. This material is rarely found in nature but commonly prepared in the laboratory.

Godanthi Haritāla – gypsum is a soft sulphate mineral, comprised of calcium sulphate.

Godugdha – milk.

Gokshura – *Tribulus terrestris* (puncture vine) is an important Āyurvedic herb and appears in a large number of alchemical preparations.

Gola – means rounded, or ball-like, sometimes used to refer to the Earth.

Gomeda – garnet can be found in a variety of colours, however the reddish shade predominates. The name *go-meda* means, 'likened to cow's fat'.

Gomutra – cow urine, is another popular alchemical ingredient, especially for those processes that involve iron.

Graha – means to seize or grasp, it also refers to one of the nine astrological manifestations of the planets.

Guḍūchī – *Tinospora cordifolia* or Giloy, is an important Āyurvedic herb and appears in a large number of alchemical preparations.

Guggulu – *Commiphora wightii* or Indian bdellium-tree is a highly prized tree that produces a medicinal resin. The use of this resin is very old, being mentioned in some of the Vedas.

Gunja – *Abrus precatorius* or crab-eyed pea, appears in the category of Upavisha (mildly toxic plant). This useful legume appears in a number of alchemical remedies.

Guru – planet Jupiter.

Guru Ratna, a stone sacred to the planet Jupiter, typically yellow sapphire or topaz.

Hālāhala – original poison, released from the milky ocean during Sagar Manthan.

Harid ratna – alternate name for emerald.

Haridra – *Curcuma longa*, better known as turmeric.

Harītakī – *Terminalia Chebula*, also known as *Chebulic myrobalan*, is one of three fruits used in the well-known Āyurvedic formula, Triphala.

Hastā – means 'the hand'.

Hastā Rekha Shāstra – *Shāstra* means knowledge and *Hastā* means the hand or palm. *Rekha* means 'line' or, in this case, 'lines' upon the palm.

Hingola – alternate name for Hiṅgula (cinnabar).

Hingu – *Ferula foetida*, also known as devils dung, is a popular Āyurvedic herb, known for its powerful digestive enhancing properties. The raw resin requires some purification prior to use. Its smell is unmistakable.

Hiṅgula HgS – cinnabar, the name refers to the colour of this mineral, specifically matching the deep red plumage of a parrot.

Hiraka – diamond.

Hiraka Bhasma – the alchemical ash of a diamond.

Hṛdayam – all things relating to the heart, or that which nourishes and heals the heart.

Indra – Vedic thunder god and lord of heaven, some draw parallels to the Greek god Zeus, or the Roman god Jupiter.

Indranila – alternate name for blue sapphire.

Jala – water

Jalamahābhūta – in five element theory, the water element.

Jalukā – that which resides with water, i.e. leech.

Jalukā Shodhana – process whereby a leech is either prepared for medical use or just after medical application. Shodhana means to clean, prepare or make suitable for human use/consumption.

Jayapāla – *Croton tiglium*, also known as purging croton. Croton seeds are used in a number of alchemical operations, they have a strong purgation effect when taken internally.

Jyestha – one of the twenty-seven Lunar asterisms or lunar mansions.

Jyotish – the ancient art of Vedic astrology, based upon the sidereal zodiac.

Kaharuba – fossilised tree amber.

Kajjali – black mercuric sulphide (or HgS) obtained by triturating purified Pārada (mercury) and Gandhaka (sulphur) together until united into a fine, smooth matt black powder.

Kajjali Bandha – the act of binding HgS into a stable form such as *Pottali*, small solidified balls of mercuric sulphide.

Kamala – lotus. All parts of the lotus plant are useful in Rasa Shāstra, its seeds are particularly useful in the processing of gemstones into alchemical remedies.

Kampilla – *Mallotus philippinensis*, also known as Kamala (not to be confused with lotus), produces a red seed pod, the covering of which showers a red powdery coating which can be used medicinally.

Kanchanara – *Bauhinia variegate* (also camel foot tree) is another popular Āyurvedic herb. The powdered bark of this tree is known to have excellent astringent, scraping properties.

Kanji – fermented or soured rice water, is used in a large number of alchemical purification practices, due largely to its availability and ease of making, i.e. watery rice water is simply allowed to ferment for a few days.

Kanksi – Another popular name for alum (aluminium sulphate).

Kankushta – an Uparasa, as yet unidentified material. Himalayan rhubarb (*Rheum emodi*) or Malabar tamarind (*Garcinia Cambogia*) have been suggested as likely candidates.

Kansya – the alloy bronze, typically mixed at a ratio of Cu: 8 parts, Sn: 2 parts.

Kanta Loha – iron extracted from magnetic iron ore.

Kanta Pashana – magnetic iron ore, also known as lodestone.

Kapardika – typically cowries (marine gastropod mollusk shells), are collected and used in alchemical remedies for all manner of digestive aliments. Yellow shells are therapeutically thought to be the best variety.

Kapha – one of three Dosha (or humors), used to define a constitutional type in Āyurveda. Kapha relates to the water element.

Kapota Puṭa – dove, or very light puṭa, approximately eight cow dung cakes in volume.

Kapura – another name for camphor, extracted from the tree *Camphora officinarum*.

Karaveera – Upavisha plant Indian oleander (Nerium oleander).

Karawella – *Momordica charantia* or bitter gourd, used in a number of alchemical operations.

Karma Rekha – one's fate line on the palm. It is associated with the middle finger and planet Saturn. It is studied to assess ones destiny.

Kasakasa – another popular name for the opium poppy (*Papaver somniferum*).

Kaṣāya – Sanskrit for astringency or the astringent principle; the elements of air and earth predominate to desiccate and bind.

Kasisa – ferrous sulphide. In the modern era FeS is almost 100% synthetic, however, it can be prepared alchemically, or found naturally occurring in minerals such as pyrites.

Kastūrī – musk deer caudal gland, extracted from the male deer.

Kaṭu – Sanskrit for pungency, the elements fire and air predominate, producing heat and burning.

Kāyachikitsa – Āyurvedic therapy administered through internal medicines.

Ketu – the southern lunar node or the body of Rāhu (minus his head). Ketu attained planetary status, lording comets, meteors and sunspots.

Ketu Ratna – cats eye (chrysoberyl), the gemstone most associated with the southern lunar node.

Khalwa Yantra – pestle and mortar.

Khaskhas – another popular name for the opium poppy (*Papaver somniferum*).

Kidaram – elephant yam (*Amorphophallus paeoniifolius*), used in the Amṛītakarana of copper.

Kithul jaggery – date palm sugar harvested from *Caryota urens*. This dark brown jaggery is considered the best medicinal grade.

Krishna Mrigam – blackbuck, is a species of antelope (*Antilope cervicapra*) found in India, Pakistan and Nepal.

Krishna Sarpa Masi Bhasma – a Rasa Shāstra preparation, Bhasma made from the whole body of a black cobra.

Krittika – one of twenty-seven lunar asterisms, ruled by the fire deity Agnī. Its symbol is a razor or bladed weapon.

Kuchala – the Upavisha plant *Nux vomica*, also known as quaker button.

Kuja – planet Mars, also known as Mangala, the auspicious one.

Kuja Ratna – the gemstone of planet Mars; is red coral (Pravala) or his secondary gemstone, red spinel, also known as balas ruby.

Kūkai – also known as Kōbō Daishi, was the founder of Shingon Buddhism in 9th century Japan.

Kukkuta Puṭa – earthen pit about ten inches square or the 'hop' of a rooster; *Kukkuta* means 'rooster'.

Kukkutanda tvak – eggshell.

Kulatha – horse gram bean, *Macrotyloma uniflorum*, a legume rich in iron and used in many alchemical purification techniques.

Kumārī – aloe vera gel.

Kūpīpākwa – specially wrapped glass bottle, usually coloured glass. The cotton cloth and clay wrappings help the glass withstand high temperatures.

Kushmanda – ash pumpkin or wax gourd, *Benincasa hispida*, is a popular type of gourd used in a number of alchemical purification techniques.

Kwatha – the term used for Āyurvedic medicinal decoctions.

Laghu – lightness, little or small.

Laghu Puṭa – earthen pit about fifteen inches square, these 'light' puṭa are used when processing sensitive materials, such as mercurials.

Laksha – the reddish resin extracted from the cocoons of *Laccifer lacca*, or lac beetle.

Lakshmi – Indian wealth goddess, consort to Lord Vishnu.

Langali – flame lily, *Gloriosa superba*, also known as Raktapushpa.

Lauha – alternate name for iron.

Lavaṇa – salt or the salty taste.

Lavaṇavarga – important table of salts used in Āyurveda and Rasa Shāstra.

Lepas – medicated pastes, applied externally to wounds.

Loha – iron (Fe).

Loha Bhasma – black iron oxide attained via alchemical processing.

Lohasiddha – the art of transmuting lower 'base' metals into precious metals such as gold and silver.

Maanikya – ruby, Al_2O_3 plus chromium.

Madura – rusted iron, typically one hundred years in age.

Majjā Dhātu – nervous system tissue, typically gauged by the quality of the eyes, i.e. whiteness, brightness etc.

Makara Dwaja – mercurial preparation, red crystals of mercury prepared in a glass bottle (kūpī). Chemical formula is HgS.

Maha Puṭa – the largest earthen pit required for puṭa. *Maha* means 'great' or 'grand'.

Mala – natural bodily wastes, i.e. sweat, urine, faeces, oils etc.

Māmśa – muscle tissue and sinew, as outlined in Sapta Dhātu (seven bodily tissues).

Manah śhilā – ruby arsenic, or arsenic sulphide As_4S_4.

Mangalya – alternative name for red lead, Pb_3O_4.

Mārana – to calcine or 'kill' a material, i.e. not able to assume its previous state.

Matsya moti – literally, 'fish pearls' typically extracted from the eyes of large fish. In some cases, it is claimed large calcite-like pearls have been found in the bodies of beached whales.

Mayūr Piccha – the 'eyed' tail feathers of the male peacock.

Mayūr Piccha Bhasma – ash obtained from the eyed tail feathers of the male peacock.

Medas Dhātu – fatty tissue, as outlined in Sapta Dhātu (seven bodily tissues).

Mgrashirsha – one of twenty-seven Nakshatra (lunar mansions) that the Moon frequents during its monthly course. Mgrashirsha means the head of the deer and is associated with Lord Shiva as he is in the Kalāpurusha-mandala, i.e. the constellation Orion, specifically the star *Lambda Orionis*.

Mhurta – previously selected auspicious moment to undertake an action, or a period of 48 minutes, associated with the moon's passage through the zodiac.

Mleccha – Vedic term for foreigner, barbarian or invader.

Mohini – Lord Vishnu in the guise of a temptress, designed to deter the demons from partaking in the Amṛīta ceremony.

Moti – another name for pearls.

Mrddara Śrnga – yellow lead oxide, also known as Litharge, PbO.

Mrga Śrnga – deer horn.

Mukta – another name for pearls.

Mukta Bhasma – alchemical ash prepared from calcined pearls.

Mukta Pisti – alchemical powder prepared from ground pearl in rosewater.

Mukta Shukti – mother-of-pearl shell.

Mula – means 'the root'.

Musa – earthenware crucible, designed to endure very high temperatures.

Nāga – typically, the metal lead, but may also refer to snakes.

Nāga Bhasma – lead oxide prepared form calcined lead.

Nāga Folk – legendary beings that inhabit the underworld, appearing as half-human, half-snake figures.

Nāga Moti – pearls obtained from the heads of hooded cobra.

Nāgapashana – serpentine stone, a type of magnesium silicate.

Nāga Sindoora – red lead, or lead tetroxide; is a commonly prepared artificial pigment.

Narasara – ammonium chloride NH_4Cl.

Nasya – nasally administered Āyurvedic medicines, usually oil-based.

Naukakruti kharal – iconical boat-shaped mortar, commonly cut from black granite.

Nava Ratna – nine gemstones associated with the nine planets in Jyotish (Vedic astrology). These are arranged into a squared pattern (or Yantra) and worn to placate the negative effects of the Graha (planets).

Navavisha – nine highly toxic substances used to fix and cleanse mercury.

Netra Tarpana – an Āyurvedic therapy that bathes the eyes in warm ghee.

Nila Ratna – another name for blue sapphire.

Nilama Bhasma – alchemical ash prepared from calcined sapphires.

Nilama Pisti – alchemical powder prepared from ground sapphires in rosewater.

Nilanjana – a variation of antimony sulphide, Sb_3S_3.

Nimbu – citrus lemon.

Nirgundi – a popular Āyurvedic herb (*Vitex negundo*) used in the processing of metals.

Ojas – seen as the culmination of Sapta Dhātu (seven bodily tissues), Ojas upholds the functionality of the seven tissues and bestows immunity to the body.

Pācana – increasing appetite.

Padelon Lavaṇa – naturally occurring black salt.

Padmaraga – alternate name for rubies.

Pancagavya – five products of the cow, these are: milk, ghee, yoghurt, urine and faeces.

Pañcakarma – five purification techniques used to restore health and vitality to the old, sick or convalescing.

Pancalavaṇa – five important salts used in alchemical preparations, these are: rock salt, black salt, sea salt, ammonium salt and Romaka salt.

Pancha Loha – five important metals used in medicines and in the construction of yantra, these are: bronze, copper, lead, brass and iron.

Pañcha Mahābhūta – the five 'great' elements, of which all materials are comprised.

Panna – another name for emeralds, $Be_3Al_2(SiO_3)_6$.

Para-Ojas – the superior portion of Ojas that resides in the area around the human heart.

Pārada – the element mercury (Hg).

Pasloha – alternate name for Pancha Loha.

Patala – meaning 'subterranean' or the name assigned to the capital city of the Nāga Folk.

Pippali Grtha – a medicated ghee based around its main ingredient, Pippali (long pepper).

Pirojaka – alternate name for turquoise.

Pisti – ultra fine powder.

Pitta Dosha – one of three Dosha (or humors), used to define one's constitutional type in Āyurveda. Pitta relates to the fire element.

Pittala – brass, an alloy prepared from copper and zinc.

Prakriti – relates to primal matter.

Prana – vital breath, causing bodily animation.

Pravala – another name for red coral.

Pravala Pisti – red coral ground in rosewater.

Prithvi – the name for the Earth, or solid, stable matter.

Puja – rituals performed by Hindus in homage to one or more deities.

Punarvasu – one of twenty seven Nakshatra (lunar mansions) that the Moon frequents during its monthly course. Punarvasu (*Alpha Geminorum*) is one of two wealth-bringing stars in the constellation of Gemini.

Puranas – meaning 'in ancient times', are a collection of thirty-six religious texts written in Sanskrit and in story form cataloguing the history of the universe.

Pushpa – means flowering, or fragrant.

Pushparaga – another name for topaz, Al_2SiO.

Pushparaga Bhasma – alchemical ash prepared from calcined topaz.

Puṭa – means to envelop and heat (cook).

Rāhū – one of the nine Graha (planets); Rāhū is the north lunar nodal point.

Rāhū Ratna – gemstone sacred to Rāhū, in this case, garnet/hessonite.

Rajata – another name for silver, Ag.

Rajata Bhasma – alchemical ash prepared from calcined silver.

Rajavarta – another name for lapis lazuli.

Rakta – red coloured, usually with reference to blood.

Rakta Dhātu – blood, as a bodily tissue.

Raktamokshana – therapeutic bloodletting.

Raktika – crab-eyed beans or jequirity (*Abrus precatorius*).

Rasa Dhātu – one of seven bodily tissues in Āyurveda.

Rasa Maanikya – alchemical preparation that uses roasted arsenic to produce a material that resembles rubies.

Rasa Parpati – a flattened disc of HgS, prepared from Kajjali.

Rasa Sindoora – red mercuric crystals prepared by sublimating HgS in a glass bottle known as a kūpī.

Rasaka – another name for smithsonite, a naturally occurring form of zinc sulphide (ZnS).

Rasasiddha – one highly adept in the art of alchemical transformations.

Rasāyana – a word with many interpretations, but within the context of Rasa Shāstra, *Rasa* = mercury, and *Āyana* = pathway or 'way of mercury'. In the context of Āyurveda, it is a rejuvenation therapy/medicine.

Ravi Ratna – the gemstone most associated with the sun, in this case a Sūryakanta (sunstone).

Reetha – Indian soapnut (*Sapindus emarginatus*).

Rig Veda – the *Rigveda* is the oldest known Vedic Sanskrit text, it is one of the four sacred canonical Hindu texts.

Rishi – ancient blessed or wise ones, keepers of wisdom and knowledge.

Rohini – one of twenty-seven Nakshatra (lunar mansions) that the Moon frequents during its monthly course. Rohini (*Alpha Taurii*) was the Moon's favourite star in the zodiac.

Rudra – wrathful emanation of Lord Shiva associated with destruction, rain, storms and high winds.

Sadharana Rasa – common, or those materials between Uparasa and Dhātu, the metals.

Sagar Manthan – an ancient Vedic tale which recounts 'the churning of the milky ocean,' in order to extract the elixir of eternal life, known as Amṛīta.

Saindhava Lavaṇa – rock salt.

Sama Veda – one of the four sacred canonical Hindu texts.

Sambar Lavaṇa – sambar salt; is a chemically unique type of salt recovered from a dried lake in Rajasthan.

Saṁpuṭa – the art of closing and wrapping a crucible, prior to cooking.

Saṁskāra – necessary actions, to be undertaken.

Samudra Lavaṇa – sea salt.

Samudra Phena – cuttlefish bone.

Sangeyasab – alternate name for jade, ie: jadite – $NaAlSi_2O_6$. or its substitute, nephrite: $Ca2(Mg, Fe)_5Si_8O_{22}(OH)_2$.

Sanskrit – an ancient Indo-Aryan language that has a rich historical and cultural significance. As one of the earliest known languages as well as the liturgical language of Hinduism, it has had a profound influence on the language and culture of the Indian subcontinent.

Sapta Dhātu – the seven bodily tissues or channels, according to Āyurveda.

Sarpa-visha – snake venom, most specifically cobra venom.

Sasyaka – copper sulphate CuS or copper sulphate in its naturally occurring form, such as copper pyrite. When roasted, this material can appear slightly bluish and is more commonly known as bornite or peacock ore.

Shad-Rasa – the six tastes used in Āyurvedic medicine.

Shalya Tantra – Āyurvedic surgery and surgical techniques.

Shani – planet Saturn.

Shani Ratna – the gemstone most associated with planet Saturn, in this case blue sapphire, blue amethyst, or blue john (Derbyshire Spar).

Shankha – conch shell.

Shankha Moti – pearls obtained from Shankha shells.

Sharaava – lidded crucibles, usually pot shaped.

Shatāvari – a popular Rasāyana and female tonic herb used in Āyurvedic medicine, Latin name *Asparagus racemosus*.

Shilajit – curious black bitumen/asphaltum excreted from certain types of rocks in the Altai and Himalayan mountain range, also known as Adrija.

Shiva – one of three principle deities in Hinduism, known as *Trimurti*, Shiva: *he that Destroys* is attributed with dissolution. Within the context of this work, Lord Shiva is the deity principally evoked in Rasa Shāstra.

Shiva Ratna – clear quartz or rock crystal, SiO2. Quartz is sacred to the Hindu deity Lord Shiva.

Shodhana – to cleanse, purify or make ready for assimilation.

Shrávana – one of twenty-seven Nakshatra (lunar mansions) that the Moon frequents during its monthly course. Shrávana (*alpha Aquilae*) or Altair was known as *Vāsudeva*, the 'light of god'; indeed this is the principle star of Vishnu.

Shukra Dhātu – the seventh bodily tissue, primarily thought to relate to reproductive tissue, resistance and immunity.

Shukra Ratna – the gemstone most associated with planet Venus, in this case diamond or white sapphire.

Siddhi – special powers awakened in the body through special forms of mediation or divine/alchemical drugs.

Sindoora – literally means 'reddish colour', but can be used to refer to the red oxides of mercury, as well as red lead.

Sneha Basti – oil enemas.

Snuhi – white latex exudate from the *Euphorbia* (cactus) plant species. This highly alkalising material is used in a number of Rasa preparations or the manufacture of Ksharasutra, (medicinal thread).

Somā – principally the Moon, or one of three primary forces in nature, i.e. *Somā* = water, *Tejas* = fire and *Vāyu* = wind. Also those herbs that have a rejuvenating quality.

Somā Ratna – the gemstone most associated with the Moon, in this case, a pearl or moonstone.

Sphatika – another name for Shiva Ratna (rock crystal) or crystals of alum.

Śrnga – horns, of either deer or cattle.

Sukti – mother-of-pearl.

Suran – elephant foot yam, used in certain Amṛītakarana processings.

Sūrya – the Sun.

Sūryakanta – a gemstone associated with the Sun, in this case, literally sunstone.

Susrutha – one of the three great ancient classics of Āyurvedic wisdom. His text, *Susrutha Saṃhitā*, remains an authoritative text on surgical procedures and equipment.

Suvarnamakshika – golden coloured ore of copper and sulphur, principally, copper pyrite.

Svedana – the act of sweating, to reduce via steaming.

Swarna – alternate name of gold (Au).

Swarna Bhasma – gold ash, or the oxide of gold metal.

Swarna Gaireka – a soft reddish iron ore containing tiny flecks of golden coloured ore.

Swarna Maksika – copper pyrite CuS

Swarna Maksika Bhasma – oxide, ash of copper pyrite.

Tabasheer – another name for bamboo manna or bamboo camphor.

Tailam – oil, usually sesame oil.

Tailavarga – recommended oils in the Āyurvedic pharmacy.

Takana – borax,

Takra – curd, usually buffalo.

Taleeq – another name for Rakta Moksha (bloodletting/leeching).

Tamaragarbha – alternate name for copper sulphate.

Tamra – alternate name for copper Cu (metal).

Tamra Bhasma – alchemically prepared copper oxide.

Tarksya – alternate name for emerald.

Tejas – one of three primary energetics, some of the attributes of Tejas are illumination, warmth, clarity and purity. It relates to the fire element and Agni.

Tibb – (Tibb-e-Unani) Greco-arabic medicine system, originating in Greece and influenced by Islamic medicine and Āyurveda.

Tikshna Loha – alternate name for wrought iron.

Tikta – the bitter taste, one of five tastes used therapeutically in Āyurveda.

Tinkar – another name of sodium borate.

Tri-Loha – a mixture of gold, silver and copper.

Trikatu – popular Āyurvedic formula. *Tri*=three and *Katu*=pungent; the three herbs are black pepper, long pepper and ginger.

Trinakanta – alternate name for amber (fossilised tree resins).

Triphala – Trikatu, a popular Āyurvedic formula. *Tri*=three and *phala*=fruits; the three herbs are Amalaki, Bibhītaki and Harītakī.

Tuttha – copper (II) sulphate, usually found in the form of bright blue crystals and artificial in origin.

Uparasa – secondary (mineral) drugs used in connection with Pārada (mercury).

Upavisha – secondary herbal drugs used in connection with Pārada (mercury).

Usna – the principle of heating.

Vaccha – a very popular Āyurvedic herb, *Acorus calamus*.

Vahni Garbha – alternate name for sunstone.

Vahnijara – another name for Agnijara (ambergris).

Vahniloha – alternate name for the alloy bronze.

Vaiduryam – alternate name for cat's eye or chrysoberyl, $Be_3Al_2Si_6O_{18}$.

Vaikrānta – generally associated with the mineral tourmaline. The name *Vaikrānta* also appears in Maha Rasa, with some connection to the mineral, fluorite. Alchemically, tourmaline tends to be the mineral most frequently associated with Vaikrānta.

Vājīkarana – one of eight therapeutic branches in Āyurveda. Vājīkarana is usually associated with the promotion of fertility and aphrodisiacs.

Vajra – alternate name for diamond.

Vālukā Yantra – or 'sand bath', is a device or technique used to help regulate temperature by heat distribution, this enables glassware etc. to withstand prolonged high temperatures.

Vamana – therapeutic vomiting, used to remove excess Kapha Dosha from the tissues.

Vanga – the metal tin (Sn).

Varaha – means boar.

Varaha Moti – pearls found at the root of boar tusks.

Varaha Puṭa – an earthen pit used in high temperature cooking whose dimensions are based roughly on the dimensions of an indentation left by a sleeping boar.

Varna – caste, one's social status or social group, i.e. priest, warrior, merchant or servant.

Varta Loha – alternate name for an alloy prepared from five metals.

Vāstu – the art of laying out a construction in accordance with subtle energy principles as well as practical considerations of the terrain. Vāstu employs direction, astrology, water, timing and propitiation.

Vāsuki – one of two important and powerful serpent kings. Vasuki was famed for his diadem, which held a legendary gemstone referred to as *Nāgamani* (serpent gemstone).

Vāta Dosha – one of three Dosha (or humors), used to define a constitutional type in Āyurveda. Vāta relates to the air element.

Vatsanābha – aconite, a popular poisonous plant used in Āyurveda as a potent remedy, and in alchemy as a plant capable of fixing (steadying) mercury.

Vatsanāga – alternate name for aconite.

Vāyu – one of three primary and universal energetics. The attributes of Vāyu are movement, coolness, dryness and changeability. *Vāyu* also relates to the air element and Vāta Dosha.

Vedas – the four ancient and sacred texts of Hinduism.

Viadya – a physician, instructor or teacher.

Vijaya – cannabis (*Cannabis sattiva*).

Vimala – iron pyrite FeS2.

Visha – poisons

Vishachikitsa – alternate name for Agada Tantra, Vishachikitsa is the art of applying poison as both antidote and remedy.

Vishatinduka – alternate name for quaker buttons (*Nux vomica*).

Vishnu – one of three important Hindu deities known as *Trimurti*; Vishnu: *he that Pervades*, is attributed with preservation.

Vrischikavisha – poison harvested from the scorpion.

Yagya – fire ceremony, popular in astrological ceremonies to undo the effects of negative planetary karmas.

Yajur Veda – one of the four holy books focusing on Mantra for worship.

Yasada – alternate name for zinc, Zn.

Yoga vāhin – appropriate vehicle for a deity, medicine etc.

Index

abalone 305, 309
abdominal distension 65
Abhicāra 37
Abhrakagrāsamāṇa 213
Abhyaṅga 70
aconite (Vatsanābha) 196, 296, 338, 342, 343
  as Visha (poison) 196, 338, 339, 442
  Bachhnāga 338
  benefits 343
  cultivation 339
  formula: Jvara Mrityunjaya Rasa 380
  in classical Chinese medicine 340, 341
  medicinal background 338
  purification 342
  storage of Vatsanābha powder 342
  treatment for snake and scorpion bites 339
  Vishakalpa 339
Adhahpatana Yantra 249
Adhas 138
Adrija 256
Agada Tantra 54, 109
Agastya 109
  as Pañchatikta 132
Agastya (South Indian sage) 461
agate (Akika) 194
  formula: Akika Bhasma 373
Agni (inner fire/digestive capacity) 61
Agni Pariksha (trial by fire) 223
Akash (æther) 46, 47
Akasha Gaṅgā (Milky Way) 37
Alakshmi 31
al-Bīrūnī 139
alchemical preparations 157
  handmade 167, 168
  seven qualitative tests 157, 158
alchemical tradition x
alchemy 37, 38, 43, 177, 209
alloys 95, 99, 103, 161
aloe gel (Kumārī) 170, 210, 320
  as Bhavana 169, 170
alum (Kanksi) 191, 399
  as Uparasa 191
Āmalakī 385
  anti-cancer properties 385
Āma (undigested waste) 60
ambergris (Agnijara) 191, 406
  as Sadharana Rasa 191
amber (Kaharuba) 194, 428
  as Uparatna 194
Amlavarga (essential sour drugs) 132
Amlavetasa *see* Amlavarga
ammonium chloride (Narasara) 191, 275, 276, 405
  as Sadharana Rasa 191
  counteracting poisons 275
  health benefits 276
  three Doshas 275
Amṛīta (divine nectar) 29–34, 300
  as nectar of immortality 31
Amṛītakarana 174, 283, 388
analogy of fertile soil 66
Andhkari 206
aṅgulī (measurement of development) 143
animal products 305, 434–441
  as medicines 305
añjali 56
Anjana sannibha 223
Anna Lepa 87
anti-ageing 224
anti-cancer herbs 384–386
  41 herbs 386
  Rasāyana drugs 384
  Upavisha 385
antimony sulphide (Anjana) 191, 401
  as Uparasa 191
Anupāna (vehicle of delivery) 65, 183–186
  apple juice 172
  betel leaf 231
  dairy 184, 185
  honey 183
  human urine 113

kithul jaggery 185, 186
lime juice 186
long pepper 323
Triphala decoction 185
water 186
Anuvāsana Basti 73, 74
Apara-Ojas 55–57, 59
Apaurusheya (conceived by divine mind) ix
apple juice (Rubb-e-seb) 172
as Bhavana 169, 172
apprentices 141, 142
competency and reliability 141–143
Hastā Rekha Shāstra (Vedic palmistry) 144, 145, 147, 149
Shat-Samudrik Shāstra (study of the body) 143
Apsaras (celestial nymphs) 83
Ariloha (metal's enemy) 99
Ariloha metals 100
aromatic ginger (Gandhamoolaka) 385
anti-cancer properties 385
arsenic disulphide (Manah shila) 191, 401
arsenic trioxide (Gauri pashana) 191, 404
as Sadharana Rasa 191
arsenic trisulphide (Haritāla) 191, 262, 265, 400
as Uparasa 191
formula: Rasa Maanikya 367
Haritāla purification, analysis 265
health benefits 266
purification 262, 264
Rasa Maanikya 266–269
asbestos (Kauseyasma) 195, 433
Ashastra 77
Ashokan pillar, Delhi 94
Ashtasaṁskāra 209, 210, 213
Ashwin Twins 83
Asiatic pennywort (Gotu Kola) 385
anti-cancer properties 385
ass milk 133
as Dugdhavarga 133
asta-chakra (eight-sided disc) 145
in Hastā Rekha Shāstra 145
Astāñga Hṛdayam 121, 122
Astāñga Sangraha 122
Asthi 55
Asthi (bone) 58, 193
Asthi Dhātu 462
astringent drugs (Ksharashtaka) 133
astringent taste 62
astrology 37, 96, 136, 144
Jyotish (Vedic Astrology) 96, 128, 214
planetary influence on taste 60
Rāhu and Ketu 99, 121
Rasashala 129
relationship between planets and metals 96–99
Yantra 470
Asura-Bhāga (demonic realm) 138
Asuras (demons) 29
Atharva Veda ix
Atman (divine mind) ix
Ātreya Smith, Vaidya 40
aurification 207, 208
Avaleha 82
Āvāpa 101, 102
Āvarita (nets) 145
in Hastā Rekha Shāstra 145
Āyurveda 45–48, 53, 55, 59, 66
alchemical tradition (Rasa Shāstra) x, 38
Anupāna 183
Doshic cycle 49
Eight Branches of 53, 54
five elements (Pañcha Mahābhūta) 46, 47
historical x, 81
Prakriti 46
rejuvenation of the body (Rasāyana) 3, 45
Ṣaḍkiryakalas (six disease stages) 48
Sapta Dhātu 55
therapeutics 63–65
three Doshas 45, 46
Agada Tantra 54
Bhūtavidyā 53
Kaumara Bhritya 53
Kāyachikitsa 53
Rasāyana 53
Shalya Tantra 54
Susrutha School of Medicine 54
Vājīkarana 54
Āyurvedic drugs 385
anti-cancer properties 385, 386
Ayuryog project 208, 216

B16 melanoma 383
Bhallātaka medicine 383
Bāhyadruti 213
Bala (mallow decoction) 293
Bala (physical strength) 417
Bala Roga, *see* Kaumara Bhritya
Banyan tree 134
    as Kshara 134
Basti (enema) 73, 74
    Anuvāsana Basti 74
    Nirūha Basti 74
    reducing Vāta 73
bezoar stone (Gorochana) 195
Bhasma (alchemical ash) 22, 157, 158
    storage 157, 158
    testing 158
Bhavana 123
    mediums used 168, 169
    types of 169
Bheda (disease progression) 49
Bhedi 206
bhilawan nut/marking nut (Bhallātaka) 197, 343
    Agni (digestive) benefits 343
    anti-cancer properties 385
    as Tailavarga 133
    as Visha 343
    cooking/preparing oil 346
    dangers of prolonged use 349
    dermotological benefits 344, 349
    effects of Bhallātaka oil 346
    formula: Bhallātaka Rasāyana 381, 382
    health benefits 349
    in Rasa medicine 386
    oil extraction – Patala Yantra 344, 345
    purification 347, 348
    rejuvenation properties 345
Bhuiamla 385
    anti-cancer properties 385
Bhūta (spirits) 53, 57
Bhūtavijja, *see* Bhūtavidyā
Bibhītaki oil 121
Biopharm Leeches 80
bismuthinite, *see* bismuth sulphide (Chapala)
bismuth sulphide (Chapala) 189, 190, 205, 206, 395
bitter gourd (*Momordica charantia*) 132
    as Pañchatikta 132
bitter taste 62
bitumen (Shilajit) 41, 190, 244, 256, 260, 393
    analysis 260
    appearance 257
    grades 257
    health benefits 261
    origin 257
    purification methods 258, 259
black salt (Viḍa) 273
    as Pancalavana 134
    health benefits 273
black salt (Viḍa) artificial 273
blacksmithing 94
blood 51, 97
blood circulation 63
blue sapphire (Nilama) 193
    association with Saturn 299
    extending life 299
    formula: Nava Ratna Raja Mrganka Rasa 374–76
    health benefits 299
    Nilama purification 297, 298
    Nilama purification, analysis 299
borax 134
    as Pancamitra 134
boron 255
bovine drugs (Pancagavya) 133
Bṛāhmi Gṛtha (medicated ghee) xvi, 323, 327
    facilitating Prana (life force) 330
    health benefits 330
    importance of lipids 330
    preparation 327–329
    study aid 327
Brassica juncea, *see* mustard seed oil
brass (Pittala) 97, 192
Brihaspati 330
Bṛṃhaṇa therapies 64, 85
Bronze Age 93
bronze (Kansya) 97, 192
Budha (Mercury diety) 214
Budha (Mercury planet/deity) 97
    in Hastā Rekha Shāstra 147
Budh Rekha (line of Mercury) 150
buffalo milk 133
    as Dugdhavarga 133
butter 185
    as Anupāna 185

cakrika (edged disc) 34
cakrika (flat disks of ground material) 102, 153
calamine (Rasaka) 165, 395
calcite (Surama sapheda) 196
camel foot tree (Kanchanara) 385
    anti-cancer properties 385
camel milk 133
    as Dugdhavarga 133
cane sugar 133
    as Madhuratraya 133
Caraka Saṃhitā x, 53, 81, 122, 259
    Caraka (author) 53
    Cikitsāsathānam 81, 82
Cāraṇa 213
castor oil (Eranda) 171
    in Bhavana 169, 171
catarrh 61
cedar wood bark (Devadaru) 385
    anti-cancer properties 385
centipede (Wu Gong) 111, 114
Ceylon champaca (Rukmal) 386
    anti-cancer properties 386
Chaitra (March/April) 72
chalk (Khatika) 196
Chandra Puṭa 174
Chauhan, Partap (Dr) 42
chickpea 132
    as Amlavarga 132
China root (Chopchini) 386
    anti-cancer properties 386
Chinese alchemy 235
chrysoberyl (Vaiduryam) 193
cinnabar (Hiṅgula) 189, 192, 231
    as Raktavarga 133
clotted blood 62
cobra venom 196
coconut oil 133
    as Tailavarga 133
Colombo 1
colon 47
conch shell (Shankha) 195
    in Hastā Rekha Shāstra 145
constellations 138
constipation 62, 65
copper pyrite (Swarna Maksika) 190, 391
copper sulphate (Sasyaka) 190, 394
copper (Tamra) 97, 98, 104, 192, 280, 281
    Amṛītakarana 283
    health benefits 284
    storage 283
    Tamra purification 282, 283
coral (Pravala) 193
cow and buffalo dung (Vanopala) 175, 176
cowrie shell (Kapardika) 191
cow's milk 133
    as Dugdhavarga 133
crab-eyed peas (Gunja) 134
    anti-cancer properties 385
    as Pancamitra 134
croton (Jayapāla) 133, 197
    as Tailavarga 133
crown flower (Arka) 196
    anti-cancer properties 385
    as Kshara 134
    as Ksharashtaka 133
    as Upavisha 196
    as Visha 444
cuttlefish bone (Samudra Phena) 195
Cyavana Prāśa 82, 84

Dadima 385
    anti-cancer properties 385
Dagmar Wujastyk, Professor 216, 217
Damaru Yantra 159
Damodar Joshi, Dr 252
deer horn (Mrga Śrnga) 195, 306
    deer horn Bhasma analysis 307
    for heart disease and swelling 308
    purification – brown Bhasma 306
    purification – white Bhasma 307
deer musk (Kastūrī) 195
Dehasiddhi (physical transformation) 44, 209
Deva-Bhāga (godly realm) 138
Devas (gods) 29
Ḍhālana (quenching) 101, 103
Dhanvantari 31
Dhanvantari School of Pañcakarma 76
Dhattura 63, 196, 333, 334
    anti-cancer properties 385
    association with Lord Shiva 333, 336
    association with the Moon and Somā elixir 336
    asthma relief 335
    calming/narcotic properties 336

dosage, control and dangers 336
formula: Brihat Jvarankusha Rasa 381
health benefits 337
husk removal 337
oil from seeds 334, 335
Pañcakarma therapy 335
Dhātu (Vedic metal) 192, 193
alloy 192, 193, 409–417
Dhātu materials 409–417
metal 192
Dhoomagati (evaporation of mercury) 210
Dhwankshi (discolouration) 206
diamond (Hiraka) 193, 300
association with Venus 303
benefits to immune system 303
Hiraka Bhasma, analysis 302
purification of Hiraka Bhasma 301, 302
Sagar Manthan 300
diarrhoea 66
Dīksha Rekha (line of Jupiter) 150
Dinacharya 84
Dīpana (stimulating) 211, 212
doctrine of signatures 319
Dolā-yantra 219, 258
Dompe 9–12, 18
bus journey 9–11
factory location and production 12, 18
ferry ride 11
Dosha (body regulators) 45–47, 50
interaction of 47
Kapha Dosha 46
Pañcha Mahābhūta (five great elements) 46, 47
Pitta Dosha 46
Prakriti (individual constitution) 46
Ṣaḍkiryakalas (six stages of disease) 48
Vāta Dosha 46
Doshic cycle and disease process 49
Bheda (progression) 49
Prakopa (aggravation) 49
Prasara (diffusion) 49
Sanchaya (increase) 49
Sthana Samsraya (relocation) 49
Vyakti (Symptomatic) 49
Draarvi 206
Dugdhavarga (essential types of milk) 133
dung (Gomaya) 175
East Himalayan fir (Taalisa) 386
anti-cancer properties 386
eggshell (Kukkutanda) 195
elephant foot yam xvi
elephant milk 133
as Dugdhavarga 133
emerald (Tarksya) 193

faeces 133
as Pancagavya 133
false daisy (Bhringaraj) 169
anti-cancer properties 385
Bhavana 169
Feng Shui 138
ferrous sulphate (Kasisa) 191, 398
fish or fish's tail (Matsya) 145
in Hastā Rekha Shāstra 145
five-leafed chaste tree (Nirgundi) 385
anti-cancer properties 385
Five-liquid process 104
flame lily (Langali) xvi, 196, 349
abortifacient and pregnancy risks 352
Agni (digestive) benefits 352
anti-cancer properties 386
anticoagulant 350
anti-inflammatory 351
growing conditions 350
nutraliser for scorpion and insect bites 352
purification 351
sciatica relief 352
toxicity 350
fluorite (Vaikrānta) 190, 391
Four Pillars of Treatment 88

Gaja Puṭa 174, 180
galena (Nilanjana) 191, 402
gallbladder 51
Gandhaka Rasāyana 363
Ganesha 330
Garbhadruti (enclosed liquefaction) 213
garnet (Gomeda) 193
Garuda Purana 317
Gati (unpredictable movements) 209
gemstones 293, 294
Bala 293, 294
curative power 294
origins 293, 294

Yagya 293
ghee 117, 323
Agni (digestive) benefits 324, 326, 330
as Anupāna 184
as Madhuratraya 133
as Pancagavya 133
as Rasāyana 325
clarifying process 323
health benefits 326, 327
preparation 325, 326
Puja 324
Giri (mineral impurity in Pārada) 205
goat bone (Ajasthi) 196, 439
goat's milk 133
as Dugdhavarga 133
Goddess Lakshmi 31
gold (Swarna) 97
in Makara Dwaja 15, 16
granite 162, 163
grape decoction (Draksa) 385
anti-cancer properties 385
Guru Chandrika 150
Guru (Jupiter/Jupiter deity) 60
in Hastā Rekha Shāstra 148
gypsum (Godanthi Haritāla) 196

Hālāhala 30
Hansagati (loss of mercury) 210
Harima Fudoki 234
Harītakī 386
anti-cancer properties 386
Hastā Rekha Shāstra (Vedic Palmistry) 144, 146–151
astrological interpretations 150, 151
Chinha (symbols) 145
Line of Enemies (Maṅgal Rekha) 149
links to Jyotish 144, 145
Parvata 144
Rāhu and Ketu 150
Rekha (lines on the hand) 144–147, 151
heart-leaved moonseed (Guduchi) 385
anti-cancer properties 385
as Pañchatikta 132
hellebore (Kutki) 385
anti-cancer properties 385
herbal remedies 383
anti-cancer properties 383
Hiṅgula (cinnabar) 192, 232, 233, 237
formula: Hiṅguleśwara Rasa 363
health benefits 238
in classical Chinese medicine 237
Ko Hung 235
Kushta Sangraf 239–241
Liao family 235
purification of Hiṅgula 236, 237
honey 133, 134
as Anupāna 183
as Madhuratraya 133
as Pancamitra 134
Hṛdaya Rekha (heart line) 150
human milk 133
as Dugdhavarga 133

Indian almond (Kshudrabeeja) 385
anti-cancer properties 385
Indian aspen (Ankenda) 386
anti-cancer properties 386
Indian bdellium (Guggulu) 197, 353
anti-cancer properties 357, 385
anti-inflammatory/arthritic 357
antioxidant 357
appearance 353
as Pancamitra 134
chemical constituents 353
collecting/obtaining Guggulu 353
dermatological benefits 357
grades and varieties 354
growing regions 353
health benefits 357
Mitra-Pañchaka, use and preparation 358
purification and additional procedure 354–356
risks and contraindications 358
thyroid regulation 357
Triphala decoction 355
use in formulas 356
Indian coral tree (Parijata) 385
anti-cancer properties 385
Indian ginseng (Ashwagandha) 385
anti-cancer properties 385
Indian liquorice (Gunja) 197
Indian madder (Manjistha) 385
anti-cancer properties 385
as Raktavarga 133

Indian oleander (Karaveera) 197
as Kshara 134
Indian sorrel 132
as Amlapañchaka 132
Indra (Vedic thunder god) 52, 293
insomnia 50
iron (Loha) 97, 98, 105, 106, 192
subcategories 198
iron pyrite (Vimala) 190, 392

jade (Sangeyasab) 194
jaggery 133, 134
as Anupāna 185
as Madhuratraya 133
as Pancamitra 134
Jalamahābhūta (water element) 59
Jala pariksha (floating) 223
Jalgati (mercury dissolution) 210
Jalukā (medical leeches) 77, 79
Jāraṇa (mercury amalgamation, digestion) 213
Jeevgati (loss of mercury) 210
Jew's stone (Badarasma) 195, 431
Jingū kōgō 234
Jiva Institute, Faridabad 42
joint pain 65
Jujube fruit 132
as Amlapañchaka 132
as Amlavarga 132
Jupiter, see Guru (Jupiter/Jupiter deity) 136
Jyotish: The Art of Vedic Astrology xv
Jyotish (Vedic astrology) xv, 136
Hastā Rekha Shāstra 144
Muhurta 136, 137
Yantra 131, 470

Kajjalabhas 223
Kalpataru 31
Kamadhenu 31
Kapha Dosha 46, 49, 58, 59
attributes of 52
Kapikacchu 386
anti-cancer properties 386
Kapota Puṭa 180
Karanda (essential sour drug) 132
as Amlavarga 132
Kaustubha Jewel 31
Kāyachikitsa 53
Keerthi Rekha (Sun line) 150
Kelani Gaṅgā River 11
ferry ride 11
Keralīya Pañcakarma 85
Anna Lepa 87
Piṇḍa Sweda 86
Pizichhil/Kāya Seka 86
Śirodhara/Dhārā Karma 85
Śiro Lepa 87
Khadira (reddening drug) 133
as Raktavarga 133
Khalwa Yantra (pestle and mortar) 160
types of 160–165
kithul jaggery 185
as Anupāna 185, 186
Kokichi Mikimoto 309
Kokum (essential sour drug) 132
as Amlapañchaka 132
Krāmaṇa (empowering mercury penetration) 213
Kuja (Mars Planet/deity) 60, 97
in Hastā Rekha Shāstra 147
Kukkuta Puṭa 174, 180, 283
Kumaratunga, Chandrika Bandaranaike 383
Kundalini Shakti 121
kūpī 14
heating stages 17
sand bath (thermal regulation) 16
Vat 69 bottle 14
kūpīpākwa 224
Kurma 30
Kushta Sangraf (Unani medicine) 239
as Rasāyana 239
benefits 239, 241
preparation method 239, 240
preparation of Kusta Sangraf, analysis 240, 241
Kwatha (Arjuna) 133
as Raktavarga 133
Kwatha (herbal decoctions) 82

lac beetle resin (Laksha) 195
Laghu Puṭa 180
Laṅghana therapies 64
lapis lazuli (Rajavarta) 194
formula: Rajavarta Rasa 366
lead monoxide (Mrddara Śrnga) 192

lead (Nāga) 97–99, 192
lead tetroxide (Nāga Sindoora) 194
lemon (Nimbu) 132
  as Amlavarga 132
  as Bhavana 169
Lepas (medicated balms) 82, 87
leukaemia L-1210 383
  Bhallātaka as medicine for 383
limb paralysis 65
lime 132
  as Amlavarga 132
limestone (Sehunda) 196
liquefaction 213
liquorice (Yastimadu) 386
  as ingredient for Rasa medicine 386
liver 51
lodestone (Kanta Pashana) 195
Loha (iron) 192
Loha Pariksha 223
Lohasiddhi (transmutation of metals) 44
lotus flower (Padma) 145
  in Hastā Rekha Shāstra 145
lotus seed (Kamala) 169
  as Bhavana 169, 171

Madagascar periwinkle (Nyantara) 386
  anti-cancer properties 386
Mādhyamakakārikā 38
Maha Puṭa 174
Maha Rasa 190, 390
  Maha Rasa materials 390–396
  mineral 190
  plant 190
Majjā (animal fats, ghee) 65
Majjā Dhātu (nervous system) 248
malabar tamarind (Kankusta) 132, 189, 191, 403
  as Amlavarga 132
Malagati 210
mala (natural body waste) 62, 205
Malkari (tri-Dosha disturbance) 206
Maṅgal Rekha (Line of Enemies) 149
Mārana (calcination) 173
  Puṭa 173
  Saṁpuṭa 173
  Sharaava Saṁpuṭa 173
Mardana (trituration) 210, 212
Mars 98

mechanisation 166
Medas Dhātu (fat tissue) 65
Menakā 83, 84
mercurial drugs 213, 214
  Apathyas 213, 214
  Pathyas 213
mercuric oxide (Giri Sindoora) 192
mercuric sulphide (Kajjali) 15, 190, 222, 223, 389
  benefits 224
  Kajjali Bandha 222
  Makara Dwaja 15
  preparation of 15, 16, 21
  Rasa Sindoora 15
mercury 97, 190, 201, 205, 206, 209, 389
  Abhraka (mica) 213
  amalgamation 213
  ancient sources 217, 218
  appearance 202
  Ashtasaṁskāra (eight rejuvenation practices) 209, 213
  as Rasāyana 190
  caste/status 95
  cinnabar (sulphide form) 202
  combining 203
  dangers 241, 242
  extraction from cinnabar 231
  five types 204, 205
  health risks 242, 243
  Hiṅgula/cinnabar (mercurial ore) 160, 233–238
  impurities 205, 206
  in gold production 202, 203
  in Rasa Shastra 15
  Kajjal (mercury combined with sulphur) 159, 222–226
  Kushta Sangraf 239–241
  licensing 201
  liquefaction 213
  loss of Pārada 210, 211
  Makara Dwaja 227–231
  mercurial drugs 213
  mercuric sulphide 190
  mercury-based medicines 201
  natural impurities 205
  origins 204
  purification of mercury 219, 220, 222
  purification of Pārada 212

reducing toxicity 242–244
Saṁskāra 213
synonyms, 108 names of Lord Shiva 44
transmutation 207, 213
Mercury (planet) 97, 214
Budha 214
Jyotish 214
Rasāyana 44
metal-based medicines 279
metal (Loha) 95
metallurgy 91–94
early metallurgy 91, 93
metals 95
Ariloha metals 100
aurification 207, 208
caste/status 95, 96
categories of metal 95
metallic immune booster 107
metal reduction 99–106
metals and planets, relationship 96–99
transmutation 207, 209
Mgrashirsha (November/December) 72
mica (Abhraka) 113, 190, 213, 367, 372, 378, 390
milk (Godugdha) 133, 170, 184
as Anupāna 184
as Bhavana 170
as Madhuratraya 133
as Pancagavya 133
Milk hedge (Snuhi) 196
mineral-based medicines 247
misra (impure metal/alloy) 95
Miśraka 205
Mohini 33
monkey face fruit (Kampilla) 191, 404
Moon (Chandra) 34, 37, 60, 97
Chandra Puṭa 174
in Hastā Rekha Shāstra 145, 147
moonstone (Chandrakanta) 194
Moringa (essential bitter drug) 132
as Pañchatikta 132
mountain knotgrass (Gorakshganja) 385
anti-cancer properties 385
Mrddara Śrnga (lead monoxide) 192
Mrga Śrnga (deer horn) 195
formula: Mrga Śrnga White Bhasma 378
Murchana (mercury fixing) 335
Mūrchā (thickening) 210, 212
muscovite, see mica (Abhraka) 190
mustard seed oil 133
as Tailavarga 133

Nāḍī Sveda Yantra 71
Nāga beings 38, 39, 121
Nāga (lead) 192, 206
Naga Pashana 3
Nāgārjuna 37–39
Nāgā beings 38
Rasasiddha 39
Rasendramaṅgala 39
the *Mādhyamakakārikā* 38
nakshatra (stars of less than six points) 145
in Hastā Rekha Shāstra 145
names of Lord Shiva 44
Nasya (nasally administered medicine) 75
Bṛṃhaṇa 75
Śamana 75
Virechana 75
Naukakruti kharal 163
Navavisha (nine toxic substances) 109
Neem 132
as Pañchatikta 132
Netra Tarpana 39
Niripola clinic 384
41 anti-cancer herbs 385, 386
Nirodha (reviving) 211, 212
Nirūha Basti 73, 74
Nirvāpa (method of metal reduction) 101
Nischandratvam (loss of mercurial lustre) 223
Niyama (mercury fixing) 211, 212
nutgrass juice (Musta) 386
anti-cancer properties 386

Ojas 55, 59
Apara-Ojas 55
Ojas and Sapta Dhātu 55
Para-Ojas 55
Prana 55
oleander (Karaveera) 134
opium poppy (Khasabeeja) 197
oral tradition ix, x

Pācana (appitite increasing) 70
Palasha (astringent drug) 133
as Ksharashtaka 133

Palika Yantra 289
Pañcakarma (purification therapies) 66, 72
Keralīya Pañcakarma 85
Nasya (nasal medication) 75
Nirūha Basti and Anuvāsana Basti 73
Paścatakarma 80
Raktamokshana (medical blood-letting) 76, 77
Rasāyanadi 81
Saṁśodhana regimen 69
Vamana (therapeutic vomiting) 72
Virechana (purgation) 73
Pancalavaṇa (five salts method) 105, 106
Pañcha Mahābhūta (five great elements) 46, 47
Akash (æther) 47, 50
attributes of 47
Jala (water) 47
Prithvi (earth) 47
Tejas (fire) 47
Vāyu (air) 47, 50
Pañchamrittika (auspicious clay mixture) 132
Pañchatikta (essential bitter drugs) 132
Pārada (mercury) 97, 190, 201, 204–206, 211, 212
effects of contamination 205
five types of 205
impurities 205, 206
loss of 210
metal impurities 206
natural impurities 205
types of Pārada 205
Para-Ojas 55
Parvata (projections on the palm) 144
Patala Yantra 335
Pātana (eliminating lead from mercury) 211, 212
peacock feather (Mayūr piccha) 195
as an antidote 323
Bhasma storage 320
formula: Maheshvara Dhupa 379, 380
health benefits 323
Maha-Mayuri 318
neutralising toxins 319
peacock feather Bhasma, analysis 320, 321
purification of peacock feather Bhasma 319, 320
Rasa Jala Nidhi recipe for extracting copper from peacock feathers 321–323
pearl farming 310
pearl (Mukta) 193
aphrodisiac 313
culturing 310
eight sacred types 317, 318
formula: Kanchanadi Rasa 377
Garuda Purana 317
grading system table 312
health benefits 316
Kokichi Mikimoto 309
marine cultured pearls 310
marine vs. freshwater 312
medicinal-grade pearls 311
nacre 312
naturally occurring pearls 310
natural vs fake, identification 311
pearl Bhasma, analysis 315, 316
pearl farming 309, 310
purification of pearl Bhasma 315
purification (Pisti) 314
rejuvenation qualities 313
sources (table) 318
peepal tree resin (Lac) 133
as Raktavarga 133
peyā (gruel) 80
pharmacy siting and construction 127
Pisti 22
preparation of 182
Pithara Yantra 287
Pitta Dosha 46, 49, 58
attributes of 51
plant ashes 132
as Pañchamrittika 132
plant-based medicines 333
poisoning 66
poison nut (Kuchala) 197
pomegranate 132
as Amlapañchaka 132
as Amlavarga 132
powdered brick 132
as Pañchamrittika 132
Pracchānakarma (blood-letting incisions) 77
Practical Āyurveda 40
Pradhanakarma, *see* Pañcakarma (five therapies)
Prakopa (aggravation of Doshas) 49
Prakriti (constitution of an individual) 46
Prana (breath of life) 55, 330

Prasara (diffusion of Doshas) 49
prickly chaff flower (Apamarga) 133
as Ksharashtaka 133
Prithvi (the Earth/stable matter) 47
public sanitation x
Puja 324
Punarnava (Mountain ebony decoction) 385
anti-cancer properties 385
pungent taste 62
Punya Rekha (line of fortune) 150
Puranas 29
Vishnu Purana 29
Purbeck granite 163
purgation therapy 72
puṭa (earthern pit) 173, 180
mārana 173
saṁpuṭa 173
Sharaava Saṁpuṭa 173
types of 173–175
Puṭa paka (temperature) 176–181
puṭa (to cook) 173
Puti (impure) 95, 98, 285, 288

quartz (Sphatika) 194
formula: Bhutankusa Rasa 371, 372

Rāhu and Ketu 34, 35, 97, 99, 121
Asleshā-bhava 121
in Hastā Rekha Shāstra 149
Kaliya 121
Raktamokshana (medical blood-letting) 76–80
Ashastra 77
Jalukā 77
saliva of leech 78
Sashastra 77
types of medical leech 78
Raṃjana (dyeing/colouring) 213
Rasa (blood-red Pārada) 205
Rasa formulae 361–382
Rasahṛdayatantra 210
Rasa Jala Nidhi 118, 321
Rasa Maanikya 266–269, 367
analysis 269
formula 367
preparation 266–269
Rasa materials 190–197
alloy 192
animal 191, 195, 196
Dhātu 192, 193
Maha Rasa 190
metal 190, 192
mineral 190–192, 194–196
plant 191, 196, 197
precious gemstone 193
Ratna 193
Sadharana 191, 192
salt 191
semi-precious gemstone 194
Sudhā Varga 196
Uparasa 190, 191
Uparatna 194
Upavisha 196, 197
Visha 196
Rasa medicines 157, 158, 222, 227
41 herbs 385, 386
anti-cancer herbs 384–386
for B16 melanoma 383
for leukaemia L-1210 383
herbal 383
Kajjal (mercury combined with sulphur) 159
Niripola clinic 384
use of Upavisha 383
Rasa Parpati 206, 225
benefits 226
preparation method 225, 226
preparing Kajjali 225
Rasaprakāśasudhākara 207
Rasashala 127, 136
celestial considerations 136
central floor mandala 128
classical floor plan 129
Eight Cardinal Point diagram 138
influence of Sarpa (snakes) 130, 131
modern floor plan, Rajagirya Āyurvedic Hospital 135
modern interpretations 134–136
Muhurta 136, 137
pharmacy essentials 132, 133
pharmacy siting and construction 127, 128, 130, 136
Sarpa Bhaya Hara Asseeyaa Yantra 131
Sthir Vāstu diagram 130
Rasa Shāstra x

Dehasiddhi 44
Lohasiddhi 44
Science of mercury 43
Rasa Shāstra in therapeutic application 383
Rasa Shāstra, The Art of Vedic Alchemy, volumes 1–4 42
Rasasiddha 39
Rasa, synonym for mercury 43, 201, 205
Rasa (taste) 59
Rasāyanadi 81
Rasāyana drugs 384
anti-cancer herbs 384
Niripola clinic 384
test results 386
Rasāyana (life-extender/rejuvenator) 4, 44, 45, 53
administration of 81
Netra Tarpana 39
Rasendra (blackish-coloured Pārada) 205
Rasendramaṅgala 39
Rasendra Sāra Saṃgraha 100
Ratna (precious gems) 193
precious gemstone 193
Ratna materials 418–423
raw milk 169
as Bhavana 169
red agate (akika) 294, 296
analysis 296
as Uparatna 194, 424
benefits 296
EDX analysis 296
formula 373
health benefits 296
purification 295, 296
Unani Tibb 294
reddening drugs (Raktavarga) 133
red Indian leadwort (Rakta chitrak) 386
anti-cancer properties 386
red iron oxide (Gaireeka) 191, 398
as Pañchamrittika 132
red sandalwood 133
as Raktavarga 133
red sulphide of mercury (Makara Dwaja) 2, 227, 230
benefits 231
Makaradwaja (guardian of the underworld) 2
preparation method 228, 229
preparation of 15–18, 20, 21
rejuvenation 4
Rekhapurnata 223
Rig Veda ix
Rishi Cyavana 82–84
Rishis (wise men) x, 52, 53, 84
rock salt 132
as Pancalavana 134
as Pañchamrittika 132
Rohitaka 386
as ingredient for Rasa medicine 386
rosewater (Arq gulab) 172
as Bhavana 169, 172
ruby (Maanikya) 193
Rūkṣaṇa therapies 64
rust of iron (Mandura) 192

Sadharana Rasa 404
animal 191
mineral 191, 192
plant 191
Sadharana Rasa materials 404–408
salt 191
Ṣaḍkiryakalas (six stages of disease) 48
Ṣaḍupakarmas (therapeutics) 63
Bṛṃhaṇa (increasing/building) 63, 64
Laṅghana (lightening) 63, 64
Rūkṣaṇa (drying/depleting) 63, 64
Snehana (oleation/lubrication) 63, 65
Stambhana (fixing/retaining) 63
Svedana (steaming/fomentation) 63, 65
saffron 133
as Raktavarga 133
Sagar Manthan 29, 30, 32
Alakshmi 31
Amṛīta 31, 32
Asuras 29
demonic nodes 35
Devas 29
Dhanvantari 31
Goddess Lakshmi 31
Hālāhala 30, 31
Kalpataru 31
Kamadhenu 31
Kaustubha Jewel 31
Kurma 30
Lord Shiva 30, 31

Lord Vishnu 30–32, 34
Mohini 33
Rāhu and Ketu 34, 35
Sasyaka 31
Sūrya and Chandra 34
Vasuki 30
Saindhava (rock salt) 272
Sal ammoniac 134
as Pancalavana 134
salt (Lavaṇa) xvi
salts (Lavaṇavarga) 270
association with the Moon (Somā) 270
order of preference in Āyurveda 271
salt and the digestive process 270, 272
Tripatu 272
Śamanādi 84
Sama Veda ix
sambar (Romaka) 134, 276
as Pancalavana 134
aggravates Pitta Dosha 276
Saṃsarjanakram 80
Saṁskāra 213
Saṁskāras 209
Saṁśodhana 68, 69
Keralīya Pañcakarma 85
Pañcakarma (five therapies) 72
Paścatakarma 80
patients unsuited for 87, 88
Pūrvakarma 70
Samudra Lavaṇa (sea salt) 274
Vāta Dosha 274
Samudra Phena (cuttlefish bone) 195
Sanatana Dharma (eternal truth) ix
Sanchaya 49
sand bath 16
Sangeyasab (jade) 194
Sanskrit x, 43, 387
sapphire (Nilama) 193
Sapta Dhātu (seven tissues) 55, 56
Apara-Ojas (culmination of Sapta Dhātu/immunity/life force 59
Asthi (bone/adipose tissue) 58
Kedārīkulyā Nyāya (Dhātu transmission) 57
Khalekapota Nyāya (Dhātu selectivity) 57
Kṣīradadi Nyāya (Dhātu transformation) 57
Majjā (marrow) 58
Māṁsa (muscle tissue) 58
Medas (fat/adipose tissue) 58
Rakta (haemoglobin) 58
Rasa (plasma) 58
Shukra (reproductive fluids) 59
three mechanisms of Dhātu formation 57
Sāraṇa 213
Sarpa Bhaya Hara Asseeyaa Yantra 131
Sarpa (snakes) 145
in Hastā Rekha Shāstra 145
Sarpa-visha (cobra venom) 196
Sashastra 77
Sasyaka 31, 190
Sasyaka (copper pyrite) 190
Sattva (essence) 358, 359
Saturn 60
Saturn (Shani) 97
sciatica 65
scorpion (Quan Xie) 111
sea salt 134
as Pancalavana 134
Sehunda (limestone) 196
Semecarpus anacardium, *see* Bhallātaka
serpentine (Nāgapashana) 194
formula: Jawahar Mohra Bhasma 368
sesame 133
as Ksharashtaka 133
sesame oil 133
as Tailavarga 133
Sesbania grandiflora, *see* Agastya
Sevana 213
Shad-Rasa (six tastes) 169
Shilajit (bitumen) 41, 244, 256
formula: Chandraprabha Vati 365
Shalakya Tantra (treatment of eyes, ears, nose and throat) 53
Shalya Tantra (Āyurvedic surgery) 54
shamans 37
sharp poison (Tīkṣṇa) 111
shāstra (knowlege, authority) 43
Shatāvari (wild asparagus decoction) 385
anti-cancer properties 385
Shilajit (bitumen) 41, 190, 244, 256, 260, 393
analysis 260
appearance 257
composite 257
Formula: Chandraprabha Vati 364
origin 257

purification 257–259
Shingon Mikkyō 39
Shiva 30, 121
108 names of 44
Rudra 1
Sri Rudram Chamakam (prayer) 43
Shodhana (purification) 66–68
Shrávana (July/August) 72
Shrinkhala (chains) 145
in Hastā Rekha Shāstra 145
Shukra Mekhala 149
Shukra (Venus planet/deity) 97
in Hastā Rekha Shāstra 148
Siddhartha Gautama Buddha 37
Nāgārjuna 37
Siddha Tradition 111
Silk Road 218
silver (Rajata) 97, 192
Sindoorakara 99
Sirā-vyadha (venepuncture) 77
Śirovirechana 75
six actions of taste 59, 60
Amla (sour) 60
Kaṣāya (astringent) 60
Katu (pungent) 60
Lavaṇa (salty) 60
Madhur (sweet) 60
Rasa (taste) 59
requiring Jalamahābhūta (water) 59
Tikta (bitter) 60
Slakshnatvam 223
small intestine 47
Smith, Vaidya Ātreya xi
Snehana 70
Snehana therapies 65
Snuhi Ksheera 134, 386
anti-cancer properties 386
Snuhi (milk hedge) 133, 134, 196
as Kshara 134
as Ksharashtaka 133
soapstone (Dugdha Pashana) 196
sodium borate (Tankana) 195
Somā 56
Somātmaka 56
sooksmagami 270
sour drugs, 5 (Amlapañchaka) 132
sour orange (Ambuldodam) 169
as Bhavana 169
sour taste 61
Souvarcala (saltpetre) 274
artificial 275
in Āyurvedic preparations 275
spasms 65
Sri Lanka 1
equatorial positioning 8
Śrnga (animal horn) 77
Stambhana therapies 66
Sthana Samsraya 49
stomach 47, 52
Sudhā (pure) 95, 97
Sudhā Varga 196
Sugandi 386
anti-cancer properties 386
Sukti (mother-of-pearl) 195
Sulphur 397
sulphur (Gandhaka) 189, 190, 248–255
analysis 254, 255
as antidote for mercury poisoning 255
benefits 255
benefits to immune system 255
combined with mercury 249
combining with mercury 248
curative properties 249
formula: Gandhaka Rasāyana 363, 364
purification methods 250–254
use of in Rasa Shastra 15
Sūryakanta (sunstone) 194
Sun (Sūrya) 34, 37, 60, 97
in Hastā Rekha Shāstra 146
Sūrya Puṭa 174
Suran root 283
Sūrya Kṣara, *see* Souvarcala (saltpetre)
Sūrya Puṭa 174
Susrutha 54
Susrutha Saṃhitā 54, 122, 127
Susrutha School of Medicine 54
Sūta 205
Svedana 63, 65, 66, 70, 71, 210–212
Swarna Bhasma 107
sweet drugs, (Madhuratraya) 133
sweet flag (Vaccha) 386
anti-cancer properties 386
sweet taste 60

tabasheer 102
Tailam (medicinal oils) 82
Taila (oils) 65
Tailavarga (essential oils) 133
tamarind 132, 133
  as Amlapañchaka 132
  as Amlavarga 132
  as Ksharashtaka 133
Tamra Bhasma 386
  ingredient for Rasa medicine 386
Tantra (method of counteracting poison) 54
Tao Hongjing 234
Tejas (fire of intelligence) 47, 330
termite cement 132
  as Pañchamrittika 132
threadstem carpetweed (Parpataka) 386
  anti-cancer properties 386
Tilaparni 386
  anti-cancer properties 386
Tinospora cordifolia, see Guduchi
tin (Vanga) 97, 98, 192, 206
  formula: Madukadi Curna 369
  health benefits 285, 287
  impact on growth of body 285
  purification of Vanga 285–287
topaz (Pushparaga) 193
tourmaline (Vaikrānta) 194
transmutation 207
Tri-Dosha 206
Tri-Loha (Three Metals) 193
Triphala decoction 169–170, 185
  as Anupāna 185
  as Bhavana 169, 170
Trishula (Shiva's trident) 145
  in Hastā Rekha Shāstra 145
turmeric (Haridra) 385
  anti-cancer properties 385
turquoise (Pirojaka) 194

Unani 122
unmada (mental illness) 53
Uparasa (secondary minerals/mineral drugs) 190, 191
  mineral 190, 191
  plant 191
  salt 191
  Uparasa materials 397–403
Uparatna (semi-precious gemstones) 194
  semi-precious gemstones 194
  Uparatna materials 424-429
Upari (uppermost and auspicious) 138
Upavisha (semi-poisonous) materials 196, 197
  Upavisha materials 444–450
urine 133
  as Pancagavya 133
Usna (heating poison) 111
Utthāpana (washing of mercury in sour gruel) 211, 212

Vahni (burning effect of ingesting Pārada) 205
Vahni Mṛtsnā 233
Vājīkarana (aphrodisiacs) 54
Vālukā Yantra (sand bath) 14, 228, 267
Vamana (therapeutic vomiting) 72
Varaha Puṭa 174, 180
Varna (social caste) 95
Varta Loha (Five Metals) 193
Vāstu Shāstra 138
  Eight Cardinal Point diagram 139
  Sthir Vāstu diagram 130, 138
  Vāstu Purusha 130
Vasuki (serpent king) 30
Vāta Dosha 46, 49, 58
  attributes of 50, 51
  influence of Rāhu and Ketu 149
Vāyu (energetic) 47
Vedas ix, x
  Apaurusheya ix
  Atharva Veda ix
  Atman (divine mind) ix
  oral tradition ix
  Rig Veda ix
  Sama Veda ix
  Sanatana Dharma (eternal truth) ix
  Vedic era ix
  Yajur Veda ix
Vedha (transmutation) 213
Vedic alchemy 248
  Rasa Shāstra: The Art of Vedic Alchemy, volumes 1–4 42, 362
Vedic astrology (Jyotish) 96, 98, 128, 214
Venus 60
Vidyādhara yantra 231
Vijaya (cannabis indica) 197

Vikasi (pervasive poison) 111
viper (Fu She) 111
Virechana (purgation) 73
Visha and Upavisha (plant poisons) 333
Vishachikitsa 54
Vishakalpa 339
Visha (poison) 109, 113, 118, 196, 205
anti-Visha substances 114–116
caste 119
copper (Tamra) 280
effects of poisoning 111, 117
poison as antidote 112, 113, 121
purified visha 339
snake categories 120
snake venom 119–123
sources of Visha 110
treatment of poisoning 113, 117
types of Visha 111
Visha and Upavisha 333
Visha as medicine 118, 122, 123
Visha materials 442, 443
Vishnu 29, 30, 32
cakrika (edged disk) 34
Sagar Manthan 30
Vi-shrinḳhala (broken lines) 145
in Hastā Rekha Shāstra 145
vomiting as therapy 61, 66, 72
Vyakti (disease stage of Dosha) 49
Vyavayi (fast acting poison) 111

warm water 186
as Anupāna 186
water hyssop (Bṛāhmi) 385
anti-cancer properties 385
water (Jala) 47

Yagya (fire ceremony) 293
Yajur Veda ix
yantra 38
Adhahpatana Yantra 249, 250
Damaru Yantra 159
Dolā-yantra 219, 258
Nāḍī Sveda Yantra 71
Palika Yantra 289
Patala Yantra 335
Pithara Yantra 287
Tamari Pathar Yantra 163
Vālukā Yantra 14, 228, 267
Vidyādhara Yantra 231
Yantra (Jyotish/Vedic astrology) 129
Yava (astringent drug) 133
as Ksharashtaka 133
Yogarāja 207
Yoga vāhin (vehicle for medicines) 190
yoghurt 133
as Pancagavya 133

zinc (Yasada) 192, 206, 288
DNA/RNA formation 288
formula: Kushta Musallus (Unani) 370
formula: Majun Aarad Khurma 370
health benefits 291
infertility pancea 288
purification 289
purification (powdered zinc) 290
toxicity 291

About the author

Andrew Mason lives the UK and is a renowned lecturer who travels worlwide to share his knowledge of Rasa Shāstra, Jyotish (Indian astrology), and Hastā Rekha Shāstra (palmistry). Since completing his training in Āyurveda and Indian alchemy, he has dedicated his efforts to documenting the practices of ancient medical systems, with a particular focus on herbo-metallic-mineral preparations.

For individuals interested in the practical techniques of Rasa Shāstra, the author's website, www.neterapublishing.com, has four digital slideshows that provide detailed insights. Titled *Rasa Shāstra: The Art of Vedic Alchemy, volumes 1–4*, they offer a unique glimpse into this ancient practice, accompanied by rarely seen imagery from India and Sri Lanka.